Generalized Gaussian Error Calculus

Michael Grabe

Generalized Gaussian Error Calculus

With 47 Figures

Springer

Dr. rer. nat. Michael Grabe
Am Hasselteich 5
38104 Braunschweig, Germany
michael.grabe@gmx.net

ISBN 978-3-642-03304-9 e-ISBN 978-3-642-03305-6
DOI 10.1007/978-3-642-03305-6
Springer Heidelberg Dordrecht London New York

Library of Congress Control Number: 2009940174

Cover design: eStudio Calamar Steinen

Printed on acid-free paper

Springer is part of Springer Science+Business Media (www.springer.com)

To Lucy, Niklas, Finley, and Rafael

Preface

The book of nature is written in the language of mathematics
Galileo Galilei, 1623

Metrology strives to supervise the flow of the measurand's true values through consecutive, arbitrarily interlocking series of measurements. To highlight this feature the term *traceability* has been coined. Traceability is said to be achieved, given the true values of each of the physical quantities entering and leaving the measurement are localized by specified measurement uncertainties.

The classical Gaussian error calculus is known to be confined to the treatment of random errors. Hence, there is no distinction between the true value of a measurand on the one side and the expectation of the respective estimator on the other. This became apparent not until metrologists considered the effect of so-called unknown systematic errors. Unknown systematic errors are time-constant quantities unknown with respect to magnitude and sign. While random errors are treated via distribution densities, unknown systematic errors can only be assessed via intervals of estimated lengths.

Unknown systematic errors were, in fact, addressed and discussed by Gauss himself. Gauss, however, argued that it were up to the experimenter to eliminate their causes and free the measured values from their influence. Unfortunately, this is not possible. Considering the present state of measurement technique, unknown systematic errors are of an order of magnitude comparable to that of random errors and this causes the Gaussian error calculus to break down. Consequently, the metrological community needs to consider how the error calculus to-be should address the coexistence of random errors and unknown systematic errors.

In the late 1970s, a seminar entitled *On the Statement of the Measuring Uncertainty* [14] was held at the Physikalisch-Technische Bundesanstalt in Braunschweig which, regrettably enough, induced a bifurcation of error calculus: one branch attempted to save the Gaussian approach by formally randomizing unknown systematic errors and thus producing the *Guide to the*

Expression of Uncertainty in Measurement, GUM for short [15]; the other proposed a revision from scratch and issued an essentially new, generalized version of the Gaussian error calculus. This latter approach will be discussed here.

The proceeding considers time-constant unknown systematic errors to spawn biased estimators thus preventing the true values of the measurands and the expectations of the respective estimators from coinciding. Eventually, this physically founded distinction gave birth to the term traceability.

Independently of the question of how to treat unknown systematic errors, the author devotes attention to another point of interest relating to the treatment of random errors. Commonly, random errors are considered normally distributed, at least approximately, and we shall, as a matter of course, retain this assumption. The inclusion of the multidimensional model in error calculus, an obvious extension, seems as yet to have been overlooked. For this model to be deployed properly, each of the variables involved is required to hold the same number of repeated measurements. This apparently banal request turns out exceedingly beneficial: it solves the cumbersome problem error calculus traditionally suffers from when it comes to assign confidence intervals in error propagation. Renouncing the stipulation of equal numbers of repeated measurements puts experimenters beyond the validity of the distribution density of the empirical moments of second order—and this very observation causes the trouble.

The generalized Gaussian approach presented here produces reliable, easy to attain measurement uncertainties meeting the demands of traceability. At the same time, the approach features the properties of a building kit: any overall uncertainty is the sum of the contribution of random errors taken from a confidence interval as provided by Student and the contribution of unknown systematic errors, as expressed by an appropriate worst case estimation.

The book supplements the author's monograph*Measurement Uncertainties in Science and Technology* [28]. Its principles have, however, been tautened and cast into an order attaching prior-ranking importance to metrology's starting point, the traceability of physical units, physical constants and physical quantities at large. In addition, the previously discussed formalism has been extended in some respects. Finally, the procedures are now illustrated by scores of numerically based diagrams helping the reader to pursue the idea of traceability.

Braunschweig, November 2009 *Michael Grabe*

Contents

Part V Fitting of Straight Lines

Part VI Fitting of Planes

Part VIII Non-linear Fitting

Part IX Appendices

Part I

Basics of Metrology

1 True Values and Traceability

As much as we expect the laws of physics to be true, we expect the constants of physics to possess true values. Metrology starts from here, from the very grasping of true values and, following the modus operandi of physicists, their traceability to the primary standards of the system of physical units agreed on.

1.1 Metrology

Metrology aims at the entirety of experimental procedures needed to assess the true values of physical quantities and physical constants. At the same time, metrology asks for a world-wide uniformity of measuring results. Uniformity in turn premises the traceability of metrological standards to the International System of Units, the SI for short [2].

As a matter of fact, due to measuring errors, metrology is not in a position to uncover the true values of physical quantities or constants directly. All metrologists can do is to attempt to assess or to *localize* the true values of physical quantities by means of appropriately positioned intervals of reliable extensions.

1.2 Traceability

Traceability is the fundamental concept of metrology and refers to the handing over of true values in the course of a measurement. Basically, the transfer is accomplished via a comparator mediating the difference between the standard and the measurand, Fig. 1.1. Consider two measuring objects of equal physical quality, say $\beta_{0,j}$ and $\beta_{0,k}$. Traceability expresses itself through

$$\beta_{0,j} - \beta_{0,k} = x_0 \tag{1.1}$$

where x_0 designates the true indication of the comparator. Let us state:

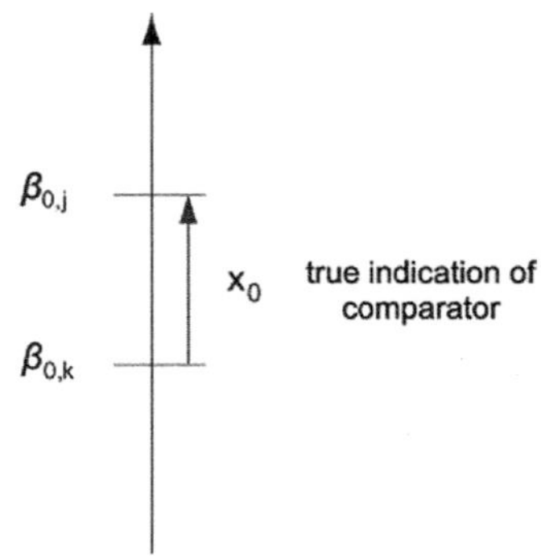

Fig. 1.1. In its most basic form traceability connects the true values of two objects, say, $\beta_{0,j}$ and $\beta_{0,k}$ with the true indication of the comparator, say, x_0

Traceability controls the transfer of the true value of one measuring object to the true value of another measuring object mediated by the true indication of a comparator countervailing the action.

Likewise, traceability is established through mathematical concatenations as functions and algorithms, say, e.g. least squares.

If accomplished, traceability controls the flow of true values all about within the widely ramified net of physical quantities. If not established, the mutual compatibility of measuring results turn out ill-defined.

1.3 Measurement Errors

Let x denote a quantity to be measured and x_0 its true value. Then, the first differs from the second due to the joint action of a random error and a time-constant unknown systematic error, Fig. 1.2.

Random errors are due to uncontrollable, irregular, statistically born microscopic processes arising during the taking up of a series of repeated measurements. A common attempt to quantify their dispersion is to allot an empirical standard deviation to them.

By contrast, unknown systematic errors are evoked during the configuration of the measuring device. There are alignments to be made, be they optical, mechanical, electrical or otherwise. Due to technical reasons, none of them will be thoroughly perfect. Rather, there will be smaller or larger residual deviations from the ideal positioning which the experimenter cannot clear up. Furthermore, the experimenter will have to account for environmental and

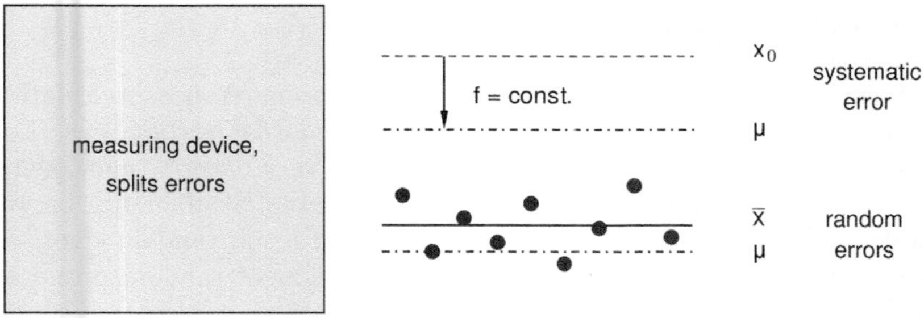

Fig. 1.2. A measuring device, operating stationarily, separates by itself random from systematic errors; x_0 true value, μ center of scattering of the series of repeated measurements, $\bar{x}$ arithmetic mean. Prior to the measurements the device somehow allocates a fixed, time-constant value f to the systematic error where $-f_{s,x} \leq f \leq f_{s,x}$

boundary conditions being controllable only with finite accuracy, mechanical or electrical switching-on effects, inherently different approaches to measure one and the same physical quantity and, if need be, to consider varying theoretical approaches. According to their nature, we address unknown systematic errors as biases. Here, the idea of non-drifting or stationarily operating measuring devices is of primary importance. Any such device separates by itself random and unknown systematic errors. When it comes to error propagation, this separation channels the flow of random and systematic errors into categorically differing branches. Obviously, an unknown systematic error shifts the bulk of measured data collectively. Unfortunately, neither the direction nor the amount of any such shift is known or ever knowable. Rather, the measurements are burdened by an unknown, non-eliminable quantity or bias for short staying visually hidden—as long as the measuring device betrays no drift.

Unknown systematic errors can only be assessed through intervals, their boundaries being taken from an analysis of the buildup, the functioning and behavior of the measuring device as a whole. From there, experimenters are in a position to specify intervals limiting their possible ranges. As has been shown elsewhere, these intervals may always be taken to be symmetric to zero [28].

After all, unknown systematic errors are considered time-constant, unknown with respect to magnitude and sign and confined by intervals symmetric to zero. Though biasing the true values of the measurands, they are not seen to affect the scattering of the random errors.

1.4 Precision and Accuracy

The term precision quantifies the random scattering of measured data. A common measure for precision is the empirical standard deviation. In addition to random errors there are biases being, by their very nature, time-constant non-statistical perturbations. Figuring out the potential difference between the estimate of the true value and the true value itself, random errors and biases enter collectively. Only the combined action of random errors and biases constitute what is called the measuring accuracy. As the Gaussian error calculus dismisses unknown systematic errors,[1] we shall have to break new ground.

1.5 Measurement Uncertainty

The measuring uncertainty expresses the measuring accuracy quantitatively – the lower the uncertainty the higher the accuracy of the measurement. Due to formal reasons, uncertainties emerge as positive quantities which afterward have to be provided with a ± sign. A measuring uncertainty should not be practiced as an end in itself, rather it should localize the position of the true value of the measurand by the smallest reliable interval – whatever size this actually may have. While high level metrology strives for extremely low uncertainties, everyday metrology gets by with uncertainties of moderate sizes depending on the particular kind of application. Remarkably enough, legal metrology addresses the properties of uncertainties as follows:

Neither the seller nor the buyer should be in a position to extricate directed advantages deploying metrological quantifications.

Case related, uncertainties may or may not carry a physical dimension.

1.6 Measuring Result

Any measurement should issue an estimate of the true value of the measurand and a measuring uncertainty expressed as

$$\begin{gathered}\text{estimator} \pm \text{uncertainty} \\ \text{estimator} - \text{uncertainty} \leq \text{true value} \leq \text{estimator} + \text{uncertainty}\,,\end{gathered} \tag{1.2}$$

[1] Just to recall: Gauss identified and discussed unknown systematic errors, however excluded them from his formalism.

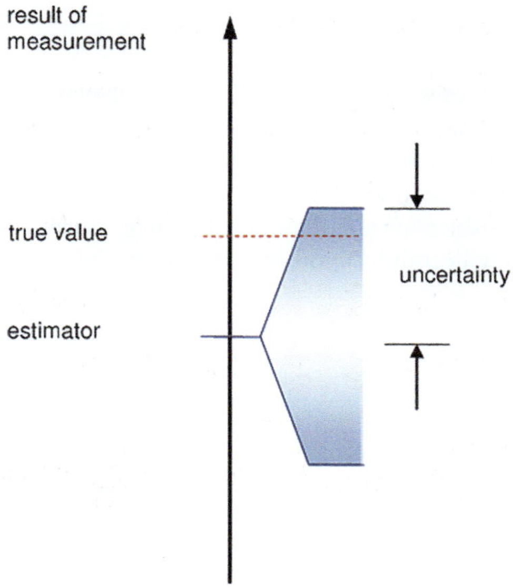

Fig. 1.3. The interval estimator ± uncertainty is required to localize the true value of the measurand

Fig. 1.3. In general, estimators happen to be mean values, the pertaining symbols carrying a bar on top. Given x designates a measurand, its mean would read $\bar{x}$. The common symbol of the measuring uncertainty is u. To establish a relationship to the measurand an appropriate subscript should be added; the final account would read

$$\bar{x} \pm u_{\bar{x}} \,. \tag{1.3}$$

Example

Let

$$(10.31 \pm 0.05)\,\mathrm{g}$$

denote the result of a weighing. Here, 10.31 g designates the estimate of the unknown true value of the mass and ± 0.05 g the estimate's uncertainty.

Uncertainties depend on the error model drawn on. This aspect asks experimenters to verify the localization properties of their uncertainties. But as nature strictly hides true values, formal checks can only be done via numerical simulations. Clearly, only a data simulation puts us in a position to firstly preset a true value and to subsequently test the localization of which via the uncertainty as issued by the (simulated) measuring process.

1.7 Rivaling Physical Approaches

Let any two rivaling physical theories be distinguished by measurement. To decide which approach is more likely to be correct, a decision is necessarily confined to the limits of the measurement accuracy. As has been emphasized repeatedly, measurement uncertainties should express the smallest intervals localizing the true values with sufficient reliability. Uncertainties being either too narrow or too wide might spoil experimenters' efforts and obscure their conclusions.

2 Models and Approaches

Asking the error model to map the physical properties of the experimental set-up abolishes the Gaussian error calculus.

2.1 Gaussian Error Model

Following Gauss [27], experimenters are faced with random errors and time-constant unknown systematic errors. The former, being due to random perturbations, fluctuate from one measurement to the next; the latter at the same time stemming from time-constant effects remain unaltered.

Gauss, as a matter of fact, dismissed unknown systematic errors arguing that it is to be left to the experimenter to get rid of them. Roughly about two centuries later, due to enhanced measuring accuracies and international measurement comparisons, this very reasoning issued a dilemma: unknown systematic errors, proving non-eliminable, caused Gauss' error calculus to break down.

To an observer, repeated measurements scatter randomly with respect to some center, say, μ. At the same time a potential unknown systematic error remains invisible. Denoting random errors by the symbol ε, the Gaussian error equation, in case of n repeated measurements, reads

$$x_l = \mu + \varepsilon_l \,; \quad l = 1, \ldots, n \,. \tag{2.1}$$

Obviously, this equation reflects what the experimenter observes but covers not necessarily the actual operating principle of the measuring device. Rather, experience urges to discern the center of scattering, μ, and the true value, x_0, of the measurand. As a rule, these quantities are separated by an unknown systematic error or a *bias*. On that score, Gauss framed the conditions of the error calculus incomplete, his formalism being doomed to failure:

As far as empirical data embed unknown systematic errors or biases the Gaussian error calculus should be considered obsolete and the same applies to other statistically borne procedures burdened by unknown systematic errors, as e.g. the analysis of variance.

2.2 Generalized Gaussian Approach

In contrast to the original Gaussian approach, the generalized Gaussian approach introduces unknown systematic errors right from the outset.

Errors of this, say, second kind probably (re)entered the stage on account of a paper issued by Eisenhart [6] in the early 1950s, pondering their consequences in regard to quoting realistic measurement uncertainties. Systematic errors, Eisenhart argued, would displace the center of the scattering of the random errors relative to the true value of the measurand. Remarkably enough, the aftermath of Eisenhart's paper did not trigger off an impetus revolutionizing error calculus. Rather, the metrological community had to wait 30 more years for an initiative to evolve. The starting point happened to be set on February 20, 1978 when a seminar was held at the Physikalisch-Technische Bundesanstalt in Braunschweig entitled *On the Statement of the Measuring Uncertainty*[1] [14]. Regrettably enough, the seminar initiated a bifurcation of error calculus giving rise to the "Guide to the Expression of Uncertainty in Measurement" [15], GUM for short, and to what the author calls "Generalized Gaussian Error Calculus" which we shall exclusively address in the following.

To this end, we confine our considerations to non-drifting measuring devices operating, as Eisenhart [6] vividly explained, *in a state of statistical control.* Such working condition reflects what may also be termed a *stationary random process.*

Akin to (2.1), we notionally decompose each of the measured values, say, $x_l;\ l = 1, \dots, n$ into three terms, the true value x_0, a specific l-dependent random error ε_l and a common, time-constant unknown systematic error f

$$x_l = x_0 + \varepsilon_l + f\,, \quad f = \text{const.}\,; \quad l = 1, \dots, n\,. \tag{2.2}$$

This equation, depicting the standard situation of metrology, marks the starting point of the generalized Gaussian error calculus. It maps the physical behavior of the measuring device suggesting the random dispersion of the data to be charged by one and the same time-constant unknown systematic error.

Further, we consider the random errors to be normally distributed and the unknown systematic error to be limited by an interval symmetric to zero

$$N(\mu, \sigma^2); \qquad -f_s \le f \le f_s\,. \tag{2.3}$$

While μ designates the center, σ stands for the spread or dispersion of the density. Assuming f to be confined to a symmetric interval does in no way limit the scope of application of our discussions. As has been shown in [28], any non-symmetric interval may easily be transformed into a symmetric one.

Due to formal reasons, we shall understand measured data as realizations of random variables. Consider, e.g., a sequence of repeated measurements

[1] Über die Angabe der Messunsicherheit.

$$x_1,\, x_2,\, \dots\,, x_n\,.$$

Here, the x_l are considered successive realizations of some random variable X,

$$X = \{x_1,\, x_2,\, \dots\,, x_n\}\,.$$

We shall assume the measured data to be normally distributed, i.e. to follow

$$p_X(x)\,\mathrm{d}x = \frac{1}{\sigma\sqrt{2\pi}}\exp\left(-\frac{(x-\mu)^2}{2\sigma^2}\right)\mathrm{d}x\,.$$

Quite obviously, the expectation of the normally distributed random variable X itself reproduces the center of scattering

$$\mu = \int\limits_{-\infty}^{\infty} x\; p_X(x)\mathrm{d}x\,.$$

Similarly, the expectation of the random variable $(X-\mu)^2$ reproduces the theoretical variance

$$\sigma^2 = \int\limits_{-\infty}^{\infty} (x-\mu)^2\, p_X(x)\mathrm{d}x\,.$$

which, remarkably enough, is in no way affected by the bias f.

To shorten the notation of expected values, we shall resort to curly brackets [28], e.g.

$$E\{X\} = \mu \quad \text{and} \quad E\{(X-\mu)^2\} = \sigma^2\,.$$

From (2.2) we draw

$$\mu = x_0 + f\,; \qquad -f_s \le f \le f_s\,. \tag{2.4}$$

The distinction between the expectation μ and the true value x_0 discloses the break down of the Gaussian error calculus.

The insertion of (2.4) into (2.2) leads us back to the Gaussian approach

$$x_l = \mu + \varepsilon_l\,; \quad l = 1, \dots, n$$

which we observe to hide the unknown systematic error f. Substituting $x_l - \mu$ for ε_l turns (2.2) into the identity

$$x_l = x_0 + (x_l - \mu) + f\,; \quad l = 1, \dots, n \tag{2.5}$$

the handling of which proves much more convenient than (2.2). Summing over l issues a formal decomposition of the arithmetic mean

$$\bar{x} = x_0 + (\bar{x} - \mu) + f \, . \tag{2.6}$$

For convenience, this notation, deliberately detached from a sum sign,[2] will be used whenever appropriate.

In the following we shall strictly adhere to a persistent distinction between theoretical parameters on the one hand and empirical estimators on the other. On this note, the counterpart of the theoretical variance is the empirical variance

$$s^2 = \frac{1}{n-1} \sum_{l=1}^{n} (x_l - \bar{x})^2 \tag{2.7}$$

where the bias f affects neither this nor that.

In case of more than one variable, the generalized Gaussian approach asks for equal numbers of repeated measurements for each of the variables considered. Otherwise, the so-called empirical covariances would get lost. This, in turn, obstructed the interaction of the empirical variances and those, as yet provisionally addressed, empirical covariances as expressed via the multivariate density of the empirical moments of second order which will be discussed in Sect. 4.2. To summarize:

The generalized Gaussian error calculus

- considers unknown systematic errors time-constant quantities
- formalizes random and systematic errors according to the physical behavior of the measuring device, i.e. strictly separated
- asks, in the multivariate case, for equal numbers of repeated measurements for each of the variables implied and, finally,
- persistently distinguishes theoretical parameters from empirical estimators.

The distinction between theoretical parameters and empirical estimators is known to go without saying. Nevertheless, it might be advantageous to address an example: Confidence ellipses may either be based on the theoretical variances and the theoretical covariance or, alternatively, on the empirical variances and the empirical covariance. Ellipses of the first kind are to be taken from the exponent of the two-dimensional normal probability density and are deployable only and only if the theoretical parameters are known—which in metrology they are not. Ellipses of the second kind, by contrast, are issued on the part of Hotelling's density and prove metrology-tailored as they get along with directly accessible empirical estimators.

Figure 2.1 visualizes the error model's scenario. The left hand diagram illustrates the theoretical density, $N(\mu, \sigma^2)$, as well as its empirical counterpart,

[2] More precisely sigma sign.

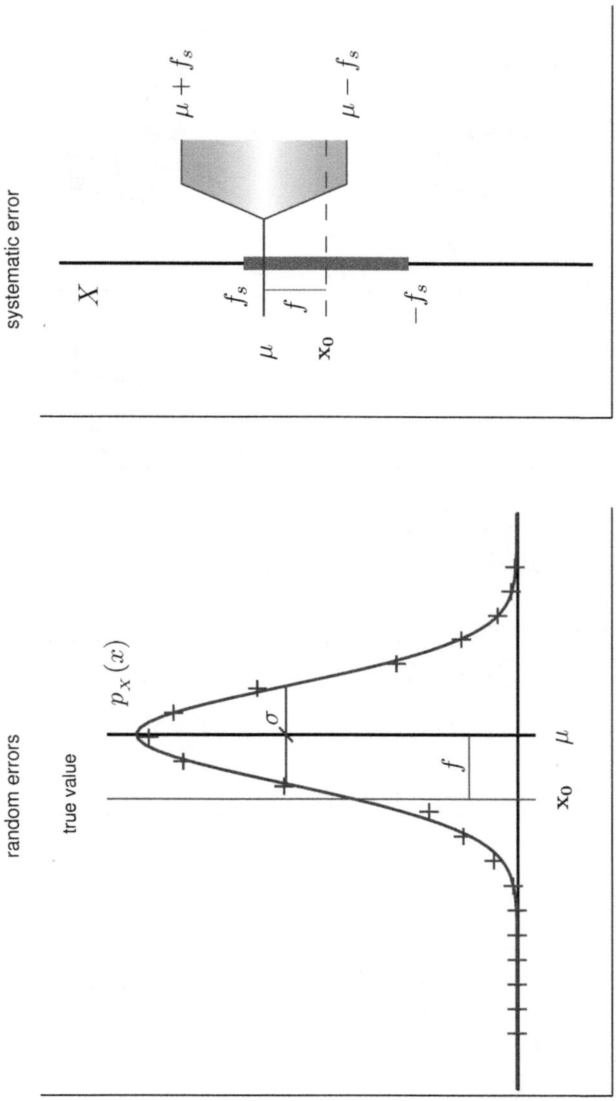

Fig. 2.1. Error model for stationarily operating devices. On the *left*: the normal density the center of which is shifted by f from x_0 to μ. On the *right*: localization of x_0 by an interval $\mu + f_s \ldots \; \mu - f_s$

$p(x_i)$, taken out of $N = 1000$ simulated data sorted into boxes of width Δx_i according to

$$p(x_i) = \frac{\Delta N_i}{N \Delta x_i} \, ; \quad i = 1, \ldots, r \, .$$

The values $p(x_i)$; $i = 1, \ldots, r$ are marked by crosses. The right hand diagram relates the quantities μ and f to the true value x_0. We ask the interval extending from $\mu - f_s$ to $\mu + f_s$ to localize the true value x_0. The diagrams recall Eisenhart's idea to separate random errors from unknown systematic errors.

2.3 Robust Testing Conditions

Data simulations prove indispensable when in comes to moot a new proposal aiming at assessing measurement uncertainties. As has been emphasized repeatedly, nature hides the true values of the measurands. Consequently, experimenters are not in a position to verify their uncertainties via empirical data. The only way out is to simulate data emulating real experiments. As here the true values are know a priori, as they are to be preset, the localization properties of uncertainties are open to be scrutinized ex post.

As practice suggests to assume random and systematic errors to range in like orders of magnitudes we shall tie down numerical simulations to this premise. Just for the purposes of data simulation—and only to this end—we notionally assign a rectangular distribution density to f,

$$p(f) = \frac{1}{2f_s}\,; \qquad -f_s \le f \le f_s\,. \tag{2.8}$$

The postulated density formally produces a theoretical variance

$$\sigma_f^2 = f_s^2/3\,. \tag{2.9}$$

The random number generator which, according to our assumptions, produces normally distributed errors is to be provided with a suitable theoretical standard deviation σ_ε. Setting arbitrarily

$$\sigma_f = \sigma_\varepsilon \tag{2.10}$$

issues an error interval

$$-f_s \le f \le f_s\,, \qquad f_s = \sqrt{3}\,\sigma_\varepsilon\,. \tag{2.11}$$

Obviously, this appears a reasonable choice in order to keep random and systematic errors in comparable sizes. To emphasize, (2.8) only refers to data simulations. Experimenters can neither choose nor manipulate f. Rather, unable to judge the actual value of a particular unknown systematic error, experimenters have no other choice but to rely on intervals expressed in the form of $-f_s \le f \le f_s$, the limits $\pm f_s$ of which stemming from the mode of operation of the measuring device [28].

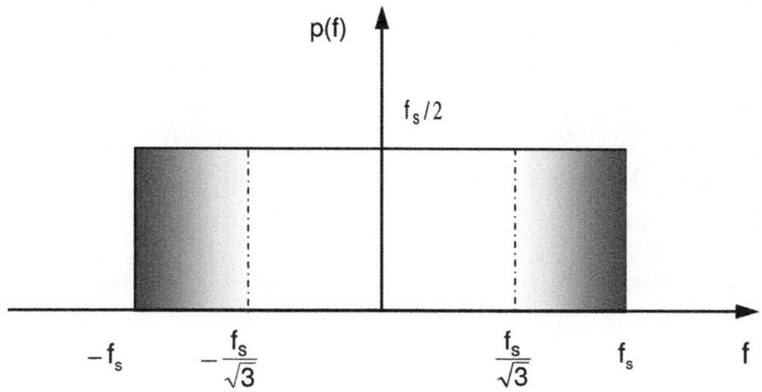

Fig. 2.2. Data simulations refer to values of f taken either out of the *left* or the *right* shaded region $-f_s \le f \le -f_s/\sqrt{3}$ or $f_s/\sqrt{3} \le f \le f_s$

For the purposes of data simulation, we strive to demonstrate the robustness of our approach. Thus, we deliberately provoke critical situations letting f assume values out of either of the two ranges

$$-f_s \le f \le -f_s/\sqrt{3} \quad \text{or} \quad f_s/\sqrt{3} \le f \le f_s\,, \tag{2.12}$$

Fig. 2.2. This will, the author envisages, codify robust testing conditions.

Also, robust testing conditions prove serviceable to scrutinize the localization properties of the ISO Guide. Here, the observation is, the more the unknown systematic errors tend to exhaust the limits of the intervals they are confined to, i.e. the nearer they are either to the lower or to the upper bounds, the sooner the Guide's uncertainties fail to localize the quested true values. This, if it actually happened to occur, would interrupt the flow of true values and thus the traceability of measures.

Incidentally, excluding critical measuring conditions right from the outset would degrade the outstanding rank of metrology within the natural sciences.

Ultimately, the Guide's so-called extension factor k_P turns out to be a weak, somewhat dark concept and does not seem apt to belatedly stabilize measurement uncertainties.

2.4 Linearizations

In general, error calculus is based on linearizations. The lower the uncertainties of the input data the more we are encouraged to consider linearizations admissible. Nevertheless, in individual cases verifications may stand to reason.

Under linearization the error model (2.2) produces a linear sum of three more or less bulky terms which, for convenience, will be ordered as follows: We let the first term express the respective true value, the second the influence of random errors and, finally, the third the influence of the unknown systematic errors. We shall meet this formal decomposition in any kind of error propagation, be this a functional relationship, a fitting procedure, or a system of linear equations.

The generalized Gaussian error model (2.2) and the attempted linearizations maintain a persistent separation of random errors and unknown systematic errors.

Should the magnitudes of the measurement errors rule linearizations out, coarser methods to assess measurement uncertainties had to be deployed. This, necessarily, produced coarser results.

2.5 Quiddity of Least Squares

Submitting a linear system to a least squares adjustment involves three steps:

- to verify that the linear system is readable in terms of true values,
- to design a solution vector via orthogonal projection and, finally,
- to assign uncertainties to the components of the solution vector.

True System

The inconsistencies of the linear system should solely be due to measurement errors. Let

$$\boldsymbol{A}\boldsymbol{\beta} \approx \boldsymbol{x} \tag{2.13}$$

be an inconsistent, overdetermined, linear system to be submitted to least squares. Here, $\boldsymbol{A}$ denotes an $(m \times r)$, $m > r$ matrix of real coefficients a_{ik}, rank $(\boldsymbol{A}) = r$, $\boldsymbol{x}$ an $(m \times 1)$ column vector of erroneous input data x_i and, finally, $\boldsymbol{\beta}$ an $(r \times 1)$ column vector of unknowns β_k

$$\boldsymbol{A} = \begin{pmatrix} a_{11} & a_{12} & \dots & a_{1r} \\ a_{21} & a_{22} & \dots & a_{2r} \\ \dots & \dots & \dots & \dots \\ a_{m1} & a_{m2} & \dots & a_{mr} \end{pmatrix}, \quad \boldsymbol{x} = \begin{pmatrix} x_1 \\ x_2 \\ \dots \\ x_m \end{pmatrix}, \quad \boldsymbol{\beta} = \begin{pmatrix} \beta_1 \\ \beta_2 \\ \dots \\ \beta_r \end{pmatrix}.$$

The aim is to estimate the r unknowns β_k ; $k = 1, \dots, r$. Clearly, the inconsistencies of (2.13) should relate to measurements errors and not to a physically ill-posed problem, which the method of least squares could not cure. Thus, we ask the system to be readable in terms of true values. If, instead of the erroneous data $x_i\,;\ i = 1, \dots, m$, we notionally inserted the respective true values $x_{0,i}\,;\ i = 1, \dots, m$ the linear system would read

$$\boldsymbol{A}\boldsymbol{\beta}_0 = \boldsymbol{x}_0\,. \tag{2.14}$$

Here $\boldsymbol{\beta}_0$ denotes the true solution vector with components $\beta_{0,k}\,;\ k = 1, \dots r$ and $\boldsymbol{x}_0$ the vector of the true input data $x_{0,i}\,;\ i = 1, \dots, m$ being hidden under measuring errors.

Should (2.14) not happen to apply, the adjustment attempted to treat a physically ill-defined problem. Remarkably enough, even then the method of least squares would yield a "solution vector," given rank $(\boldsymbol{A}) = r$. So there is nothing else for it but to state that the method of least squares per se does not let experimenters off the hook.

In the following, we shall always assume (2.14) to apply. Though there are more equations than unknowns, we may formally solve (2.14) for $\boldsymbol{\beta}_0$,

$$\boldsymbol{\beta}_0 = \boldsymbol{B}^{\mathrm{T}}\boldsymbol{x}_0\,; \qquad \boldsymbol{B} = \boldsymbol{A}(\boldsymbol{A}^{\mathrm{T}}\boldsymbol{A})^{-1}\,. \tag{2.15}$$

The flow of true values or maintenance of traceability is illustrated in Fig. 2.3. Let us once more refer to (2.14) which formally equates the vector $\boldsymbol{x}_0$ of true values with a linear combination of the column vectors of the matrix $\boldsymbol{A}$. We observe that $\boldsymbol{x}_0$ lies *in* the column space of the matrix $\boldsymbol{A}$. This is to be confronted with (2.13). There, due to measuring errors, the vector $\boldsymbol{x}$ of observations lies *outside* the column space of the matrix $\boldsymbol{A}$.

Orthogonal Projection

The method of least squares is just a trick: The vector $\boldsymbol{x}$, thrown out of the column space of $\boldsymbol{A}$, is simply projected back thus rendering the inconsistent system (2.13) solvable by brute force. Whether or not this procedural method is wise will be discussed soon. For the moment, suffice it to state that the projection per se is in no way in a position to sense any properties of the measurement errors of $\boldsymbol{x}$.

The orthogonal projection is brought about by a projection operator

$$\boldsymbol{P} = \boldsymbol{A}(\boldsymbol{A}^{\mathrm{T}}\boldsymbol{A})^{-1}\boldsymbol{A}^{\mathrm{T}}\,. \tag{2.16}$$

The operator formally produces a mathematically consistent system

$$\boldsymbol{A}\,\bar{\boldsymbol{\beta}} = \boldsymbol{P}\boldsymbol{x} \tag{2.17}$$

the least squares solution vector of which is

$$\bar{\boldsymbol{\beta}} = \boldsymbol{B}^{\mathrm{T}}\boldsymbol{x}\,; \qquad \boldsymbol{B} = \boldsymbol{A}(\boldsymbol{A}^{\mathrm{T}}\boldsymbol{A})^{-1}\,. \tag{2.18}$$

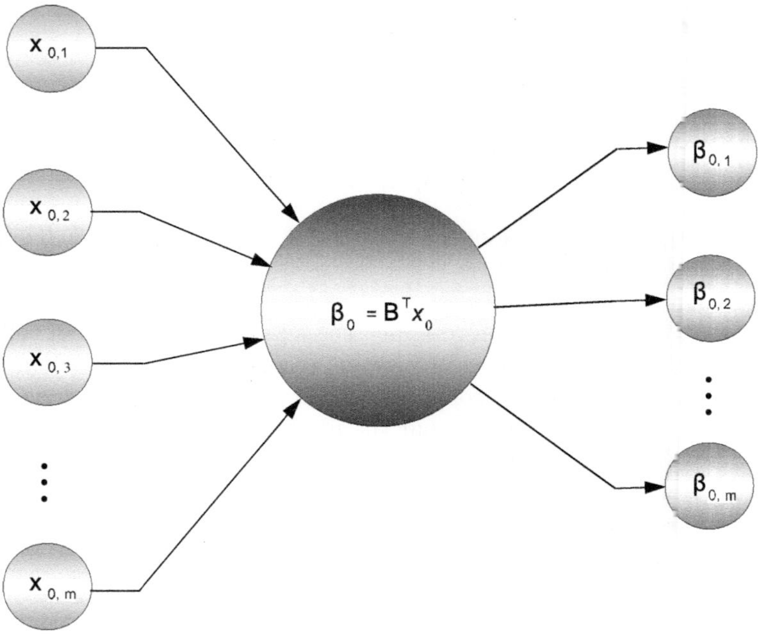

Fig. 2.3. Traceability in least squares

Remarkable enough, the vector $\bar{\boldsymbol{\beta}}$ is linear in the input data. In a sense, the input gets weighted by the coefficients of the matrix $\boldsymbol{B}$. The orthogonal projection minimizes the sum of squared residuals [28]

$$\bar{Q} = \left(\boldsymbol{x} - \boldsymbol{A}\bar{\boldsymbol{\beta}}\right)^{\mathrm{T}} \left(\boldsymbol{x} - \boldsymbol{A}\bar{\boldsymbol{\beta}}\right) . \tag{2.19}$$

In case the input data $x_1, x_2, \ldots, x_m$ are free of systematic errors and the random errors are uncorrelated and come from the same parent distribution $N(\mu, \sigma^2)$, the minimized sum of squared residuals provides an estimate

$$s^2 = \frac{\bar{Q}}{m - r} ; \qquad E\{S^2\} = \sigma^2 \tag{2.20}$$

of the theoretical variance σ^2.

Assignment of Uncertainties

Let us ask: Is it really wise to rely on a solution vector basically conjured from a cross-grained linear system? Conceding the method of least squares is a trick, what eventually matters is the question whether or not the uncertainties of the components $\bar{\beta}_k$; $k = 1, \ldots, r$ of the solution vector are apt to localize the true values $\beta_{0,k}$; $k = 1, \ldots, r$. Let the $u_{\bar{\beta}_k}$ denote the uncertainties of the $\bar{\beta}_k$. Then, the adjustment's result should comply with

$$\bar{\beta}_k - u_{\bar{\beta}_k} \leq \beta_{0,k} \leq \bar{\beta}_k + u_{\bar{\beta}_k} \, ; \quad k = 1, \ldots, r \, . \tag{2.21}$$

Remarkably enough, the error model (2.2) guarantees this very outcome hence proving Gauss' trick successful. However, the observation that Gauss confined himself to the error equation (2.1) while we, by contrast, rely on (2.2) renders things a bit more intricate.

For the time being let us observe: the narrower the intervals (2.21), the more information is available. However, the intervals' length do not only depend on the properties of the input data but also on the structure of the design matrix $\boldsymbol{A}$. Given the experimenter is in a position to choose differing matrices, he accordingly encounters differing uncertainties. Moreover, he may deploy weighting factors in order to boost the influence of those input data he considers more accurate and to reduce the influence of other input data he judges less accurate.

After all, we may not expect uncertainty statements to be unique. In any case, however, they should localize the true values.

Weighting Procedures

The classical version of the method least squares is based on the error model (2.1). In particular, as long as there are no biases, the Gauss–Markoff theorem is valid and provides weight factors. These weights are said to be optimal and produce what is called *minimum variance estimators.*

As a matter of fact, systematic errors or biases abolish the Gauss–Markoff theorem. This is why weight factors are no longer readily available. They may, however, be introduced heuristically via trial and error. But would such a proceeding maintain the localization of true values $\beta_{0,k}$; $k = 1, \ldots, r$? To recall, weight factors cause two things:

- they numerically shift estimators and
- they shrink uncertainties.

Hence, it might happen that a set of weights shifts the estimators $\bar{\beta}_k$; $k = 1, \ldots, r$ away from the true values $\beta_{0,k}$; $k = 1, \ldots, r$ while at the same time tightened uncertainties abolish the localization of the true values. Basically, in case of experimental data, we are not in a position to retrace such effects. However, deploying simulated data, delocalizations appear well detectable as has been demonstrated in [28]. To emphasize, the experimenter can never and will never be in a position to reveal whether or not the intervals (2.21) incur delocalizations.

Remarkably enough, the situation changes, given the adjustment refers to the error model (2.2) mapping the physical properties of stationary experiments. We then observe:

Within the assumptions of the error model referred to, the localizations of true values are maintained under any kind of weighting. In particular, weights may be chosen by trial and error in order to let uncertainties breathe in a way left to the judgement of the experimenter.

In practice, the experimenter will choose a first set of weights, somehow drawn from the uncertainties of the input data. Inspecting the newly produced uncertainties, he may deploy a second set of more or less similar weights thus being in a position to control uncertainties either selectively or flatly at least within certain ranges. In a sense, this procedure may be seen to replace the Gauss–Markoff theorem.

2.6 Analysis of Variance

Considering the physical effects unknown systematic errors inflict on series of repeated measurements, the analysis of variance is doomed to run into a dead end. Clearly, the empirical variances to be fed into the analysis of

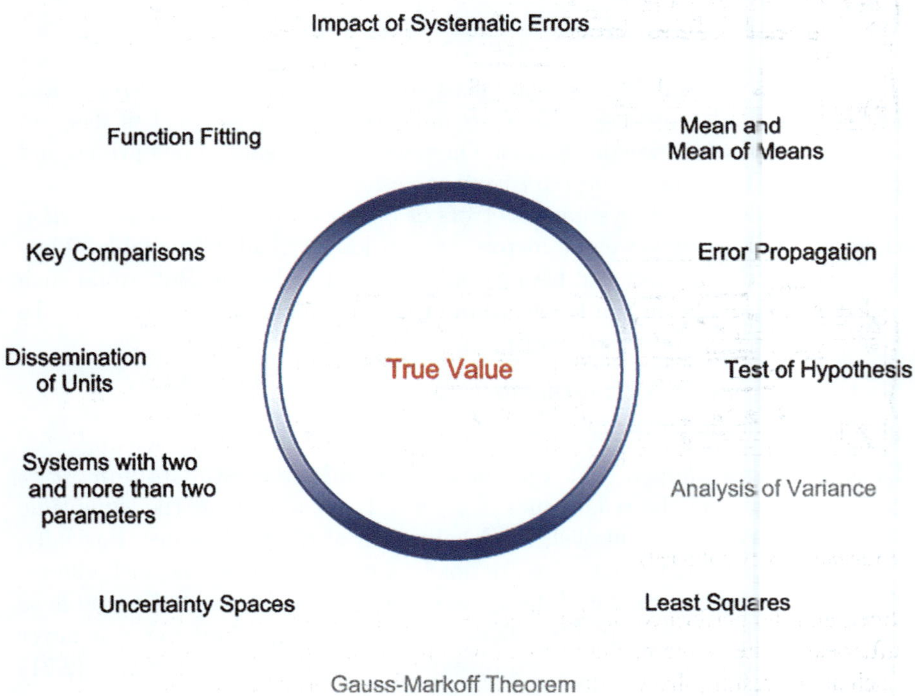

Fig. 2.4. Road map toward a new error calculus. The procedures printed in *dark* are to be revised, those in *light* become obsolete

variance are never less than zero. Unknown systematic errors, in contrast, by their very nature being either negative or positive, cannot be extracted from the input data and then somehow re-fed as positive variances. This is why empirical data burdened by unknown systematic errors may not be submitted to the analysis of variance. – Supposedly, this classical and sophisticated tool of data analysis happened to be conceived while unknown systematic errors lay hidden under a veil of oblivion.

2.7 Road Map

Biased empirical data, and as a rule we should expect empirical data to be biased, induce a breakdown of the majority of procedures of classical error calculus. While most of them appear to be reparable, at least two of them, the Gauss–Markoff theorem and the analysis of variance are doomed to failure. Certainly, both aspects, the repair of those which prove reparable and the final breakdown of those which are not, are expected to provoke extensive discussions, Fig. 2.4.

Part II

Generalized Gaussian Error Calculus

3 The New Uncertainties

The generalized Gaussian error calculus expresses the influence of random errors and unknown systematic errors by means of confidence intervals and worst-case estimations, respectively. These are the basic building blocks the new approach relies on.

3.1 Gaussian Versus Generalized Gaussian Approach

The Gaussian error calculus addresses the expected values of unbiased estimators, in the simplest case the expectation μ of the arithmetic mean $\bar{x}$. The Gaussian approach takes the expectations to be the true values of the measurands.

By contrast, the generalized Gaussian approach considers biased estimators thus introducing a distinction between expectations and true values:

Gaussian focus	–	expected values of unbiased estimators
Generalized Gaussian focus	–	true values of the measurands

3.2 Uncertainty and True Value

Let there be n repeated measurements of some measurand. The simplest least squares estimator of the true value x_0 is known to be the arithmetic mean

$$\bar{x} = \frac{1}{n}\sum_{l=1}^{n} x_l \,. \tag{3.1}$$

The presence of systematic errors causes the mean to be biased. To that effect, the uncertainty $u_{\bar{x}}$ should be apt to localize the true value x_0 of the measurand,

$$\bar{x} - u_{\bar{x}} \le x_0 \le \bar{x} + u_{\bar{x}} \,.$$

In general, arithmetic means aiming at one and the same physical quantity and coming from different laboratories will not coincide. This, per se,

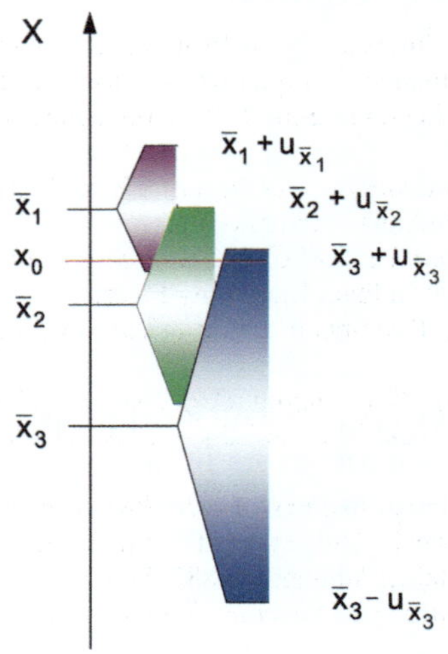

Fig. 3.1. Arithmetic means $\bar{x}_1, \bar{x}_2, \ldots$ aiming at the same physical quantity, say, x_0 and measured in different laboratories are not required to coincide. However, the intersection of their uncertainties should localize the true value of the measurand

is acceptable. However, we expect their uncertainties to overlap and their intersection to localize the true value of the measurand, Fig. 3.1.

3.3 Designing Uncertainties

From now on, whenever appropriate, we shall provide the systematic error f and the expected value μ with indices. For example, instead of (2.2) and (2.4) we shall write

$$x_l = x_0 + \varepsilon_l + f_x$$

and

$$\mu_x = x_0 + f_x \, ; \qquad -f_{s,x} \le f_x \le f_{s,x} \, ,$$

respectively. Given the measuring device works stationarily, the empirical variance

$$s_x^2 = \frac{1}{n-1} \sum_{l=1}^{n} (x_l - \bar{x})^2 \tag{3.2}$$

proves useful to quantify the scattering of random errors. Obviously, the difference

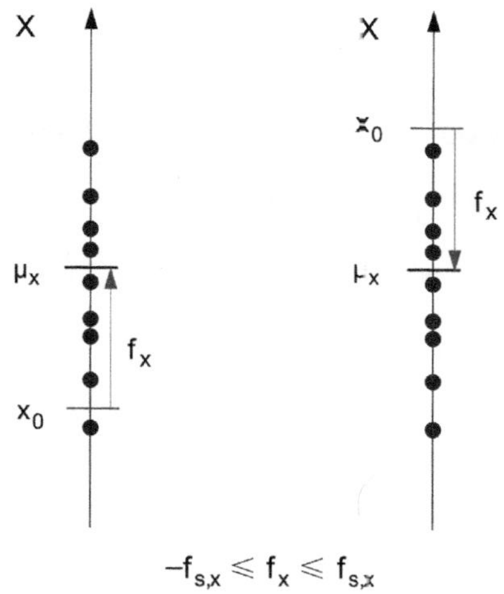

Fig. 3.2. The measured data do not scatter with respect to the true value x_0 of the measurand, they rather scatter throughout a neighborhood of the parameter $\mu_x = x_0 + f_x$, where f_x denotes the actual value of the systematic error. As this value is unknown with respect to magnitude and sign, the uncertainty has to make allowance for $x_0 \leq \mu_x$ as well as for $x_0 \geq \mu_x$

$$x_l - \bar{x} = \varepsilon_l - \frac{1}{n}\sum_{l=1}^{n} \epsilon_l \tag{3.3}$$

eliminates the bias f_x. As the bias is limited to an interval symmetric to zero [28]

$$-f_{s,x} \leq f_x \leq f_{s,x} \tag{3.4}$$

the localization of the true value x_0 appears plausible if the uncertainty $u_{\bar{x}}$ of the arithmetic mean $\bar{x}$ is the sum of the empirical standard deviation s_x and the worst case estimation $f_{s,x}$ of f_x following

$$u_{\bar{x}} = s_x + f_{s,x}\,, \tag{3.5}$$

Fig. 3.2. The idea of confidence intervals suggests to localize the position of the unknown expectation

$$E\{\bar{X}\} = \mu_x \tag{3.6}$$

via

$$\bar{x} - \frac{t_P(\nu)}{\sqrt{n}}\, s_x \leq \mu_x \leq \bar{x} + \frac{t_P(\nu)}{\sqrt{n}}\, s_x\,, \tag{3.7}$$

where $t_P(\nu)$ denotes the Student-factor of degrees of freedom $\nu = n - 1$.

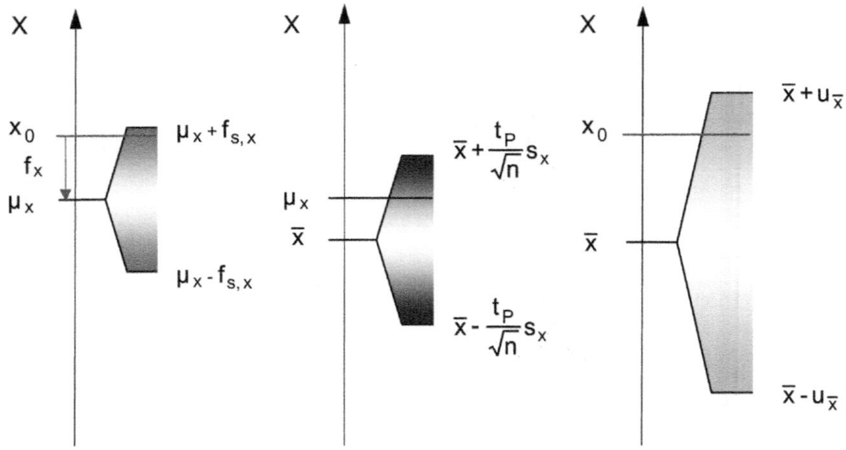

Fig. 3.3. The intervals $\mu_x \pm f_{s,x}$ and $\bar{x} \pm u_{\bar{x}}$ are meant to localize the true value x_0 of the measurand; no probability is given. The interval $\bar{x} \pm t_P/\sqrt{n}\ s_x$ is expected to localize the parameter μ_x with probability P

Indeed, we may expect this interval to localize the parameter μ_x with probability or at confidence level P. In general, we shall put $P = 95\%$. Then, instead of (3.5), the uncertainty interval takes the form

$$u_{\bar{x}} = \frac{t_P(\nu)}{\sqrt{n}}\, s_x + f_{s,x}\,. \tag{3.8}$$

After all, the measuring result

$$\bar{x} \pm u_{\bar{x}} \quad \text{or} \quad \bar{x} - u_{\bar{x}} \le x_0 \le \bar{x} + u_{\bar{x}} \tag{3.9}$$

is meant to localize the true value x_0, Fig. 3.3.

Often two or more than two means aiming at the same physical quantity are averaged to produce, say, a grand mean. Here, the argument is that any metrological information at hand should be rendered useful. On the other hand, averaging a given set of arithmetic means might seem a contradiction in terms: Why not take the mean the uncertainty of which is minimal and dismiss the others?

In a sense, this question aims at the heart of metrology: the true values of the measurands are unknown. Selecting the mean with the smallest uncertainty might be wrong and possibly even momentous. Ultimately, it is our ignorance about true values which suggests pooling procedures however different the uncertainties of the means might be. Pooling stands to reason at least as long as the respective uncertainties mutually overlap. On the other hand, pooling of means might lure us into trouble, if either

- one interval misses or even several intervals miss the common true value or
- the respective true values differ due to physical reasons.

In the latter case, the uncertainty of the grand mean would imply the static deviation between varying physical properties, say, inherently differing true values. Suchlike deviations, however, defy error calculus.

Pooling appears advisable if and only if the true values of the means coincide.

If metrology's dominant question aims at the localization of true values, how can we be sure that the quotation (3.9) localizes the measurand's true value x_0?

3.4 Quasi Safeness

While the confidence interval (3.7) proposes a probability, the quotation of the overall uncertainty (3.9) does not. Nevertheless, we might at least wish to qualify somehow the localization of the true value x_0.

In the first instance random errors, emerging from real experiments, appear to lack the outermost tails of the theoretical normal density $N(\mu_x, \sigma_x)$. This in turn means that we have hardly to expect (3.7) to miss the parameter μ_x. Furthermore, as we have brought to bear the unknown systematic error f_x via a worst case estimation, we may expect the interval (3.9) to localize the true value x_0 *quasi-safe*—given everything has gone well.

4 Treatment of Random Errors

The multidimensional normal model covers the behavior of a set of random variables related to different measuring devices. We shall, in particular, refer to the density of the empirical variances and empirical covariances, commonly summed up as empirical moments of second order.

4.1 Well-Defined Measuring Conditions

By approximation, we shall consider random errors normally distributed. This assumption suggests to deploy the multidimensional normal model. There is, however, a technical problem inasmuch as experimenters are used to deciding freely on the numbers of repeated measurements. By contrast, the multidimensional normal model premises equal numbers of repeated measurements.

Commonly, measuring results are not expected to depend significantly on the numbers of repeated measurements. Consider just one variable: Given n is not extremely small, then, whether there are n or $n+1$ repeated measurements will hardly induce a different scenario.[1] For two variables, say, X and Y with n_x and n_y repeated measurements, we may either have $n_x \neq n_y$ or $n_x = n_y$. Reducing the first case to the latter will barely issue a significantly different result, given n_x and n_y do not differ abnormally.

To emphasize, only $n_x = n_y$ clears the way for the multidimensional normal model and this is why we shall call $n_x = n_y$ *well-defined measuring conditions*.

Consider two series of repeated measurements

$$x_1, x_2, \ldots, x_n \quad \text{and} \quad y_1, y_2, \ldots, y_n \tag{4.1}$$

with arithmetic means

$$\bar{x} = \frac{1}{n}\sum_{l=1}^{n} x_l\,, \qquad \bar{y} = \frac{1}{n}\sum_{l=1}^{n} y_l \tag{4.2}$$

[1] On the other hand, for practical reasons, n cannot be made arbitrarily large as the experimental set up might sometime start to drift.

and moments of second order

$$s_x^2 = \frac{1}{n-1}\sum_{l=1}^{n}(x_l - \bar{x})^2\,, \quad s_y^2 = \frac{1}{n-1}\sum_{l=1}^{n}(y_l - \bar{y})^2$$
$$s_{xy} = \frac{1}{n-1}\sum_{l=1}^{n}(x_l - \bar{x})(y_l - \bar{y})\,. \tag{4.3}$$

The empirical variances and the empirical covariance are well to be distinguished from their theoretical counterparts, the theoretical variances, and the theoretical covariance. Denoting by S_x^2, S_y^2, S_{xy} random variables with realizations s_x^2, s_y^2, s_{xy}, we have

$$E\{S_x^2\} = \sigma_x^2\,, \quad E\{S_y^2\} = \sigma_y^2\,, \quad E\{S_{xy}\} = \sigma_{xy}\,.$$

4.2 Multidimensional Normal Model

The joint density of the random variables $\bar{X}, \bar{Y}, S_x^2, S_y^2, S_{xy}$ factorizes into a density p_1 of the arithmetic means and a density p_2 of the empirical moments of second order [26]

$$p(\bar{x}, \bar{y}, s_x^2, s_{xy}, s_y^2) = p_1(\bar{x}, \bar{y})\, p_2(s_x^2, s_{xy}, s_y^2)\,. \tag{4.4}$$

Remarkably enough, the density

$$p_2(s_x^2, s_{xy}, s_y^2) = \frac{(n-1)^{n-1}}{4\pi\Gamma(n-2)|\boldsymbol{\sigma}|^{(n-1)/2}}\left[s_x^2 s_y^2 - s_{xy}^2\right]^{(n-4)/2}$$
$$\times \exp\left(-\frac{n-1}{2|\boldsymbol{\sigma}|}h(s_x^2, s_{xy}, s_y^2)\right)\,; \tag{4.5}$$
$$h(s_x^2, s_{xy}, s_y^2) = \sigma_y^2 s_x^2 - 2\sigma_{xy}s_{xy} + \sigma_x^2 s_y^2$$

applies to dependent as well as to independent random variables $\bar{X}, \bar{Y}$. In case of independence, we have $\sigma_{xy} = 0$. Nevertheless, even then $p_2(s_x^2, s_{xy}, s_y^2)$ does not factorize.

Experimenters are used to answering the question of how to interpret an empirical covariance, given its expectation vanishes, by declaring it dispensable. On the other hand, this obviously conflicts with (4.5) which tells us that the empirical covariance S_{xy} is demonstrably an integral part of the density and that its neglect would mutilate the statistical coherence of the three random variables S_x^2, S_y^2, S_{xy}. In particular, there should exist a smooth transition from correlated to weakly correlated and finally to uncorrelated variables, implying $\sigma_{xy} \to 0$. Regardless of this, the empirical covariance s_{xy}, being sample dependent, would continue to take values out of the interval $-s_x s_y \le s_{xy} \le s_x s_y$. Ultimately, this is the message transferred by the density of the moments of second order.

To make the point as clear as possible: Given the variables considered are known to be independent, he who wants to exclude empirical covariances right from the outset is entitled to do so. As will be shown however, the formal consequences thereof will turn out unfavorable.

On the other hand, accepting the properties of the density of the moments of second order asks experimenters to always consider equal numbers of repeated measurements. Clearly, unequal numbers would prevent the formalizing of empirical covariances.

Concatenated measurands should always be subjected to the same number of repeated measurements thus allowing the formalization of empirical covariances.

4.3 Permutation of Repeated Measurements

After all, there is still a point of concern. Given the random variables are known to be independent, the sequences of the x_i and y_i appear permutable, each permutation producing a different empirical covariance. But which one is valid?

Obviously, under the premise of independence, each permutation is equally well admissible. As an example, consider the sequences

$$x_1,\ x_2,\ x_3,\ x_4,\ x_5,\ x_6,\ x_7,\ x_8,\ x_9,\ x_{10}$$
$$y_1,\ y_2,\ y_3,\ y_4,\ y_5,\ y_6,\ y_7,\ y_8,\ y_9,\ y_{10}$$

and

$$x_1,\ x_2,\ x_3,\ x_4,\ x_5,\ x_6,\ x_7,\ x_8,\ x_9,\ x_{10}$$
$$y_7,\ y_4,\ y_6,\ y_8,\ y_2,\ y_{10},\ y_3,\ y_5,\ y_1,\ y_9\ .$$

While permutations do not affect the empirical variances s_x^2, s_y^2, the empirical covariance s_{xy} will in general take different values. But this is quite in order and does in no way affect the consequences of using (4.5). Rather, the ambiguousness of the empirical covariance in case of independent random variables reflects the basic operating principle of the density per se.

Considering empirical covariances will be shown to generalize Student's confidence intervals to multivariate error propagation. Both in the one-dimensional and the multidimensional case, the length of confidence intervals are sample dependent.

5 Treatment of Systematic Errors

Unknown systematic errors are proposed to be propagated via worst-case estimations. This proceeding expresses our state of knowledge and meets the concerns of metrology.

5.1 Repercussion of Biases

Averages are seen to reduce the influence of measuring errors. In the first place, averages shall abate the fluctuations of random errors. The question as to how averages affect time-constant systematic errors depends on the particular situation. Simple arithmetic means leave them unaffected, within least squares estimators, however, they get reduced.

Let us reconsider the arithmetic mean (3.1). Inserting the error equation (2.2), we recognize the impact of the bias f_x,

$$\bar{x} = x_0 + \frac{1}{n}\sum_{l=1}^{n} \epsilon_l + f_x \,. \tag{5.1}$$

The arithmetic mean $\bar{x}$ is biased with respect to the true value x_0. This observation insistently asks us to shape the measuring uncertainty appropriately.

In Sect. 6.3 we shall consider a weighted mean $\bar{\beta}$ of some individual means $\bar{x}_i\,;\; i = 1, \ldots, m$,

$$\bar{\beta} = \sum_{i=1}^{m} w_i\,\bar{x}_i\,, \qquad \sum_{i=1}^{m} w_i = 1\,,$$

the $w_i > 0$ denoting so-called weight factors. The systematic errors f_i of the individual means, confined to intervals

$$-f_{s,i} \le f_i \le f_{s,i}\,; \qquad i = 1, \ldots, m\,,$$

issue an uncertainty component

$$f_{s,\bar{\beta}} = \sum_{i=1}^{m} w_i f_{s,i}\,. \tag{5.2}$$

Indeed, $f_{s,\bar{\beta}}$ will hardly overestimate the influence of systematic errors. Just to have an example, let us put $m = 3$ and $w_1 = w_2 = w_3$. This issues

$$f_{s,\bar{\beta}} = (f_{s,1} + f_{s,2} + f_{s,3})/3\,. \tag{5.3}$$

Metrology's Achilles' heel anchors in the traceability of units, calibration chains, individual standards, physical constants and measures in a wider sense. Hence, we emphasize: The attempt to implement traceability should consider biases.

5.2 Uniqueness of Worst-Case Assessments

Worst-case estimations provide a natural basis to assess unknown systematic errors or biases. However, the proceeding should be handled with care as we strive for unique quotations. This is why we should never propagate worst-case estimations themselves since this might produce ambiguous results. To assess the overall influence of a whole lot of biases, the triangle inequality is the appropriate tool. However, prior to its becoming operative, those biases which occur more than once should be factored out.

Not until then should the triangle inequality be applied; this observation strictly excludes the propagation of worst-case estimations. But this is the only precaution we have to consider in order to ensure unique and meaningful measurement uncertainties.

Part III

Error Propagation

6 Means and Means of Means

We shall confine ourselves to unweighted and weighted arithmetic means as issued by the method of least squares.

6.1 Arithmetic Mean

Let us consider n independent repeated measurements

$$x_1, x_2, \ldots, x_n$$

with formal decompositions

$$x_l = x_0 + (x_l - \mu_x) + f_x\,, \quad -f_{s,x} \le f_x \le f_{s,x}\,; \qquad l = 1, \ldots, n\,.$$

The measurements produce a mean

$$\bar{x} = \frac{1}{n}\sum_{l=1}^{n} x_l \tag{6.1}$$

and an empirical standard deviation

$$s_x^2 = \frac{1}{n-1}\sum_{l=1}^{n}(x_l - \bar{x})^2\,. \tag{6.2}$$

As shown in Appendix H, Student's t exists in two versions

$$T(n-1) = \frac{X - \mu_x}{S_x} \quad \text{and} \quad T(n-1) = \frac{\bar{X} - \mu_x}{S_x/\sqrt{n}}\,.$$

Correspondingly, the inequality

$$-t_P \le T \le t_P$$

issues confidence intervals

$$x_l - t_P(n-1)\, s_x \le \mu_x \le x_l + t_P(n-1)\, s_x\,; \quad l = 1, \ldots, n \tag{6.3}$$

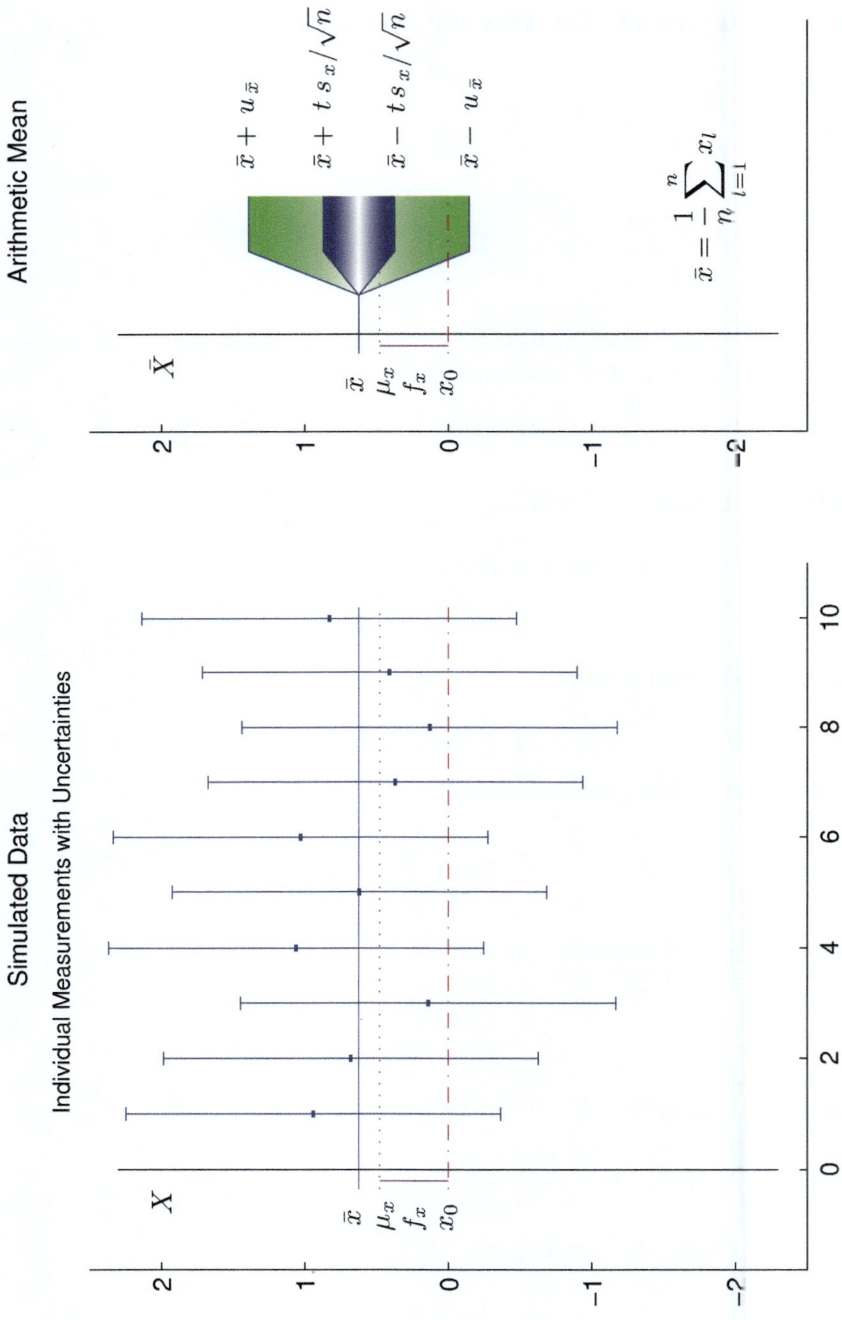

Fig. 6.1. Biased arithmetic mean $\bar{x}$ with uncertainty $u_{\bar{x}}$

and

$$\bar{x} - \frac{t_P(n-1)}{\sqrt{n}}\, s_x \le \mu_x \le \bar{x} + \frac{t_P(n-1)}{\sqrt{n}}\, s_x\,. \tag{6.4}$$

In equal measure, both inequalities localize the expected value μ_x, the first with respect to the n input data and the second with respect to the arithmetic mean. To localize the true value x_0 we may either refer to

$$x_l \pm u_x\,, \quad u_x = t_P(n-1)\, s_x + f_{s,x}\,; \quad l = 1, \ldots, n \tag{6.5}$$

or

$$\bar{x} \pm u_{\bar{x}}\,, \quad u_{\bar{x}} = \frac{t_P(n-1)}{\sqrt{n}}\, s_x + f_{s,x}\,. \tag{6.6}$$

Data Simulation

Figure 6.1 depicts $n = 10$ repeated measurements with $f_x = \text{const}$. We observe the biased arithmetic mean, the confidence interval expressing the influence of random errors and localizing the expectation μ_x and, finally, the uncertainty $u_{\bar{x}}$ of the mean $\bar{x}$ encompassing the random errors and the unknown systematic error, likewise superimposed to every x_l. To recall, the true value x_0 is only known due to data simulation—in reality, it remains hidden.

6.2 Extravagated Averages

Means to be averaged should relate to identically true values. Disregarding this pre-condition means to attempt to adjust an inconsistent linear system. This, in fact, violates the basic idea of the method of least squares.

Measurement uncertainties are thought to express measurement errors, but are hardly apt to quantify static disparities between physically varying constants.

6.3 Mean of Means

The pooling of a set of means is directed by the question of whether the individuals carry equal weight.

Unweighted Grand Mean

Given m arithmetic means and uncertainties

$$\bar{x}_i = \frac{1}{n}\sum_{l=1}^{n} x_{il}\,, \quad u_{\bar{x}_i} = \frac{t_P\,(n-1)}{\sqrt{n}}\,s_i + f_{s,i}\,; \quad i = 1\,,\ldots\,,\,m\,.$$

We assume the $\bar{x}_i \pm u_{\bar{x}_i}$ to localize the common true value $x_0 = x_{0,1} = \ldots = x_{0,m}$. It stands to reason to ask for a grand mean $\bar{\beta}$ and its uncertainty $u_{\bar{\beta}}$.

To this end we submit the inconsistent linear system

$$\begin{pmatrix} 1 \\ 1 \\ \ldots \\ 1 \end{pmatrix} \beta \approx \begin{pmatrix} \bar{x}_1 \\ \bar{x}_2 \\ \ldots \\ \bar{x}_m \end{pmatrix}, \qquad \boldsymbol{a}\beta \approx \bar{\boldsymbol{x}} \tag{6.7}$$

to least squares letting

$$\boldsymbol{a} = (1 \quad 1 \quad \ldots \quad 1)^T$$

denote an auxiliary vector. Multiplying (6.7) on the left by a^T produces the grand mean

$$\bar{\beta} = \frac{1}{m}\sum_{i=1}^{m} \bar{x}_i\,. \tag{6.8}$$

The itemization

$$\begin{array}{ccccccc} x_{11},\; x_{12}, & \ldots\,, x_{1l} & \ldots\,, x_{1n} & \Rightarrow & \bar{x}_1 & = & \sum_{l=1}^{n} x_{1l} \\ & \downarrow & & & & & \\ x_{21},\; x_{22}, & \ldots\,, x_{2l} & \ldots\,, x_{2n} & \Rightarrow & \bar{x}_2 & = & \sum_{l=1}^{n} x_{2l} \\ & \downarrow & & & & & \\ \ldots\;\ldots & \ldots\;\ldots & \ldots\;\ldots & \ldots & & & \ldots \\ & \downarrow & & & & & \\ x_{m1},\, x_{m2}, & \ldots\,, x_{ml} & \ldots\,, x_{mn} & \Rightarrow & \bar{x}_m & = & \sum_{l=1}^{n} x_{ml} \end{array} \tag{6.9}$$

suggests to introduce n grand means

$$\bar{\beta}_l = \frac{1}{m}\sum_{i=1}^{m} x_{il}\,; \qquad l = 1,\ldots,n \tag{6.10}$$

based on the respective l-th measurements. Of course, the $\bar{\beta}_l$ lead us back to

$$\bar{\beta} = \frac{1}{n}\sum_{l=1}^{n}\bar{\beta}_l \,. \tag{6.11}$$

The error equations

$$x_{il} = x_{0,i} + (x_{il} - \mu_i) + f_i\,, \qquad \bar{x}_i = x_{0,i} + (\bar{x}_i - \mu_i) + f_i$$

produce

$$\bar{\beta}_l = \frac{1}{m}\sum_{i=1}^{m}\left[\, x_{0,i} + (x_{il} - \mu_i) + f_i \,\right], \qquad \bar{\beta} = \frac{1}{m}\sum_{i=1}^{m}\left[\, x_{0,i} + (\bar{x}_i - \mu_i) + f_i \,\right].$$

Subtraction yields the n differences

$$\bar{\beta}_l - \bar{\beta} = \frac{1}{m}\sum_{i=1}^{m}(x_{il} - \bar{x}_i)\,; \qquad l = 1,\ldots,n\,. \tag{6.12}$$

From this we take the empirical variance of the $\bar{\beta}_l$ with respect to the grand means $\bar{\beta}$,

$$\begin{aligned} s_{\bar{\beta}}^2 &= \frac{1}{n-1}\sum_{l=1}^{n}\left(\bar{\beta}_l - \bar{\beta}\right)^2 \\ &= \frac{1}{m^2(n-1)}\sum_{l=1}^{n}\left[\sum_{i=1}^{m}(x_{il} - \bar{x}_i)\right]\left[\sum_{j=1}^{m}(x_{jl} - \bar{x}_j)\right] \\ &= \frac{1}{m^2}\sum_{i,j}^{m} s_{ij} \end{aligned} \tag{6.13}$$

in which the

$$s_{ij} = \frac{1}{n-1}\sum_{l=1}^{n}(x_{il} - \bar{x}_i)(x_{jl} - \bar{x}_j)\,; \quad i,j = 1,\ldots,m\,; \qquad s_{ii} \equiv s_i^2$$

denote the empirical variances and covariances of the input data. After all, putting

$$\boldsymbol{s} = \begin{pmatrix} s_{11} & s_{12} & \ldots & s_{1m} \\ s_{21} & s_{22} & \ldots & s_{2m} \\ \ldots & \ldots & \ldots & \ldots \\ s_{m1} & s_{m2} & \ldots & s_{mm} \end{pmatrix}, \qquad s_{ii} \equiv s_i^2 \tag{6.14}$$

relation (6.13) takes the form

$$s_{\bar{\beta}}^2 = \frac{1}{m^2}\,\boldsymbol{a}^{\mathrm{T}}\boldsymbol{s}\,\boldsymbol{a}\,. \tag{6.15}$$

The expected value

$$E\{\bar{\beta}\} = \mu_{\bar{\beta}} = x_0 + \frac{1}{m}\sum_{i=1}^{m} f_i \tag{6.16}$$

issues the propagated systematic error

$$f_{\bar{\beta}} = \frac{1}{m}\sum_{i=1}^{m} f_i \tag{6.17}$$

with worst-case estimation

$$f_{s,\bar{\beta}} = \frac{1}{m}\sum_{i=1}^{m} f_{s,i}\,. \tag{6.18}$$

This relation, averaging systematic errors, affirms that an increasing number m of means does not entail a surge of the propagated systematic error $f_{s,\bar{\beta}}$ as discussed in (5.3).

The confidence interval

$$\bar{\beta} - \frac{t_P(n-1)}{\sqrt{n}}\,s_{\bar{\beta}} \;\le\; \mu_{\bar{\beta}} \;\le\; \bar{\beta} + \frac{t_P(n-1)}{\sqrt{n}}\,s_{\bar{\beta}} \tag{6.19}$$

localizes the expected value $\mu_{\bar{\beta}}$ with probability P. Hence, the overall uncertainty of the grand mean $\bar{\beta}$ turns out to be

$$u_{\bar{\beta}} = \frac{t_P(n-1)}{\sqrt{n}}\,s_{\bar{\beta}} + f_{s,\bar{\beta}}\,. \tag{6.20}$$

Weighted Grand Mean

To find a weighted grand mean, we left multiply (6.7) by a weight matrix. We choose suchlike matrices to be diagonal

$$\boldsymbol{G} = \operatorname{diag}\left\{\, g_1,\, g_2,\, \ldots\,,\, g_m \,\right\}\,; \quad g_i = 1/u_{\bar{x}_i}\,. \tag{6.21}$$

Submitting the weighted system

$$\boldsymbol{G}\boldsymbol{a}\beta \approx \boldsymbol{G}\bar{\boldsymbol{x}}. \tag{6.22}$$

to least squares yields

$$\bar{\beta} = \sum_{i=1}^{m} w_i \bar{x}_i. \tag{6.23}$$

For convenience, we gather the weights

$$w_i = \frac{g_i^2}{\sum\limits_{i=1}^{m} g_i^2}\,, \qquad \sum_{i=1}^{m} w_i = 1$$

within an auxiliary vector

$$\boldsymbol{w} = (w_1 \quad w_2 \quad \ldots \quad w_m)^T .$$

This suggests to formalize the empirical variance according to

$$s_{\bar\beta}^2 = \boldsymbol{w}^{\mathrm{T}} \boldsymbol{s}\, \boldsymbol{w}\,. \tag{6.24}$$

Finally, we consider the worst-case estimation of the propagated systematic error

$$f_{s,\bar\beta} = \sum_{i=1}^{m} w_i f_{s,i}. \tag{6.25}$$

Hence, the overall uncertainty of the weighted grand mean is given by

$$u_{\bar\beta} = \frac{t_P(n-1)}{\sqrt{n}}\, s_{\bar\beta} + f_{s,\bar\beta}\,, \quad \bar\beta - u_{\bar\beta} \le x_0 \le \bar\beta + u_{\bar\beta}\,. \tag{6.26}$$

As is known, biases invalidate the Gauss–Markov theorem. This is why a unique weight matrix does not exist.[1] The stipulation $g_i = 1/u_{\bar{x}_i}$ may be considered a start and subsequently be varied by trial and error. As (6.26) suggests, the experimenter is entitled to this proceeding as any weight matrix safeguards the localization of the true value x_0—be the final uncertainty favorable or not.

Ultimately, it is up to the experimenter to decide which measurements to boost and which to belittle. The more weight he assigns to the mean with the smallest uncertainty, the more will the grand mean's uncertainty resemble this very uncertainty. Thus, the uncertainty of the grand mean is free to float between the smallest and the largest uncertainty of the set of means considered. Whatever choice is taken, the final uncertainty will localize the true value x_0—given this has applied to each of the results $\bar{x}_i \pm u_{\bar{x}_i}\,;\; i = 1, \ldots, m$.

To experience the weighting property of (6.25), let us add another example and put $u_{\bar{x}_1} = a$, $u_{\bar{x}_2} = 2a$, $u_{\bar{x}_3} = 3a$ so that $w_1 = 36/49$, $w_2 = 9/49$, $w_1 = 4/49$. Hence,

[1] Even if the classical error calculus were applicable, the ideal Gauss–Markov weight matrix did not exist as the theorem refers to theoretical variances and theoretical covariances which, at least in metrology, prove inaccessible. Strictly speaking, the theorem was applicable at no time.

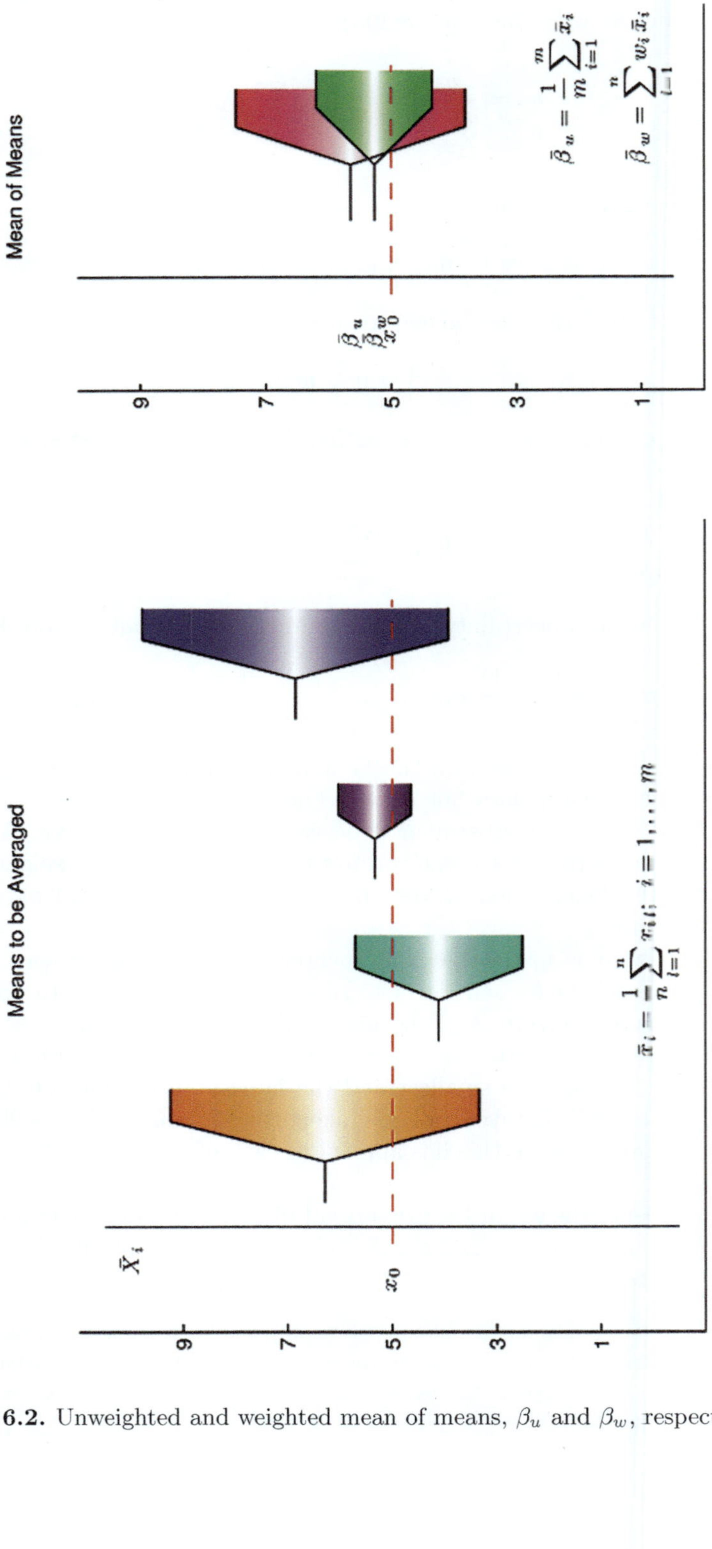

Fig. 6.2. Unweighted and weighted mean of means, β_u and β_w, respectively

$$f_{s,\bar{\beta}} = (36 f_{s,1} + 9 f_{s,2} + 4 f_{s,3}) / 49 \,.$$

The example discloses the weight factors $w_i;\ i = 1, \ldots, m$ to drive the propagated systematic error $f_{s,\bar{\beta}}$ into a reasonable order of magnitude however large m may be.

Data Simulation

Figure 6.2 pools $m = 4$ arithmetic means under the premise of a given common true value x_0. We observe the unweighted and the weighted grand mean, $\bar{\beta}_u$ and $\bar{\beta}_w$, respectively.

6.4 Individual Mean Versus Grand Mean

Let us elaborate the uncertainties $u_{\bar{d}_i}$ of the differences

$$\bar{d}_i = \bar{x}_i - \bar{\beta}\,; \quad i = 1, \ldots, m \tag{6.27}$$

between any of the individual means $\bar{x}_i;\ i = 1, \ldots, m$ and the grand mean $\bar{\beta}$. Denoting by $d_{0,i}$ the true values of the $\bar{d}_i$, we are looking for

$$|\, \bar{d}_i - d_{0,i} \,| \le u_{\bar{d}_i}\,; \quad i = 1, \ldots, m \,.$$

As the $d_{0,i}$ vanish, we should observe

$$|\, \bar{x}_i - \bar{\beta} \,| \le u_{\bar{d}_i}; \quad i = 1, \ldots, m \,. \tag{6.28}$$

Inserting the error equations

$$\bar{x}_i = x_{0,i} + (\bar{x}_i - \mu_i) + f_i\,; \quad i = 1, \ldots, m$$

into (6.27) produces

$$\bar{d}_i = x_{0,i} + (\bar{x}_i - \mu_i) + f_i - \sum_{j=1}^{m} w_j \left[x_{0,j} + (\bar{x}_j - \mu_j) + f_j \right] \tag{6.29}$$

which firstly reissues

$$d_{i,0} = 0\,; \quad i = 1, \ldots, m \,. \tag{6.30}$$

Reverting to individual measurements x_{il} we find

$$\bar{d}_i = \frac{1}{n} \sum_{l=1}^{n} \left[(x_{il} - \mu_i) - \sum_{j=1}^{m} w_j (x_{jl} - \mu_j) \right] + f_i - \sum_{j=1}^{m} w_j f_j$$

so that

$$\bar{d}_{il} = (x_{il} - \mu_i) - \sum_{j=1}^{m} w_j (x_{jl} - \mu_j) + f_i - \sum_{j=1}^{m} w_j f_j \,; \quad l = 1, \ldots\ n \tag{6.31}$$

and

$$\bar{d}_i = \frac{1}{n} \sum_{l=1}^{n} \bar{d}_{il} \,. \tag{6.32}$$

The differences

$$\bar{d}_{il} - \bar{d}_i = (x_{il} - \bar{x}_i) - \sum_{j=1}^{m} w_j (x_{jl} - \bar{x}_j)$$

suggest the empirical variance

$$s_{\bar{d}_{il}}^2 = \frac{1}{n-1} \sum_{l=1}^{n} (\bar{d}_{il} - \bar{d}_i)^2$$

$$= \frac{1}{n-1} \sum_{l=1}^{n} \left[(x_{il} - \bar{x}_i)^2 - 2(x_{il} - \bar{x}_i) \sum_{j=1}^{m} w_j (x_{jl} - \bar{x}_j) + \left(\sum_{j=1}^{m} w_j (x_{jl} - \bar{x}_j) \right)^2 \right]$$

which we turn into

$$s_{\bar{d}_{il}}^2 = s_i^2 - 2 \sum_{j=1}^{m} w_j \, s_{ij} + \boldsymbol{w}^{\mathrm{T}} \boldsymbol{s} \, \boldsymbol{w} \,. \tag{6.33}$$

Relation (6.29) issues the propagated systematic error

$$f_{\bar{d}_i} = (1 - w_i) f_i \ - \sum_{j=1\ j \neq i}^{m} w_j f_j \tag{6.34}$$

with worst-case estimation

$$f_{s,\bar{d}_i} = (1 - w_i) f_{s,i} \ + \sum_{j=1\ j \neq i}^{m} w_j f_{s,j} \,. \tag{6.35}$$

Here, we refer to Sect. 5.2 where we addressed the need to keep worst-case assessments unique, i.e. to factor out those biases which occur more than once.

Adding $\pm w_i f_{s,i}$ on the right-hand side of (6.35) we arrive at

$$f_{s,\bar{d}_i} = (1 - 2w_i) f_{s,i} \ + \sum_{j=1}^{m} w_j f_{s,j} \tag{6.36}$$

Hence, the uncertainty of the difference (6.27) takes the form

$$u_{\bar{d}_i} = \frac{t_P(n-1)}{\sqrt{n}} \sqrt{s_i^2 - 2\sum_{j=1}^{m} w_j\, s_{ij} + \boldsymbol{w}^{\mathrm{T}}\boldsymbol{s}\,\boldsymbol{w}} \tag{6.37}$$

$$+(1-2w_i)f_{s,i} + \sum_{j=1}^{m} w_j f_{s,j}$$

Data Simulation

To recall, the common true value x_0 is unknown. Data simulation, however, gives us the chance to preset the true value and thus to scrutinize the properties of uncertainties.

Let us consider a set of means $\bar{x}_i \pm u_{\bar{x}_i};\ i = 1,\dots,m$ and assume they localize a common true value x_0. We firstly compare the means $\bar{x}_i;\ i = 1,\dots,m$ directly. The results are shown in the upper diagrams of Fig. 6.3. We then address (6.28), putting however

$$\bar{x}_i - u_{\bar{d}_i} \le \bar{\beta} \le \bar{x}_i + u_{\bar{d}_i};\quad i = 1,\dots,m\,.$$

The lower diagrams confront the $\bar{x}_i \pm u_{\bar{d}_i};\ i = 1,\dots,m$ with $\bar{\beta}$.

Though the grand mean promises more stability than the individual means, the uncertainties $u_{\bar{d}_i}$ prove intricate and somewhat sensitive constructions as they are interdependent. For an illustration, let us consider two further examples, Figs. 6.4 and 6.5.

Even if the results $\bar{x}_i \pm u_{\bar{x}_i};\ i = 1,\dots,m$ mutually overlap, it may happen that one mean or even more means abstain from localizing the true value x_0. As an example, we let $\bar{x}_i \pm u_{\bar{x}_i};\ i = 2,3$ fail to localize x_0. Clearly, the experimenter himself could not notice that and if he resorted to the grand mean, in order to dispel pending doubts, he unfortunately might not detect the given irregularities, Fig. 6.4.

Ultimately, trial and error leads us to Fig. 6.5 in which the same two means fail to localize x_0. But as further data have been altered, now two of the $u_{\bar{d}_i}$ point to irregularities. Remarkably enough, while $\bar{x}_i \pm u_{\bar{x}_i};\ i = 2,3$ fail to localize x_0, $u_{\bar{d}_i};\ i = 1,5$ suggest a review.

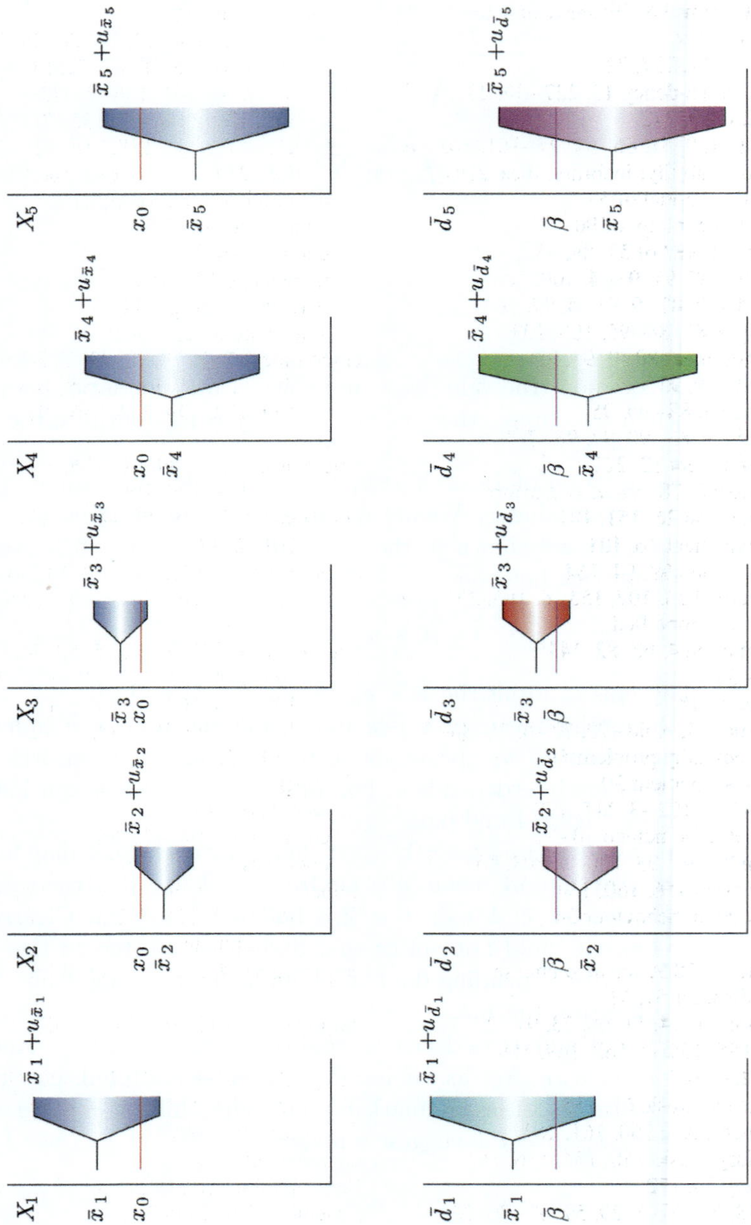

Fig. 6.3. Check of consistency of results $\bar{x}_i \pm u_{\bar{x}_i}$; $i = 1, \ldots, 5$; x_0 common true value; $\bar{\beta}$ grand mean, $u_{\bar{d}_i}$ uncertainty of $\bar{x}_i - \beta$

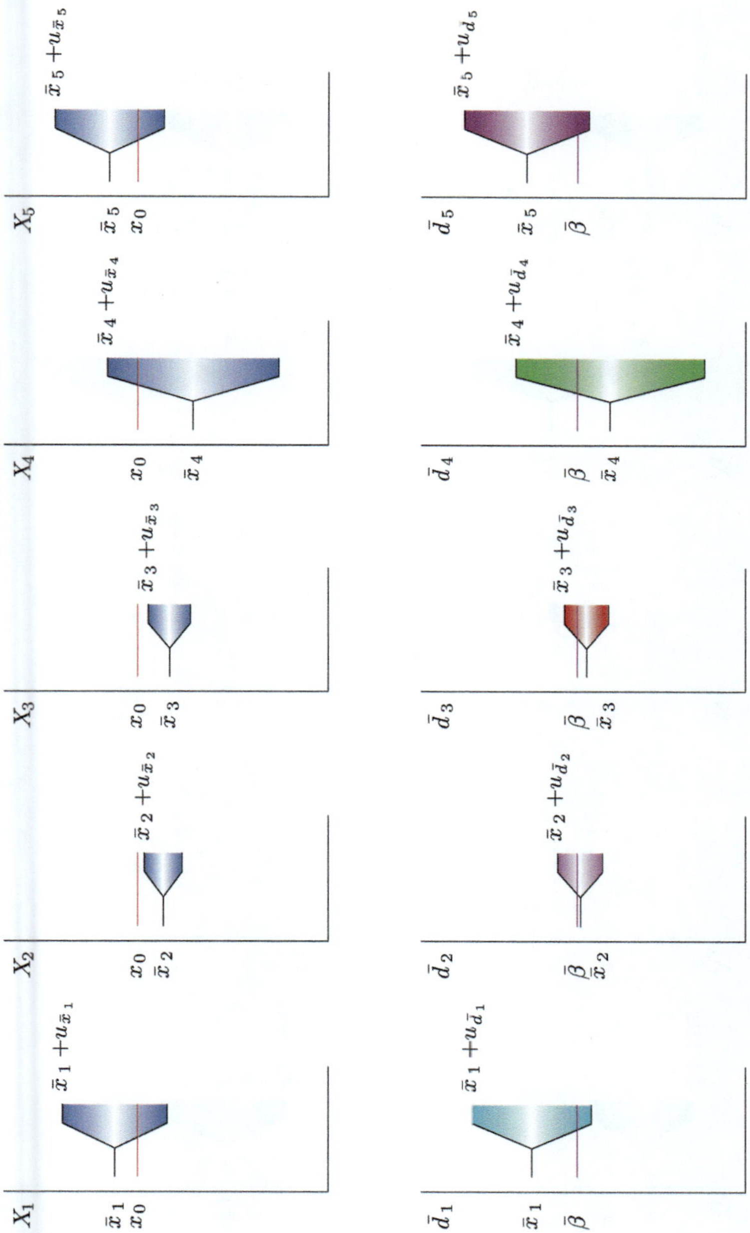

Fig. 6.4. Though $\bar{x}_i \pm u_{\bar{x}_i}$; $i = 2, 3$ fail to localize x_0, none of the uncertainties $u_{\bar{d}_i}$; $i = 1, \ldots, 5$ beckons inconsistency

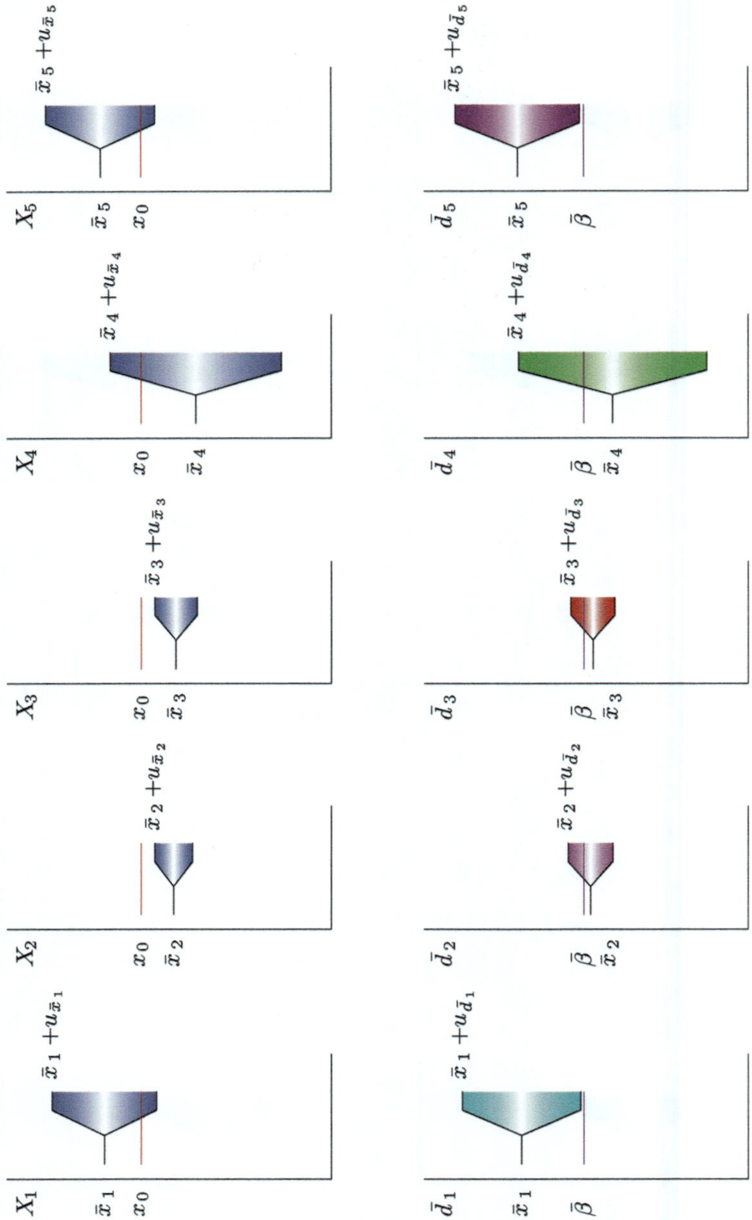

Fig. 6.5. Under additional modifications, the same means $\bar{x}_i \pm u_{\bar{x}_i}$ $i = 2, 3$ continue to fail to localize x_0. Now, however, the uncertainties $u_{\bar{d}_1}$ and $u_{\bar{d}_5}$ beckon inconsistency

7 Functions of Erroneous Variables

To assess the influence of measurement errors propagated via functions, we confine ourselves to linearized series expansions. From there, we expect the functions to behave sufficiently smooth throughout a neighborhood of the expansion point. Also, the measuring errors themselves should not exceed reasonable sizes.

7.1 One Variable

Consider a measuring result

$$\bar{x} \pm u_{\bar{x}}\,, \qquad u_{\bar{x}} = \frac{t_P(n-1)}{\sqrt{n}}\, s_x + f_{s,x} \tag{7.1}$$

and a function $\Phi(x)$. The uncertainty $u_{\bar{\Phi}}$ of $\Phi(\bar{x})$ is obviously given by

$$u_{\Phi} = \left| \frac{\mathrm{d}\Phi}{\mathrm{d}\bar{x}} \right| u_{\bar{x}}\,.$$

Let us take this result to explore the toolkit of the new error calculus.

Series Expansions

In terms of the error equations

$$x_l = x_0 + (x_l - \mu_x) + f_x\,; \qquad l = 1, \ldots, n$$

$$\bar{x} = x_0 + (\bar{x} - \mu_x) + f_x\,; \quad -f_{s,x} \le f_x \le f_{s,x}$$

we expand $\Phi(x)$ throughout a neighborhood of the true value x_0, firstly with respect to the n points $x_1, x_2, \ldots, x_n$

$$\Phi(x_l) = \Phi(x_0) + \frac{\mathrm{d}\Phi}{\mathrm{d}x_0}(x_l - \mu_x) + \frac{\mathrm{d}\Phi}{\mathrm{d}x_0} f_x + \cdots\,; \quad l = 1, 2, \cdots, n$$

and secondly with respect to the particular realization $\bar{x}$ of the notionally defined random variable $\bar{X}$

$$\Phi(\bar{x}) = \Phi(x_0) + \frac{\mathrm{d}\Phi}{\mathrm{d}x_0}\,(\bar{x} - \mu_x) + \frac{\mathrm{d}\Phi}{\mathrm{d}x_0} f_x + \cdots$$

By approximation, we assume that

- the expansions may be linearized,
- a derivative $\mathrm{d}\Phi/\mathrm{d}\bar{x}$ may be substituted for the derivative $\mathrm{d}\Phi/\mathrm{d}x_0$,
- $\mathrm{d}\Phi/\mathrm{d}\bar{x}$ may be considered constant.

Just to simplify the notation, we additionally provide the truncated expansions with equality signs which is, of course, incorrect. Thus we have

$$\Phi(x_l) = \Phi(x_0) + \frac{\mathrm{d}\Phi}{\mathrm{d}\bar{x}}(x_l - \mu_x) + \frac{\mathrm{d}\Phi}{\mathrm{d}\bar{x}} f_x\,; \qquad l = 1, 2, \ldots, n \tag{7.2}$$

and

$$\Phi(\bar{x}) = \Phi(x_0) + \frac{\mathrm{d}\Phi}{\mathrm{d}\bar{x}}\,(\bar{x} - \mu_x) + \frac{\mathrm{d}\Phi}{\mathrm{d}\bar{x}} f_x\,. \tag{7.3}$$

Subtracting (7.3) from (7.2) produces

$$\Phi(x_l) - \Phi(\bar{x}) = \frac{\mathrm{d}\Phi}{\mathrm{d}\bar{x}}(x_l - \bar{x})\,. \tag{7.4}$$

Summing this over l we find

$$\Phi(\bar{x}) = \frac{1}{n}\sum_{l=1}^{n} \Phi(x_l)\,. \tag{7.5}$$

Random Errors

Relation (7.4) provides the empirical variance of the $\Phi(x_l)$ with respect to the mean $\Phi(\bar{x})$

$$s_\Phi^2 = \frac{1}{n-1}\sum_{l=1}^{n}\left[\,\Phi(x_l) - \Phi(\bar{x})\,\right]^2 = \left(\frac{\mathrm{d}\Phi}{\mathrm{d}\bar{x}}\right)^2 s_x^2\,. \tag{7.6}$$

Systematic Error

Relation (7.3) issues the propagated systematic error

$$f_\Phi = \frac{\mathrm{d}\Phi}{\mathrm{d}\bar{x}} f_x\,, \qquad -f_{s,x} \le f_x \le f_{s,x}\,. \tag{7.7}$$

Confidence Interval and Overall Uncertainty

Let us formally associate random variables X and S_Φ with the measured data $x_1, x_2, \ldots, x_n$ and the empirical variance s_Φ.

Then, the expected value of (7.3) takes the form

$$\mu_\Phi = E\left\{\Phi(\bar{X})\right\} = \Phi(x_0) + \frac{\mathrm{d}\Phi}{\mathrm{d}\bar{x}} f_x \,. \tag{7.8}$$

Given the x_l are independent and normally distributed, this also applies to the $\Phi(x_l)$; $l = 1, \ldots, n$ as given in (7.2). Then, in view of (7.5), we are in a position to define Student's

$$T(n-1) = \frac{\Phi(\bar{X}) - \mu_\Phi}{S_\Phi/\sqrt{n}} \,. \tag{7.9}$$

Thus, we may expect the confidence interval

$$\Phi(\bar{X}) - \frac{t_P(n-1)}{\sqrt{n}} S_\Phi \le \mu_\Phi \le \Phi(\bar{X}) + \frac{t_P(n-1)}{\sqrt{n}} S_\Phi \tag{7.10}$$

to localize the unknown parameter μ_Φ with probability P. A realization of (7.10) is, of course,

$$\Phi(\bar{x}) - \frac{t_P(n-1)}{\sqrt{n}} s_\Phi \le \mu_\Phi \le \Phi(\bar{x}) + \frac{t_P(n-1)}{\sqrt{n}} s_\Phi \,.$$

Still, we have to assess (7.7). As we stick to worst-case estimations we find

$$f_{s,\Phi} = \left|\frac{\mathrm{d}\Phi}{\mathrm{d}\bar{x}}\right| f_{s,x} \,, \quad -f_{s,\Phi} \le f_\Phi \le f_{s,\Phi} \,. \tag{7.11}$$

Hence, the overall uncertainty is given by

$$u_{\bar{\Phi}} = \left|\frac{\mathrm{d}\Phi}{\mathrm{d}\bar{x}}\right| \left[\frac{t_P(n-1)}{\sqrt{n}} s_x + f_{s,x}\right] = \left|\frac{\mathrm{d}\Phi}{\mathrm{d}\bar{x}}\right| u_{\bar{x}} \,. \tag{7.12}$$

The quantity $u_{\bar{x}}$ obviously matches (7.1). The final measuring result reads

$$\Phi(\bar{x}) \pm u_{\bar{\Phi}} \tag{7.13}$$

meaning

$$\Phi(\bar{x}) - u_{\bar{\Phi}} \le \Phi(x_0) \le \Phi(\bar{x}) + u_{\bar{\Phi}} \,.$$

The magnitude of the derivative

$$\left|\frac{\mathrm{d}\Phi}{\mathrm{d}\bar{x}}\right|$$

has modified the magnitude of the initial uncertainty $u_{\bar{x}}$—though the derivative per se is not connected to the measurements.

7.2 Two Variables

Let us consider two series of measurements

$$x_1,\, x_2,\quad \ldots \quad ,x_{n_1} \qquad \text{and} \qquad y_1,\, y_2,\quad \ldots \quad ,y_{n_2} \tag{7.14}$$

and some function $\Phi(x,y)$.

The Gaussian error calculus tacitly imputes $n_1 \neq n_2$. Though this choice cannot be forbidden, it excludes the exploitation of the multidimensional normal model as discussed in Sect. 4.1. From there, we shall assume $n_1 = n_2 = n$. Hence, the problem is to find $u_{\bar{\Phi}}$ from

$$x_1,\, x_2,\quad \ldots \quad ,x_n \qquad \text{and} \qquad y_1,\, y_2,\quad \ldots \quad ,y_n\,. \tag{7.15}$$

For simplicity, we might prefer the two series to be independent. Clearly, were they dependent, they came in pairs

$$(x_1,y_1),\, (x_2,y_2),\quad \ldots \quad ,(x_n,y_n)\,. \tag{7.16}$$

But fancy the dependence gets steadily weaker and weaker. Then, in the end, we expect the uncertainty due to the data (7.16) to coincide with the uncertainty due to the data (7.15). Let us elaborate the uncertainty in case of dependence and ponder subsequently about the transition to independence.

The means and uncertainties of the input data are

$$\begin{aligned} \bar{x} &= \frac{1}{n}\sum_{l=1}^{n} x_l\,, \qquad & u_{\bar{x}} &= \frac{t_P(n-1)}{\sqrt{n}}\,s_x + f_{s,x} \\ \bar{y} &= \frac{1}{n}\sum_{l=1}^{n} y_l\,, \qquad & u_{\bar{y}} &= \frac{t_P(n-1)}{\sqrt{n}}\,s_y + f_{s,y}\,. \end{aligned} \tag{7.17}$$

Series Expansions

With a view to the error equations

$$\begin{aligned} x_l &= x_0 + (x_l - \mu_x) + f_x, \qquad & y_l &= y_0 + (y_l - \mu_y) + f_y\,; \quad l = 1,\ldots,n \\ \bar{x} &= x_0 + (\bar{x} - \mu_x) + f_x, \qquad & \bar{y} &= y_0 + (\bar{y} - \mu_y) + f_y \end{aligned}$$

we consider the expansion of $\Phi(x,y)$ throughout a neighborhood of the point (x_0, y_0), firstly with reference to the n pairs (x_l, y_l); $l = 1,\ldots,n$

$$\begin{aligned} \Phi(x_l,y_l) = \Phi(x_0,y_0) &+ \frac{\partial \Phi}{\partial x_0}(x_l - \mu_{\bar{x}}) + \frac{\partial \Phi}{\partial y_0}(y_l - \mu_{\bar{y}}) \\ &+ \frac{\partial \Phi}{\partial x_0} f_x + \frac{\partial \Phi}{\partial y_0} f_y + \cdots\,; \quad l = 1,\,\ldots\,,n \end{aligned}$$

and secondly with respect to the sample means $\bar{x}, \bar{y}$

$$\Phi(\bar{x}, \bar{y}) = \Phi(x_0, y_0) + \frac{\partial \Phi}{\partial x_0}(\bar{x} - \mu_{\bar{x}}) + \frac{\partial \Phi}{\partial y_0}(\bar{y} - \mu_{\bar{y}})$$
$$+ \frac{\partial \Phi}{\partial x_0} f_x + \frac{\partial \Phi}{\partial y_0} f_y + \cdots .$$

By approximation, we assume that

- the expansions may be linearized via truncation,
- the derivatives $\partial \Phi / \partial \bar{x}$ and $\partial \Phi / \partial \bar{y}$ may be substituted for the derivatives $\partial \Phi / \partial x_0$ and $\partial \Phi / \partial y_0$,
- $\partial \Phi / \partial \bar{x}$ and $\partial \Phi / \partial \bar{y}$ may be considered constant.

Just to simplify the notation, we provide the truncated expansions with equality signs

$$\Phi(x_l, y_l) = \Phi(x_0, y_0) + \frac{\partial \Phi}{\partial \bar{x}}(x_l - \mu_{\bar{x}}) + \frac{\partial \Phi}{\partial \bar{y}}(y_l - \mu_{\bar{y}})$$
$$+ \frac{\partial \Phi}{\partial \bar{x}} f_x + \frac{\partial \Phi}{\partial \bar{y}} f_y ; \quad l = 1, \ldots, n \tag{7.18}$$

and

$$\Phi(\bar{x}, \bar{y}) = \Phi(x_0, y_0) + \frac{\partial \Phi}{\partial \bar{x}}(\bar{x} - \mu_{\bar{x}}) + \frac{\partial \Phi}{\partial \bar{y}}(\bar{y} - \mu_{\bar{y}})$$
$$+ \frac{\partial \Phi}{\partial \bar{x}} f_x + \frac{\partial \Phi}{\partial \bar{y}} f_y ; \quad l = 1, \ldots, n . \tag{7.19}$$

Subtracting (7.19) from (7.18) produces

$$\Phi(x_l, y_l) - \Phi(\bar{x}, \bar{y}) = \frac{\partial \Phi}{\partial \bar{x}}(x_l - \bar{x}) + \frac{\partial \Phi}{\partial \bar{y}}(y_l - \bar{y}) ; \quad l = 1, \ldots, n . \tag{7.20}$$

Summing over l we have

$$\bar{\Phi} = \frac{1}{n} \sum_{l=1}^{n} \Phi(x_l, y_l) . \tag{7.21}$$

Random Errors

Relation (7.20) produces the empirical variance of the $\Phi(x_l, y_l)$ with respect to the mean value $\Phi(\bar{x}, \bar{y})$

$$s_{\Phi}^2 = \frac{1}{n-1} \sum_{l=1}^{n} [\, \Phi(x_l, y_l) - \Phi(\bar{x}, \bar{y}) \,]^2 \, .$$

On elaborating we find

$$\begin{aligned} s_{\Phi}^2 &= \frac{1}{n-1} \sum_{l=1}^{n} \left[\left(\frac{\partial \Phi}{\partial \bar{x}} \right)^2 (x_l - \bar{x})^2 \right. \\ &\quad \left. + 2 \left(\frac{\partial \Phi}{\partial \bar{x}} \right) \left(\frac{\partial \Phi}{\partial \bar{y}} \right) (x_l - \bar{x})(y_l - \bar{y}) + \left(\frac{\partial \Phi}{\partial \bar{y}} \right)^2 (y_l - \bar{y})^2 \right] \\ &= \left(\frac{\partial \Phi}{\partial \bar{x}} \right)^2 s_x^2 + 2 \left(\frac{\partial \Phi}{\partial \bar{x}} \right) \left(\frac{\partial \Phi}{\partial \bar{y}} \right) s_{xy} + \left(\frac{\partial \Phi}{\partial \bar{y}} \right)^2 s_y^2 \, . \end{aligned} \tag{7.22}$$

Let us compress (7.22) by means of the empirical variance–covariance matrix

$$\boldsymbol{s} = \begin{pmatrix} s_{xx} & s_{xy} \\ s_{yx} & s_{yy} \end{pmatrix} ; \quad s_{xx} \equiv s_x^2, \quad s_{xy} = s_{yx}, \quad s_{yy} \equiv s_y^2 \tag{7.23}$$

and an auxiliary vector

$$\boldsymbol{b} = \begin{pmatrix} \dfrac{\partial \Phi}{\partial \bar{x}} & \dfrac{\partial \Phi}{\partial \bar{y}} \end{pmatrix}^{\mathrm{T}} . \tag{7.24}$$

This issues

$$s_{\Phi}^2 = \boldsymbol{b}^{\mathrm{T}} \boldsymbol{s} \, \boldsymbol{b} \, . \tag{7.25}$$

Systematic Errors

Again, we formally introduce random variables X and Y and assume their realizations to be the measured data x_l, y_l; $l = 1, \ldots, n$. Then, the expected value of (7.19),

$$\mu_{\Phi} = E\left\{ \Phi(\bar{X}, \bar{Y}) \right\} = \Phi(x_0, y_0) + \left[\frac{\partial \Phi}{\partial \bar{x}} f_x + \frac{\partial \Phi}{\partial \bar{y}} f_y \right] ,$$

issues the propagated systematic error

$$f_{\Phi} = \frac{\partial \Phi}{\partial \bar{x}} f_x + \frac{\partial \Phi}{\partial \bar{y}} f_y \, ; \quad -f_{s,x} \le f_x \le f_{s,x} \, , \quad -f_{s,y} \le f_y \le f_{s,y} \, . \tag{7.26}$$

Confidence Interval and Overall Uncertainty

In error calculus, it is common to consider consecutive realizations $(x_l,\, y_l)$, $(x_{l+1},\, y_{l+1})$, ... ; $l = 1, 2, \ldots, n$ of the pair of random variables (X, Y) independent. Thus, the same applies to realizations $\Phi(x_l, y_l)$, $\Phi(x_{l+1}, y_{l+1})$, ... ; $l = 1, 2, \ldots, n$ of $\Phi(X, Y)$.

However, with respect to the same l, x_l and y_l may or may not be dependent.

As X and Y are jointly normally distributed, the sequence $\Phi(x_l, y_l)$; $l = 1, 2, \ldots, n$ follows a one-dimensional normal distribution, be X and Y dependent or not [28, 30]. Furthermore, let us recall that s_Φ^2 is an unbiased estimator of

$$\begin{aligned}\sigma_\Phi^2 &= E\left\{(\Phi(X,Y) - \mu_\Phi)^2\right\} \\ &= \left(\frac{\partial\Phi}{\partial\bar{x}}\right)^2 \sigma_x^2 + 2\left(\frac{\partial\Phi}{\partial\bar{x}}\right)\left(\frac{\partial\Phi}{\partial\bar{y}}\right)\sigma_{xy} + \left(\frac{\partial\Phi}{\partial\bar{x}}\right)^2 \sigma_y^2 .\end{aligned} \tag{7.27}$$

In view of (7.21), and as successive realizations of $\Phi(X, Y)$ are considered independent, we are entitled to define a

$$\chi^2(n-1) = \frac{(n-1)S_\Phi^2}{\sigma_\Phi^2} \tag{7.28}$$

and hence Student's

$$T(n-1) = \frac{\Phi(\bar{X},\bar{Y}) - \mu_\Phi}{S_\Phi/\sqrt{n}} . \tag{7.29}$$

After all,

$$\Phi(\bar{X},\bar{Y}) - \frac{t_P(n-1)}{\sqrt{n}} S_\Phi \leq \mu_\Phi \leq \Phi(\bar{X},\bar{Y}) + \frac{t_P(n-1)}{\sqrt{n}} S_\Phi \tag{7.30}$$

specifies a confidence interval of probability P with respect to the expectation μ_Φ of the arithmetic mean $\Phi(\bar{X}, \bar{Y})$.

Combining this with the worst-case estimation of (7.26),

$$f_{s,\Phi} = \left|\frac{\partial\Phi}{\partial\bar{x}}\right| f_{s,x} + \left|\frac{\partial\Phi}{\partial\bar{y}}\right| f_{s,y} , \tag{7.31}$$

produces the final measuring result

$$\begin{aligned}&\Phi(\bar{x},\bar{y}) \pm u_{\bar{\Phi}} \\ &u_{\bar{\Phi}} = \frac{t_P(n-1)}{\sqrt{n}}\sqrt{\boldsymbol{b}^{\mathrm{T}}\boldsymbol{s}\,\boldsymbol{b}} + \left|\frac{\partial\Phi}{\partial\bar{x}}\right| f_{s,x} + \left|\frac{\partial\Phi}{\partial\bar{y}}\right| f_{s,y} .\end{aligned} \tag{7.32}$$

Remarkably enough, the formalism does not require us to distinguish between dependent and independent random variables. As has been discussed, in case of dependence the data appear in ordered pairs, while in case of independence pairing is of no relevance whatsoever, though the numerical value of the empirical covariance s_{xy} does depend on the actual kind of pairing. This, however, does in no way affect the reliability of the confidence interval (7.30). To recall, the length of confidence intervals are submitted to fluctuations as expressed by Student's distribution density, may the empirical variance imply the fluctuations of one, of two, or of more than two variables. In particular, in case of independence, any value of the empirical covariance coming out of

$$-s_x\, s_y < s_{xy} < s_x\, s_y$$

is likewise acceptable. Even purposeful modifications of the sequence of the data would not invalidate (7.30). Had we, however, mutilated the empirical variance (7.22) by dismissing the empirical covariance, our perspective to properly assess the influence of random errors would have been rendered moot. After all, we expect the interval

$$\Phi(\bar{x}, \bar{y}) - u_{\bar{\Phi}} \le \Phi(x_0, y_0) \le \Phi(\bar{x}, \bar{y}) + u_{\bar{\Phi}} \qquad (7.33)$$

to localize the true value $\Phi(x_0, y_0)$ and this result is due to our decision in favor of what we have called *well-defined measuring conditions.*

Finally, let us redirect our attention to the role of the partial derivatives. Though being in no way connected to the measurements themselves, they are nevertheless involved in the momentousness of the random and systematic errors of the respective input data. Assume, e.g., the uncertainties $u_{\bar{x}}$ and $u_{\bar{y}}$ to be equal but that $\mid \partial\Phi/\partial\bar{x} \mid$ exceeds $\mid \partial\Phi/\partial\bar{y} \mid$. Then, the contribution to the uncertainty $u_{\bar{\Phi}}$ due to the errors in x will exceed the contribution due to the errors in y.

Robust Assessment

Let us address (7.32) and consider the empirical variance

$$s_{\bar{\Phi}}^2 = \boldsymbol{b}^{\mathrm{T}} \boldsymbol{s}\, \boldsymbol{b} = \left(\frac{\partial\Phi}{\partial\bar{x}}\right)^2 s_x^2 + 2\left(\frac{\partial\Phi}{\partial\bar{x}}\right)\left(\frac{\partial\Phi}{\partial\bar{y}}\right) s_{xy} + \left(\frac{\partial\Phi}{\partial\bar{y}}\right)^2 s_y^2 \,.$$

In case of independence, we may use the upper boundary of the interval $-s_x s_y < s_{xy} < s_x s_y$ to write

$$\begin{aligned} &\left(\frac{\partial\Phi}{\partial\bar{x}}\right)^2 s_x^2 + 2\left(\frac{\partial\Phi}{\partial\bar{x}}\right)\left(\frac{\partial\Phi}{\partial\bar{y}}\right) s_{xy} + \left(\frac{\partial\Phi}{\partial\bar{y}}\right)^2 s_y^2 \\ &\le \left(\frac{\partial\Phi}{\partial\bar{x}}\right)^2 s_x^2 + 2\left|\frac{\partial\Phi}{\partial\bar{x}}\right|\left|\frac{\partial\Phi}{\partial\bar{y}}\right| s_x\, s_y + \left(\frac{\partial\Phi}{\partial\bar{y}}\right)^2 s_y^2 \qquad (7.34) \\ &= \left(\left|\frac{\partial\Phi}{\partial\bar{x}}\right| s_x + \left|\frac{\partial\Phi}{\partial\bar{y}}\right| s_y\right)^2 \end{aligned}$$

which transfers (7.32) into

$$\begin{aligned}
u_{\bar{\Phi}} &= \frac{t_P(n-1)}{\sqrt{n}} \sqrt{\left(\frac{\partial \Phi}{\partial \bar{x}}\right)^2 s_x^2 + 2\left(\frac{\partial \Phi}{\partial \bar{x}}\right)\left(\frac{\partial \Phi}{\partial \bar{y}}\right) s_{xy} + \left(\frac{\partial \Phi}{\partial \bar{y}}\right)^2 s_y^2} \\
&\quad + \left|\frac{\partial \Phi}{\partial \bar{x}}\right| f_{s,x} + \left|\frac{\partial \Phi}{\partial \bar{y}}\right| f_{s,y} \\
&\leq \left|\frac{\partial \Phi}{\partial \bar{x}}\right| \left(\frac{t_P(n-1)}{\sqrt{n}} s_x + f_{s,x}\right) + \left|\frac{\partial \Phi}{\partial \bar{y}}\right| \left(\frac{t_P(n-1)}{\sqrt{n}} s_y + f_{s,y}\right) \\
&= \left|\frac{\partial \Phi}{\partial \bar{x}}\right| u_{\bar{x}} + \left|\frac{\partial \Phi}{\partial \bar{y}}\right| u_{\bar{y}} \,.
\end{aligned}$$

We consider

$$u_{\bar{\Phi}} \leq \left|\frac{\partial \Phi}{\partial \bar{x}}\right| u_{\bar{x}} + \left|\frac{\partial \Phi}{\partial \bar{y}}\right| u_{\bar{y}} \tag{7.35}$$

a robust assessment of the uncertainty $u_{\bar{\Phi}}$. Clearly, in this context, a singular empirical variance–covariance matrix $\boldsymbol{s}$ does not matter.

7.3 More Than Two Variables

For more than two variables, little remains to be added. To set up the error propagation, we either consider m series of data each comprising the same number n of repeated measurements

$$\begin{array}{ll}
x_{11},\; x_{12},\; \ldots,\, x_{1l}\; \ldots,\; x_{1n} & \text{device}_1 \\
\downarrow & \\
x_{21},\; x_{22},\; \ldots,\, x_{2l}\; \ldots,\; x_{2n} & \text{device}_2 \\
\downarrow & \\
\ldots\;\; \ldots\;\; \ldots\;\; \ldots\;\; \ldots\;\; \ldots & \\
\downarrow & \\
x_{m1}, x_{m2},\; \ldots,\, x_{ml}\; \ldots,\; x_{mn} & \text{device}_m
\end{array}$$

or, alternatively, n data tuples

$$(x_{11}, x_{21}, \ldots, x_{m1})\,, \quad (x_{12}, x_{22}, \ldots, x_{m2})\,, \ldots$$
$$(x_{1l}, x_{2l}, \ldots, x_{ml})\,, \ldots \quad (x_{1n}, x_{2n}, \ldots, x_{mn})$$

each of which being m-dimensional. We consider the n data tuples independent. However, with respect to the same l, the $x_{1l}, x_{2l}, \ldots, x_{ml}$ may or may not be dependent. Let us assess the uncertainty of some function $\Phi(x_1, x_2, \ldots, x_n)$.

As usual, we denote by

$$x_{0,1},\, x_{0,2}\,,\dots,\, x_{0,m}$$

the true values of the measurands. The error equations read

$$\begin{aligned} x_{il} &= x_{0,i} + (x_{il} - \mu_i) + f_i\,; \qquad i = 1,\dots,m\,; \quad l = 1,\dots,n \\ \bar{x}_i &= x_{0,i} + (\bar{x}_i - \mu_i) + f_i\,; \qquad -f_{s,i} \le f_i \le f_{s,i}\,. \end{aligned}$$

Series Expansion

There are series expansion of $\Phi(x_1, x_2, \dots, x_n)$ throughout a neighborhood of the point $x_{0,1},\, x_{0,2}\,,\dots,\, x_{0,m}$, firstly, with respect to the tuples $(x_{1l}, x_{2l}, \cdots, x_{ml})\,;\ l = 1, 2, \dots, n$ producing

$$\Phi(x_{1l}, x_{2l}, \dots x_{ml}) = \Phi(x_{0,1}, x_{0,2}, \dots x_{0,m})$$
$$+ \sum_{i=1}^{m} \frac{\partial \Phi}{\partial x_{0,i}}\,(x_{il} - \mu_i) + \sum_{i=1}^{m} \frac{\partial \Phi}{\partial x_{0,i}}\, f_i + \cdots\,; \quad l = 1,\dots,n$$

and, secondly, with respect to the m sample means $\bar{x}_1, \bar{x}_2, \dots, \bar{x}_m$ yielding

$$\Phi(\bar{x}_1, \bar{x}_2, \dots \bar{x}_m) = \Phi(x_{0,1}, x_{0,2}, \dots x_{0,m})$$
$$+ \sum_{i=1}^{m} \frac{\partial \Phi}{\partial x_{0,i}}(\bar{x}_i - \mu_i) + \sum_{i=1}^{m} \frac{\partial \Phi}{\partial x_{0,i}}\, f_i + \cdots\,.$$

Substituting $\partial\Phi/\partial\bar{x}_i$ for $\partial\Phi/\partial x_{0,i}$, truncating and recasting the expansions issues

$$\Phi(x_{1l}, x_{2l}, \dots x_{ml}) = \Phi(x_{0,1}, x_{0,2}, \dots x_{0,m})$$
$$+ \sum_{i=1}^{m} \frac{\partial \Phi}{\partial \bar{x}_i}(x_{il} - \mu_i) + \sum_{i=1}^{m} \frac{\partial \Phi}{\partial \bar{x}_i}\, f_i\,; \quad l = 1,\dots,n \tag{7.36}$$

and

$$\Phi(\bar{x}_1, \bar{x}_2, \dots \bar{x}_m) = \Phi(x_{0,1}, x_{0,2}, \dots x_{0,m})$$
$$+ \sum_{i=1}^{m} \frac{\partial \Phi}{\partial \bar{x}_i}(\bar{x}_i - \mu_i) + \sum_{i=1}^{m} \frac{\partial \Phi}{\partial \bar{x}_i}\, f_i\,. \tag{7.37}$$

Subtraction of (7.37) from (7.36) produces

$$\Phi(x_{1l}, x_{2l}, \dots x_{ml}) - \Phi(\bar{x}_1, \bar{x}_2, \dots \bar{x}_m)$$
$$= \sum_{i=1}^{m} \frac{\partial \Phi}{\partial \bar{x}_i}(x_{il} - \bar{x}_i)\,; \quad l = 1,\dots,n\,. \tag{7.38}$$

Summing over l issues

$$\bar{\Phi} = \frac{1}{n}\sum_{l=1}^{n} \Phi(x_{1l}, x_{2l}, \ldots, x_{ml}) \,. \tag{7.39}$$

Random Errors

From (7.38) we draw the empirical variance

$$s_{\Phi}^2 = \sum_{i,j}^{m} \frac{\partial \Phi}{\partial \bar{x}_i}\frac{\partial \Phi}{\partial \bar{x}_j} s_{ij} = \boldsymbol{b}^{\mathrm{T}} \boldsymbol{s}\, \boldsymbol{b} \tag{7.40}$$

in which we let

$$\boldsymbol{s} = \begin{pmatrix} s_{11} & s_{12} & \ldots & s_{1m} \\ s_{21} & s_{22} & \ldots & s_{2m} \\ \ldots & \ldots & \ldots & \ldots \\ s_{m1} & s_{m2} & \ldots & s_{mm} \end{pmatrix}; \qquad s_{ii} \equiv s_i^2 \,, \quad s_{ij} = s_{ji} \tag{7.41}$$

denote the empirical variance–covariance matrix of the input data with elements

$$s_{ij} = \frac{1}{n-1}\sum_{l=1}^{n}(x_{il} - \bar{x}_i)(x_{jl} - \bar{x}_j)\,; \quad i, j = 1, \ldots, m$$

and

$$\boldsymbol{b} = \left(\frac{\partial \Phi}{\partial \bar{x}_1} \quad \frac{\partial \Phi}{\partial \bar{x}_2} \quad \cdots \quad \frac{\partial \Phi}{\partial \bar{x}_m}\right)^{\mathrm{T}} .$$

an auxiliary vector.

Systematic Errors

Reading (7.37) in random variables $\bar{X}_1, \bar{X}_2, \ldots \bar{X}_m$ yields the expected value

$$\begin{aligned} \mu_{\Phi} &= E\left\{\Phi(\bar{X}_1, \bar{X}_2, \ldots, \bar{X}_m)\right\} \\ &= \Phi(x_{0,1}, x_{0,2}, \ldots, x_{0,m}) + \sum_{i=1}^{m} \frac{\partial \Phi}{\partial \bar{x}_i} f_i \,. \end{aligned}$$

Hence, the propagated systematic error is

$$f_{\Phi} = \sum_{i=1}^{m} \frac{\partial \Phi}{\partial \bar{x}_i} f_i \,; \quad -f_{s,i} \leq f_i \leq f_{s,i} \,. \tag{7.42}$$

Confidence Interval and Overall Uncertainty

The confidence interval

$$\begin{aligned}\Phi(\bar{x}_1, \bar{x}_2, \ldots, \bar{x}_m) - \frac{t_P(n-1)}{\sqrt{n}} s_\Phi \leq \mu_\Phi \\ \leq \Phi(\bar{x}_1, \bar{x}_2, \ldots, \bar{x}_m) + \frac{t_P(n-1)}{\sqrt{n}} x_\Phi\end{aligned} \tag{7.43}$$

is expected to localize the parameter μ_Φ with probability P. Combining the worst-case estimation

$$f_{s,\Phi} = \sum_{i=1}^{m} \left| \frac{\partial \phi}{\partial \bar{x}_i} \right| f_{s,i} \tag{7.44}$$

of the propagated systematic error f_Φ with the confidence interval (7.43) for the expected value μ_Φ produces the final result

$$\begin{aligned}&\Phi(\bar{x}_1, \bar{x}_2, \ldots, \bar{x}_m) \pm u_{\bar{\Phi}} \\ &u_{\bar{\Phi}} = \frac{t_P(n-1)}{\sqrt{n}} \sqrt{\boldsymbol{b}^{\mathrm{T}} \boldsymbol{s}\, \boldsymbol{b}} + \sum_{i=1}^{m} \left| \frac{\partial \Phi}{\partial \bar{x}_i} \right| f_{s,i}\end{aligned} \tag{7.45}$$

which is meant to localize the true value $\Phi(x_{0,1}, x_{0,2}, \ldots, x_{0,m})$.

Robust Assessment

Considering two variables at a time, the respective empirical covariance may be treated just as in case of two variables. This leads to

$$u_{\bar{\Phi}} \leq \sum_{i=1}^{m} \left| \frac{\partial \Phi}{\partial \bar{x}_i} \right| u_{\bar{x}_i} \tag{7.46}$$

which, obviously, renders the consideration of empirical covariances irrelevant.

Flow of True Values

Ultimately, let us visualize the flow of true values. To have an example, we consider three true values, say, $\beta_{0,1}, \beta_{0,2}, \beta_{0,3}$, estimators $\bar{\beta}_1, \bar{\beta}_2, \bar{\beta}_3$ and uncertainties $u_{\bar{\beta}_1}, u_{\bar{\beta}_2}, u_{\bar{\beta}_3}$.

Let there be some relationship defining a fourth true value $\beta_{0,4}$ via

$$\phi(\beta_{0,1},\ \beta_{0,2},\ \beta_{0,3}) = \beta_{0,4}\,.$$

Given the intervals $\bar{\beta}_1 \pm u_{\bar{\beta}_1}, \bar{\beta}_2 \pm u_{\bar{\beta}_2}, \bar{\beta}_3 \pm u_{\bar{\beta}_3}$ localize the true values $\beta_{0,1},\ \beta_{0,2},\ \beta_{0,3}$, with respect to

$$\phi(\bar{\beta}_1,\ \bar{\beta}_2,\ \bar{\beta}_3) = \bar{\beta}_4$$

or

$$\phi(\bar{\beta}_1 \pm u_{\bar{\beta}_1}, \bar{\beta}_2 \pm u_{\bar{\beta}_2}, \bar{\beta}_3 \pm u_{\bar{\beta}_3}) \Rightarrow \bar{\beta}_4 \pm u_{\bar{\beta}_4}$$

we should have

$$\bar{\beta}_4 - u_{\bar{\beta}_4} \le \beta_{0,4} \le \bar{\beta}_4 + u_{\bar{\beta}_4}\,.$$

This is what traceability asks for. The result is illustrated in Fig. 7.1.

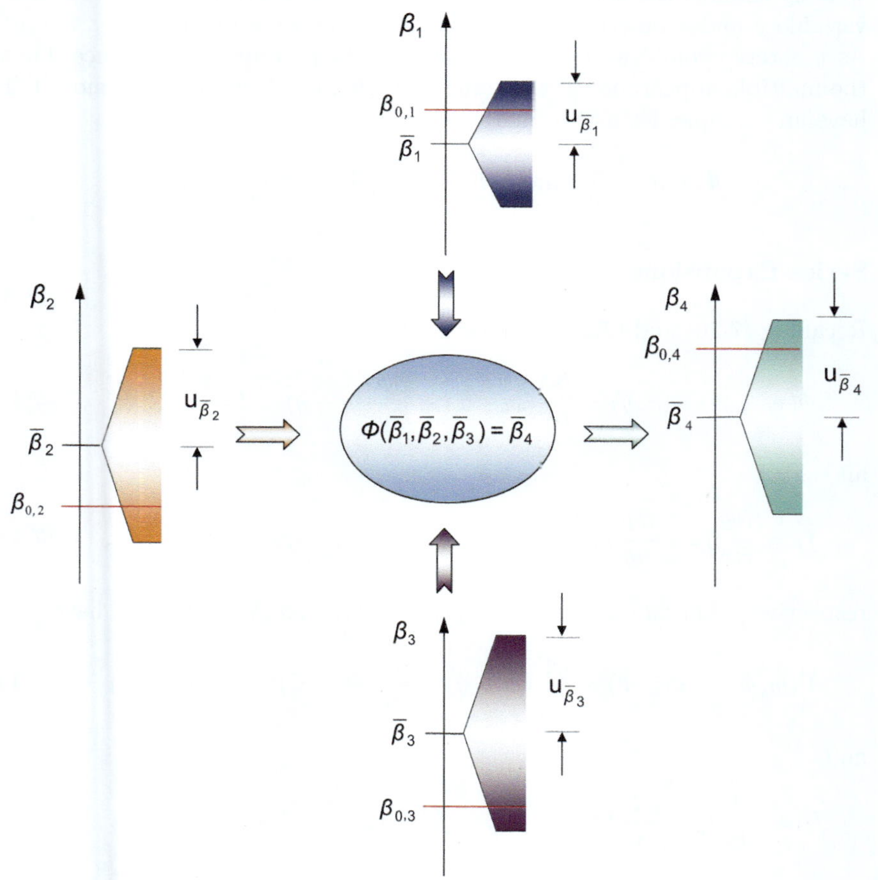

Fig. 7.1. The merging of measuring results, say, $\bar{\beta}_1,\ \bar{\beta}_2,\ \bar{\beta}_3$ to $\bar{\beta}_4 = \phi(\bar{\beta}_1,\ \bar{\beta}_2,\ \bar{\beta}_3)$ should maintain traceability

Just to extend the idea, assume there is a separate experiment allowing to measure β_4 directly. Then the uncertainties of the preceding result and the directly measured result

$$\bar{\beta}_4{}' - u_{\bar{\beta}_4'} \leq \beta_{0,4} \leq \bar{\beta}_4{}' + u_{\bar{\beta}_4'}$$

should overlap.

7.4 Concatenated Functions

More often than not, the propagation of measuring errors pervades several stages. This means that functional relationships, burdened by measuring errors, enter other functions. To map the underlying physical processes, we shall steadily channel the flow of random and systematic errors. Let us consider a variable y and a function $\Phi(x,y)$ both entering some function $\Gamma[y,\Phi(x,y)]$. As y enters twofold we expect a covariance to pop up. A covariance due to the multiple appearance of the same variable may, however, be removed. To have an example, let us consider

$$\Phi(x,y) = \frac{x}{y} \quad \text{and} \quad \Gamma[y,\Phi(x,y)] = y\exp[\Phi(x,y)]\,.$$

Series Expansions

Recalling (7.20) and (7.26) we have

$$\Phi(x_l,y_l) - \Phi(\bar{x},\bar{y}) = \frac{\partial\Phi}{\partial\bar{x}}(x_l-\bar{x}) + \frac{\partial\Phi}{\partial\bar{y}}(y_l-\bar{y})\,; \quad l=1,\ldots,n \tag{7.47}$$

and

$$f_\Phi = \frac{\partial\Phi}{\partial\bar{x}}f_x + \frac{\partial\Phi}{\partial\bar{y}}f_y\,; \quad -f_{s,x} \leq f_x \leq f_{s,x}\,, \quad -f_{s,y} \leq f_y \leq f_{s,y}\,, \tag{7.48}$$

respectively. Let us put $\Phi_l \equiv \Phi(x_l,y_l);\ l=1,\ldots,n$. With this we have

$$\Gamma(y_l,\Phi_l) - \Gamma(\bar{y},\bar{\Phi}) = \frac{\partial\Gamma}{\partial\bar{y}}(y_l-\bar{y}) + \frac{\partial\Gamma}{\partial\bar{\Phi}}(\Phi_l-\bar{\Phi})\,; \quad l=1,\ldots,n \tag{7.49}$$

and

$$f_\Gamma = \frac{\partial\Gamma}{\partial\bar{y}}f_y + \frac{\partial\Gamma}{\partial\bar{\Phi}}f_\Phi\,; \quad -f_{s,y} \leq f_y \leq f_{s,y}\,, \quad -f_{s,\Phi} \leq f_\Phi \leq f_{s,\Phi}\,. \tag{7.50}$$

Relation (7.49) is apt to produce an empirical covariance $s_{y,\Phi}$ which, however, disappears upon inserting (7.47)

$$\Gamma(y_l, \Phi_l) - \Gamma(\bar{y}, \bar{\Phi}) = \frac{\partial \Gamma}{\partial \bar{\Phi}} \frac{\partial \Phi}{\partial \bar{x}} (x_l - \bar{x}) + \left[\frac{\partial \Gamma}{\partial \bar{y}} + \frac{\partial \Gamma}{\partial \bar{\Phi}} \frac{\partial \Phi}{\partial \bar{y}} \right] (y_l - \bar{y}) \,. \tag{7.51}$$

In order to keep the propagation of systematic errors unique, we put

$$f_\Gamma = \frac{\partial \Gamma}{\partial \bar{\Phi}} \frac{\partial \Phi}{\partial \bar{x}} f_x + \left[\frac{\partial \Gamma}{\partial \bar{y}} + \frac{\partial \Gamma}{\partial \bar{\Phi}} \frac{\partial \Phi}{\partial \bar{y}} \right] f_y \,. \tag{7.52}$$

Introducing

$$b_x = \frac{\partial \Gamma}{\partial \bar{\Phi}} \frac{\partial \Phi}{\partial \bar{x}} \,, \quad b_y = \frac{\partial \Gamma}{\partial \bar{y}} + \frac{\partial \Gamma}{\partial \bar{\Phi}} \frac{\partial \Phi}{\partial \bar{y}} \qquad \text{and} \qquad \boldsymbol{b} = \begin{pmatrix} b_x \\ b_y \end{pmatrix} \tag{7.53}$$

we arrive at

$$\Gamma(y_l, \Phi_l) - \Gamma(\bar{y}, \bar{\Phi}) = b_x (x_l - \bar{x}) + b_y (y_l - \bar{y}) \tag{7.54}$$

and

$$f_\Gamma = b_x f_x + b_y f_y \,. \tag{7.55}$$

Overall Uncertainty

Obviously, the uncertainty $u_{\bar{\Gamma}}$ of $\Gamma(\bar{y}, \bar{\Phi})$ turns out as

$$\Gamma(\bar{y}, \bar{\Phi}) \pm u_{\bar{\Gamma}}$$

$$u_{\bar{\Gamma}} = \frac{t_P(n-1)}{\sqrt{n}} \sqrt{b_x^2 s_x^2 + 2 b_x b_y s_{xy} + b_y^2 s_y^2} \; + \mid b_x \mid f_{s,x} + \mid b_y \mid f_{s,y} \tag{7.56}$$

which is meant to localize the true value $\Gamma(y_0, \Phi_0)$.

Remark

Let us subject

$$f_\Gamma = \frac{\partial \Gamma}{\partial \bar{y}} f_y + \frac{\partial \Gamma}{\partial \bar{\Phi}} f_\Phi$$

to a worst-case estimation

$$f_{s,\Gamma}^* = \left| \frac{\partial \Gamma}{\partial \bar{y}} \right| f_{s,y} + \left| \frac{\partial \Gamma}{\partial \bar{\Phi}} \right| f_{s,\Phi} \,.$$

Inserting the worst-case estimation of (7.48)

$$f_{s,\Phi} = \left| \frac{\partial \Phi}{\partial \bar{x}} \right| f_{s,x} + \left| \frac{\partial \Phi}{\partial \bar{y}} \right| f_{s,y}$$

produces

$$f_{s,\Gamma}^{*} = \left| \frac{\partial \Gamma}{\partial \bar{\Phi}} \frac{\partial \Phi}{\partial \bar{x}} \right| f_{s,x} + \left[\left| \frac{\partial \Gamma}{\partial \bar{y}} \right| + \left| \frac{\partial \Gamma}{\partial \bar{\Phi}} \frac{\partial \Phi}{\partial \bar{y}} \right| \right] f_{s,y}$$

which clearly differs from the worst-case estimation of (7.52)

$$f_{s,\Gamma} = \left| \frac{\partial \Gamma}{\partial \bar{\Phi}} \frac{\partial \Phi}{\partial \bar{x}} \right| f_{s,x} + \left| \frac{\partial \Gamma}{\partial \bar{y}} + \frac{\partial \Gamma}{\partial \bar{\Phi}} \frac{\partial \Phi}{\partial \bar{y}} \right| f_{s,y} \, .$$

To stress, $f_{s,\Gamma}^{*}$, expressing an inconsistent propagation of systematic errors, conflicts with the proceeding stated in Sect. 5.2. By contrast, $f_{s,\Gamma}$ brings out a consistent proceeding. In order to keep the propagation of unknown systematic errors unique, worst-case estimations should be carried out not until those errors which occur more than once have been factored out.

7.5 Elementary Examples

Let us consider two measuring results

$$\bar{x} \pm \left(\frac{t_P(n-1)}{\sqrt{n}} s_x + f_{s,x} \right) , \qquad \bar{y} \pm \left(\frac{t_P(n-1)}{\sqrt{n}} s_y + f_{s,y} \right) \tag{7.57}$$

and see about their sum, difference, product, and quotient.

Sum

The error equations of the sum

$$\Phi(x,y) = x + y \tag{7.58}$$

read

$$\Phi(x_l, y_l) = x_0 + y_0 + (x_l - \mu_x) + (y_l - \mu_y) + f_x + f_y \, ; \quad l = 1, 2, \ldots, n$$

$$\Phi(\bar{x}, \bar{y}) = x_0 + y_0 + (\bar{x} - \mu_x) + (\bar{y} - \mu_y) + f_x + f_y \, .$$

The difference

$$\Phi(x_l, y_l) - \Phi(\bar{x}, \bar{y}) = (x_l - \bar{x}) + (y_l - \bar{y}) \, ; \quad l = 1, 2, \ldots, n \tag{7.59}$$

issues the mean

$$\bar{\Phi} = \frac{1}{n} \sum_{l=1}^{n} \Phi(x_l, y_l) \tag{7.60}$$

and the empirical variance

$$s_{\Phi}^{2} = \frac{1}{n-1} \sum_{l=1}^{n} \left[\Phi(x_l, y_l) - \Phi(\bar{x}, \bar{y})\right]^2 = s_x^2 + 2s_{xy} + s_y^2 \,. \tag{7.61}$$

The expected value of $\Phi(\bar{X}, \bar{Y})$,

$$\mu_{\Phi} = E\{\Phi(X, Y)\} = x_0 + y_0 + f_x + f_y \,, \tag{7.62}$$

discloses the propagated systematic error

$$f_{\Phi} = f_x + f_y \,, \quad -f_{s,x} \le f_x \le f_{s,x} \,, \quad -f_{s,y} \le f_x \le f_{s,y} \,. \tag{7.63}$$

Hence, we expect the interval

$$\Phi(\bar{x}, \bar{y}) \pm u_{\bar{\Phi}}$$

$$u_{\bar{\Phi}} = \frac{t_P(n-1)}{\sqrt{n}} \sqrt{s_x^2 + 2s_{xy} + s_y^2} + (f_{s,x} + f_{s,y}) \tag{7.64}$$

to localize the true value $x_0 + y_0$.

Perhaps, we might wish to pursue the variation of the length of the confidence interval

$$\Phi(\bar{x}, \bar{y}) - \frac{t_P(n-1)}{\sqrt{n}} s_{\Phi} \le \mu_{\Phi} \le \Phi(\bar{x}, \bar{y}) - \frac{t_P(n-1)}{\sqrt{n}} s_{\Phi}$$

given the empirical covariance s_{xy} runs from $-s_x s_y$ to $s_x s_y$ maintaining

$$-s_x \, s_y < s_{xy} < s_x \, s_y \,.$$

The effect of the increase of the empirical variances

$$s_{\Phi} = \sqrt{s_x^2 + 2s_{xy} + s_y^2} \,,$$

is illustrated in Fig. 7.2.

Difference

To ask for the difference of two arithmetic means, say, $\bar{x}$ and $\bar{y}$ aiming at a true value, say, z_0 is tantamount to scrutinize their mutual compatibility, Fig 7.3. Considering

$$\Phi(x, y) = x - y \tag{7.65}$$

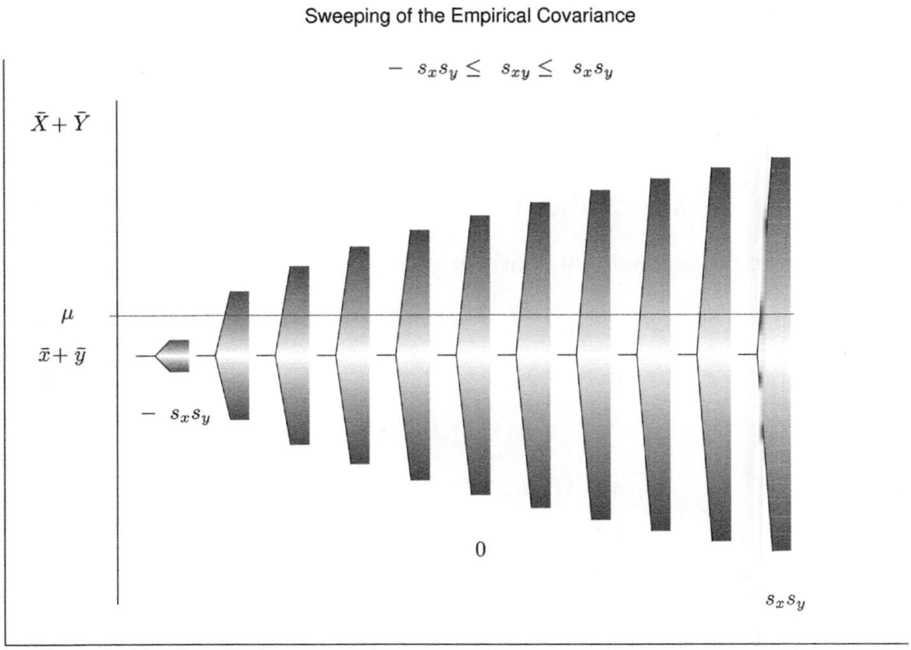

Fig. 7.2. Sweeping the empirical covariance s_{xy} through the interval $-s_x s_y \cdots s_x s_y$ alters the length of the confidence interval

the error equations

$$\Phi(x_l, y_l) = x_0 - y_0 + (x_l - \mu_x) - (y_l - \mu_y) + f_x - f_y\,; \quad l = 1.2, \ldots, n$$

$$\Phi(\bar{x}, \bar{y}) = x_0 - y_0 + (\bar{x} - \mu_x) - (\bar{y} - \mu_y) + f_x - f_y$$

produce

$$\Phi(x_l, y_l) - \Phi(\bar{x}, \bar{y}) = (x_l - \bar{x}) - (y_l - \bar{y})\,; \quad l = 1, 2, \ldots, n\,. \tag{7.66}$$

Hence, we have

$$\bar{\Phi} = \frac{1}{n} \sum_{l=1}^{n} \Phi(x_l, y_l) \tag{7.67}$$

and

$$s_{\Phi}^2 = s_x^2 - 2 s_{xy} + s_y^2\,. \tag{7.68}$$

With respect to the expectation

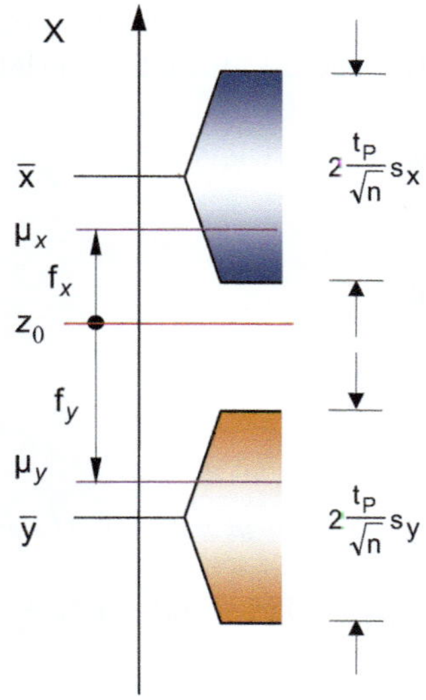

Fig. 7.3. Difference between two arithmetic means aiming at a common true value z_0

$$\mu_{\Phi} = E\{\Phi(X,Y)\} = (x_0 + f_x) - (y_0 + f_y) = \mu_x - \mu_y \tag{7.69}$$

there is a Student's

$$T(n-1) = \frac{(\bar{X} - \bar{Y}) - (\mu_x - \mu_y)}{S_{\Phi}/\sqrt{n}}\,. \tag{7.70}$$

As (7.69) issues the propagated systematic error

$$f_{\Phi} = f_x - f_y\,, \quad -f_{s,x} \le f_x \le f_{s,x}\,, \quad -f_{s,y} \le f_x \le f_{s,y} \tag{7.71}$$

the two means appear compatible given

$$|\,\bar{x} - \bar{y}\,| \le \frac{t_p(n-1)}{\sqrt{n}}\sqrt{s_x^2 - 2s_{xy} + s_y^2} + (f_{s,x} + f_{s,y})\,. \tag{7.72}$$

Condoning a singular empirical variance–covariance matrix, a coarser estimation would be

$$|\,\bar{x} - \bar{y}\,| \le u_{\bar{x}} + u_{\bar{y}}\,.$$

Round robins, as discussed in Sect. 9.2, circulate a measuring standard throughout a group of laboratories. Given the true value of the standard remains constant during the circulation, each of the participants has the chance to assess the true value of one and the same physical quantity. But then, any two uncertainties should mutually overlap.

Product

Expanding and linearizing

$$\Phi(x,y) = x\,y \tag{7.73}$$

produces

$$\Phi(x_l, y_l) = x_0 y_0 + \bar{y}(x_l - \mu_x) + \bar{x}(y_l - \mu_y) + \bar{y} f_x + \bar{x} f_y\,; \quad l = 1,2,\ldots,n$$

$$\Phi(\bar{x}, \bar{y}) = x_0 y_0 + \bar{y}(\bar{x} - \mu_x) + \bar{x}(\bar{y} - \mu_y) + \bar{y} f_x + \bar{x} f_y\,.$$

From

$$\Phi(x_l, y_l) - \Phi(\bar{x}, \bar{y}) = \bar{y}(x_l - \bar{x}) + \bar{x}(y_l - \bar{y})\,; \quad l = 1,2,\ldots,n \tag{7.74}$$

we take

$$\bar{\Phi} = \frac{1}{n}\sum_{l=1}^{n} \Phi(x_l, y_l) \tag{7.75}$$

and

$$s_\Phi^2 = \bar{y}^2 s_x^2 + 2\,\bar{x}\,\bar{y}\,s_{xy} + \bar{x}^2 s_y^2\,. \tag{7.76}$$

The expected value

$$\mu_\Phi = E\{\Phi(X,Y)\} = x_0 y_0 + \bar{y} f_x + \bar{x} f_y \tag{7.77}$$

issues the propagated systematic error

$$f_\Phi = \bar{y} f_x + \bar{x} f_y\,, \quad -f_{s,x} \le f_x \le f_{s,x}\,, \quad -f_{s,y} \le f_x \le f_{s,y}\,. \tag{7.78}$$

Hence, the interval

$$\Phi(\bar{x}, \bar{y}) \pm u_{\bar{\Phi}}$$

$$u_{\bar{\Phi}} = \frac{t_P(n-1)}{\sqrt{n}}\sqrt{\bar{y}^{\,2} s_x^2 + 2\bar{x}\bar{y}s_{xy} + \bar{x}^{\,2} s_y^2} + |\,\bar{y}\,|\,f_{s,x} + |\,\bar{x}\,|\,f_{s,y} \tag{7.79}$$

is meant to localize the true value $\Phi(x_0, y_0) = x_0 y_0$.

Quotient

Finally, let us consider

$$\Phi(x, y) = \frac{x}{y}\,. \tag{7.80}$$

Expanding and linearizing we have[1]

$$\Phi(x_l, y_l) = \frac{x_0}{y_0} + \frac{1}{\bar{y}}(x_l - \mu_x) - \frac{\bar{x}}{\bar{y}^2}(y_l - \mu_y) + \frac{1}{\bar{y}}f_x - \frac{\bar{x}}{\bar{y}^2}f_y\,; \quad l = 1, 2, \ldots, n$$

$$\Phi(\bar{x}, \bar{y}) = \frac{x_0}{y_0} + \frac{1}{\bar{y}}(\bar{x} - \mu_x) - \frac{\bar{x}}{\bar{y}^2}(\bar{y} - \mu_y) + \frac{1}{\bar{y}}f_x - \frac{\bar{x}}{\bar{y}^2}f_y\,.$$

Thus from

$$\Phi(x_l, y_l) - \Phi(\bar{x}, \bar{y}) = \frac{1}{\bar{y}}(x_l - \bar{x}) - \frac{\bar{x}}{\bar{y}^2}(y_l - \bar{y})$$

we draw

$$\bar{\Phi} = \frac{1}{n}\sum_{l=1}^{n}\Phi(x_l, y_l) \tag{7.81}$$

and

$$s_{\Phi}^2 = \frac{1}{\bar{y}^{\,2}}\,s_x^2 - 2\frac{\bar{x}}{\bar{y}^{\,3}}\,s_{xy} + \frac{\bar{x}^2}{\bar{y}^{\,4}}\,s_y^2\,. \tag{7.82}$$

The expected value

$$\mu_{\Phi} = E\{\Phi(X, Y)\} = \frac{x_0}{y_0} + \frac{1}{\bar{y}}f_x - \frac{\bar{x}}{\bar{y}^2}f_y \tag{7.83}$$

suggests a propagated systematic error

$$f_{\Phi} = \frac{1}{\bar{y}}f_x - \frac{\bar{x}}{\bar{y}^2}f_y\,, \quad -f_{s,x} \le f_x \le f_{s,x}\,, \quad -f_{s,y} \le f_x \le f_{s,y}\,. \tag{7.84}$$

After all, we consider the result

$$\Phi(\bar{x}, \bar{y}) \pm u_{\bar{\Phi}}$$

$$u_{\bar{\Phi}} = \frac{t_P(n-1)}{\sqrt{n}}\sqrt{\frac{1}{\bar{y}^{\,2}}\,s_x^2 - 2\frac{\bar{x}}{\bar{y}^{\,3}}\,s_{xy} + \frac{\bar{x}^{\,2}}{\bar{y}^{\,4}}\,s_y^2} \tag{7.85}$$

$$+\frac{1}{|\,\bar{y}\,|}\,f_{s,x} + \frac{|\,\bar{x}\,|}{|\,\bar{y}^2\,|}\,f_{s,y}$$

to localize the true value x_0/y_0.

[1]Given X and Y were standardized, normally distributed variables, the quotient $U = X/Y$ follows a Cauchy density. Then, as is known, $E\{U\}$ and $E\{U^2\}$ would not exist. We, however, confine ourselves to truncated Taylor series. In this approximation, the $\Phi(x_l, y_l)$; $l = 1, \ldots, n$ may be considered normally distributed.

7.6 Test of Hypothesis

Let us return to the difference between two arithmetic means directed at one and the same true value. Time and again, it has been emphasized that the classical error calculus is used to addressing different numbers of repeated measurements

$$\bar{x}_1 = \frac{1}{n}\sum_{l=1}^{n_1} x_{1l}\,, \qquad \bar{x}_2 = \frac{1}{n}\sum_{l=1}^{n_2} x_{2l}\,; \qquad n_1 \neq n_2\,. \tag{7.86}$$

Coherently, the empirical covariance is not to be set up simply because there are excess measurements of either of the variables. For the moment, let us disregard systematic errors and assume equal theoretical variances, so that $\mu_i = \mu$ and $\sigma_i^2 = \sigma^2$; $i = 1, 2$, respectively. Due to $n_1 \neq n_2$, the classical error calculus examines the difference between the two means through the theoretical variance of the random variable $\bar{Z} = \bar{X}_1 - \bar{X}_2$,

$$\begin{aligned}
\sigma_{\bar{z}}^2 &= E\{\left[(\bar{X}_1 - \mu) - (\bar{X}_2 - \mu)\right]^2\} \\
&= E\{(\bar{X}_1 - \mu)^2\} - 2E\{(\bar{X}_1 - \mu)(\bar{X}_2 - \mu)\} + E\{(\bar{X}_2 - \mu)^2\} \\
&= \frac{\sigma_1^2}{n_1} + \frac{\sigma_2^2}{n_2} = \sigma^2 \frac{n_1 + n_2}{n_1\, n_2}\,.
\end{aligned}$$

Thus, the random variable $\bar{Z}$ follows a

$$N\left(0,\ \sigma\sqrt{\frac{n_1 + n_2}{n_1\, n_2}}\right)$$

density. The associated standardized variable is

$$\frac{(\bar{X}_1 - \bar{X}_2)}{\sigma\sqrt{\dfrac{n_1 + n_2}{n_1\, n_2}}}\,.$$

The sum of the two χ^2-s

$$\chi^2(\nu^*) = \frac{(n_1 - 1)S_1^2 + (n_2 - 1)S_2^2}{\sigma^2}$$

with degrees of freedom

$$\nu^* = n_1 + n_2 - 2$$

suggests a Student's

$$T(\nu^*) = \frac{\dfrac{\bar{X}_1 - \bar{X}_2}{\sigma\sqrt{(n_1 + n_2)/(n_1 n_2)}}}{\sqrt{\dfrac{\chi^2(\nu^*)}{\nu^*}}}\,.$$

Written explicitly, we have

$$T(\nu^*) = \frac{\dfrac{\bar{X}_1 - \bar{X}_2}{\sigma\sqrt{(n_1 + n_2)/(n_1 n_2)}}}{\sqrt{\dfrac{(n_1 - 1)S_1^2 + (n_2 - 1)S_2^2}{\sigma^2}}}\sqrt{n_1 + n_2 - 2}\,. \tag{7.87}$$

Obviously, this statement lacks an empirical covariance. To compare (7.87) with (7.70), we put $n_1 = n_2 = n$

$$T(\nu^*) = \frac{\bar{X}_1 - \bar{X}_2}{\sqrt{S_1^2 + S_2^2}}\sqrt{n}\,. \tag{7.88}$$

Rewriting (7.70) we have

$$T(\nu) = \frac{\bar{X}_1 - \bar{X}_2}{\sqrt{S_1^2 - 2S_{12} + S_2^2}}\sqrt{n}\,. \tag{7.89}$$

While (7.88) refers to degrees of freedom $\nu^* = 2(n - 1)$, relation, (7.89) addresses degrees of freedom $\nu = (n - 1)$, where $t_P(2\nu) < t_P(\nu)$.

Both quantities possess equal statistical properties. For an illustration, we consider the respective confidence intervals

$$\begin{aligned}(\bar{X}_1 - \bar{X}_2) &- t_P(2\nu)\sqrt{S_1^2 + S_2^2}/\sqrt{n} \\ &\leq 0 \leq (\bar{X}_1 - \bar{X}_2) + t_P(2\nu)\sqrt{S_1^2 + S_2^2}/\sqrt{n}\,,\end{aligned} \tag{7.90}$$

and

$$\begin{aligned}(\bar{X}_1 - \bar{X}_2) &- t_P(\nu)/\sqrt{S_1^2 - 2S_{12} + S_2^2}\sqrt{n} \\ &\leq 0 \leq (\bar{X}_1 - \bar{X}_2) + t_P(\nu)\sqrt{S_1^2 - 2S_{12} + S_2^2}/\sqrt{n}\,.\end{aligned} \tag{7.91}$$

Interestingly enough, a dependence between the two means would promptly invalidate (7.90), but would affect (7.91) in no way.

To be more general, we now admit different expectations, $\mu_1 \neq \mu_2$, and different theoretical variances, $\sigma_1^2 \neq \sigma_2^2$, briefly touching the so-called Fisher–Behrens problem which, traditionally, refers to $n_1 \neq n_2$. To the knowledge of the author, no exact solution has ever been proposed.

However, assuming $n_1 = n_2$, (7.70) doubtlessly yields an exact

$$T(\nu) = \frac{(\bar{X}_1 - \bar{X}_2) - (\mu_1 - \mu_2)}{\sqrt{S_1^2 - 2S_{12} + S_2^2}} \sqrt{n} \,. \tag{7.92}$$

Evidently, the proceeding to ask for well-defined measuring conditions brings the empirical covariance S_{12} into play and this renders the Fisher–Behrens problem an ill-defined issue.[2]

Data Simulation

Under data simulation, (7.90) and (7.91) confirm equal statistical behavior as illustrated in Figs. 7.4 and 7.5, respectively.

[2]From a metrological point of view, the mere notation $\mu_1 \neq \mu_2$ conceals the underlying physical situation. Firstly, desisting from unknown systematic errors and assuming $\mu_1 \neq \mu_2$ presents us with a contradiction in terms. Secondly, admitting unknown systematic errors, $\mu_1 \neq \mu_2$ renders the Fischer–Behrens issue meaningless, as we are now asked to probe compatibility via (7.72), but not via a purely statistical approach.

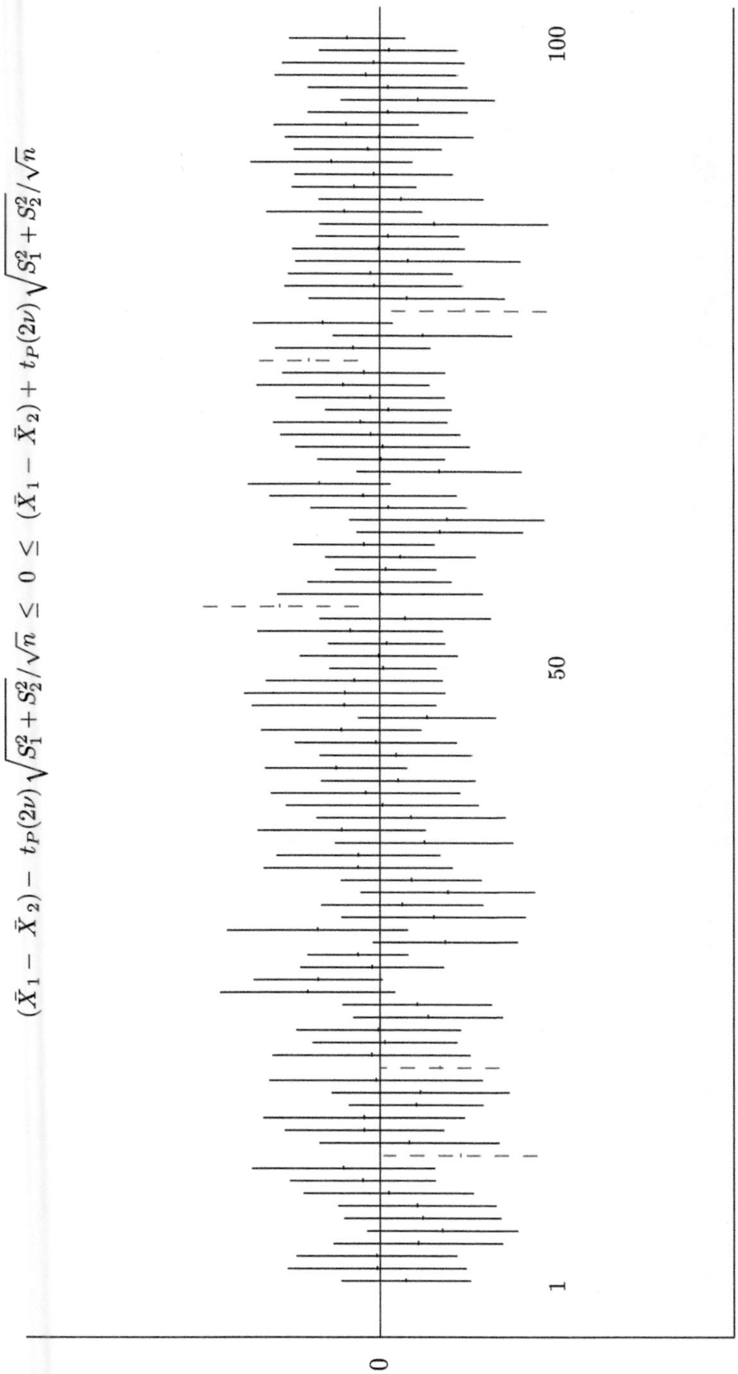

Fig. 7.4. Difference between two arithmetic means, classical approach, 100 simulations

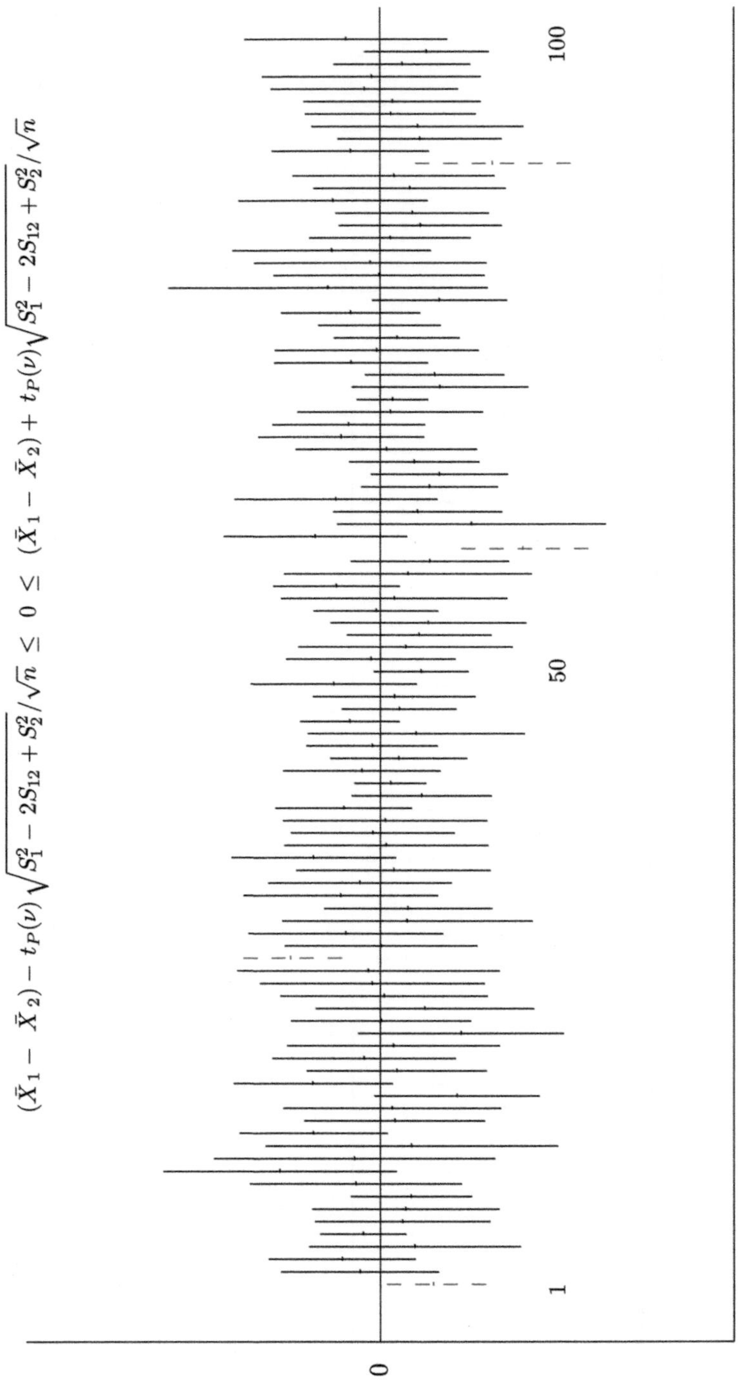

Fig. 7.5. Difference between two arithmetic means, generalized approach, 100 simulations

8 Method of Least Squares

The method of least squares controls the flow of errors via the elements of the design matrix. Hence, assuming linear systems with differing design matrices aiming at the same set of unknowns, the adjustment's uncertainties would differ even if the uncertainties of the input data were the same.

8.1 Empirical Variance–Covariance Matrix

As discussed in (2.18), the solution vector

$$\bar{\boldsymbol{\beta}} = \boldsymbol{B}^{\mathrm{T}} \bar{\boldsymbol{x}} \tag{8.1}$$

is given by the matrix

$$\boldsymbol{B} = \boldsymbol{A}(\boldsymbol{A}^{\mathrm{T}}\boldsymbol{A})^{-1} = (b_{ik}) \; ; \quad i = 1, \ldots, m \, ; \; k = 1, \ldots, r \tag{8.2}$$

and the vector of input data

$$\bar{\boldsymbol{x}} = (\bar{x}_1 \;\; \bar{x}_2 \;\; \ldots \;\; \bar{x}_m)^{\mathrm{T}} \, , \quad \bar{x}_i = \frac{1}{n} \sum_{l=1}^{n} x_{il} \, ; \quad i = 1, \ldots, m \, . \tag{8.3}$$

Inserting the formal decomposition

$$\bar{\boldsymbol{x}} = \boldsymbol{x}_0 + (\bar{\boldsymbol{x}} - \boldsymbol{\mu}) + \boldsymbol{f} \tag{8.4}$$

resolves $\bar{\boldsymbol{\beta}}$ according to

$$\bar{\boldsymbol{\beta}} = \boldsymbol{\beta}_0 + \boldsymbol{B}^{\mathrm{T}}(\bar{\boldsymbol{x}} - \boldsymbol{\mu}) + \boldsymbol{B}^{\mathrm{T}} \boldsymbol{f} \, . \tag{8.5}$$

The elements b_{ik} of the matrix $\boldsymbol{B}$ produce the components

$$\bar{\beta}_k = \sum_{i=1}^{m} b_{ik} \bar{x}_i \, ; \quad k = 1, \ldots, r \, . \tag{8.6}$$

As each of the $\bar{\beta}_k$ relies on the same set of input data, the $\bar{\beta}_k$ are necessarily dependent. The means (8.6) issue

$$\bar{\beta}_k = \sum_{i=1}^{m} b_{ik}\left[\frac{1}{n}\sum_{l=1}^{n} x_{il}\right] = \frac{1}{n}\sum_{l=1}^{n}\left[\sum_{i=1}^{m} b_{ik}x_{il}\right]$$
$$= \frac{1}{n}\sum_{l=1}^{n}\bar{\beta}_{kl}\,; \quad k = 1,\dots,r\,. \tag{8.7}$$

We consider the sums

$$\bar{\beta}_{kl} = \sum_{i=1}^{m} b_{ik}x_{il}\,; \quad k = 1,\dots,r \tag{8.8}$$

to be least squares estimators, the input data of which being the m individual measurements $x_{1l}, x_{2l}, \dots, x_{ml}$. There, each of the m means $\bar{x}_i$ contributes just the l-th measured value. To clarify the approach, we draw upon an illustration:

$$\begin{array}{cccccccc}
x_{11} & x_{12} & \dots & x_{1l} & \dots & x_{1n} & \Rightarrow & \bar{x}_1 \\
x_{21} & x_{22} & \dots & x_{2l} & \dots & x_{2n} & \Rightarrow & \bar{x}_2 \\
\dots & \dots & \dots & \dots & \dots & \dots & \dots & \dots \\
x_{m1} & x_{m2} & \dots & x_{ml} & \dots & x_{mn} & \Rightarrow & \bar{x}_m \\
\Downarrow & \Downarrow & & \Downarrow & & \Downarrow & & \Downarrow \\
\bar{\beta}_{k1} & \bar{\beta}_{k2} & & \bar{\beta}_{kl} & & \bar{\beta}_{kn} & \Rightarrow & \bar{\beta}_k
\end{array}$$

According to (8.7), the $\bar{\beta}_k$ are the arithmetic means of the n estimators $\bar{\beta}_{kl}$. The differences $\bar{\beta}_{kl} - \bar{\beta}_k$ establish the empirical variances and covariances

$$s_{\bar{\beta}_k\bar{\beta}_{k'}} = \frac{1}{n-1}\sum_{l=1}^{n}\left(\bar{\beta}_{kl} - \bar{\beta}_k\right)\left(\bar{\beta}_{k'l} - \bar{\beta}_{k'}\right); \quad k, k' = 1,\dots,r\,. \tag{8.9}$$

For convenience, we let them define the empirical variance–covariance matrix

$$\boldsymbol{s}_{\bar{\boldsymbol{\beta}}} = \begin{pmatrix} s_{\bar{\beta}_1\bar{\beta}_1} & s_{\bar{\beta}_1\bar{\beta}_2} & \dots & s_{\bar{\beta}_1\bar{\beta}_r} \\ s_{\bar{\beta}_2\bar{\beta}_1} & s_{\bar{\beta}_2\bar{\beta}_2} & \dots & s_{\bar{\beta}_2\bar{\beta}_r} \\ \dots & \dots & \dots & \dots \\ s_{\bar{\beta}_r\bar{\beta}_1} & s_{\bar{\beta}_r\bar{\beta}_2} & \dots & s_{\bar{\beta}_r\bar{\beta}_r} \end{pmatrix}, \quad s_{\bar{\beta}_k\bar{\beta}_k} \equiv s^2_{\bar{\beta}_k}\,. \tag{8.10}$$

Inserting the differences

$$\bar{\beta}_{kl} - \bar{\beta}_k = \sum_{i=1}^{m} b_{ik}\left(x_{il} - \bar{x}_i\right); \quad l = 1,\dots,n; \quad k = 1,\dots,r$$

into (8.9) we obtain

$$s_{\bar{\beta}_k \bar{\beta}_{k'}} = \frac{1}{n-1} \sum_{l=1}^{n} \left[\sum_{i=1}^{m} b_{ik} \left(x_{il} - \bar{x}_i \right) \right] \left[\sum_{j=1}^{m} b_{jk'} \left(x_{jl} - \bar{x}_j \right) \right]$$
$$= \sum_{i,j}^{m} b_{ik} b_{jk'} s_{ij} \tag{8.11}$$

in which the quantities

$$s_{ij} = \frac{1}{n-1} \sum_{l=1}^{n} \left(x_{il} - \bar{x}_i \right) \left(x_{jl} - \bar{x}_j \right) \; ; \quad i, j = 1, \ldots, m \tag{8.12}$$

designate the elements of the empirical variance–covariance matrix

$$\boldsymbol{s} = \begin{pmatrix} s_{11} & s_{12} & \ldots & s_{1m} \\ s_{21} & s_{22} & \ldots & s_{2m} \\ \ldots & \ldots & \ldots & \ldots \\ s_{m1} & s_{m2} & \ldots & s_{mm} \end{pmatrix} \tag{8.13}$$

of the input data. By means of the column vectors $\boldsymbol{b}_k$; $k = 1, \ldots, r$ of the matrix $\boldsymbol{B}$,

$$\boldsymbol{B} = \left(\boldsymbol{b}_1 \quad \boldsymbol{b}_2 \quad \cdots \quad \boldsymbol{b}_r \right) , \tag{8.14}$$

we may cast the variances and covariances (8.11) into

$$s_{\bar{\beta}_k \bar{\beta}_{k'}} = \boldsymbol{b}_k^{\mathrm{T}} \boldsymbol{s} \, \boldsymbol{b}_{k'} \; ; \quad k, \, k' = 1, \ldots, r \, . \tag{8.15}$$

Hence, (8.10) turns into

$$s_{\bar{\beta}} = \begin{pmatrix} \boldsymbol{b}_1^{\mathrm{T}} \boldsymbol{s} \, \boldsymbol{b}_1 & \boldsymbol{b}_1^{\mathrm{T}} \boldsymbol{s} \, \boldsymbol{b}_2 & \ldots & \boldsymbol{b}_1^{\mathrm{T}} \boldsymbol{s} \, \boldsymbol{b}_r \\ \boldsymbol{b}_2^{\mathrm{T}} \boldsymbol{s} \, \boldsymbol{b}_1 & \boldsymbol{b}_2^{\mathrm{T}} \boldsymbol{s} \, \boldsymbol{b}_2 & \ldots & \boldsymbol{b}_2^{\mathrm{T}} \boldsymbol{s} \, \boldsymbol{b}_r \\ \ldots & \ldots & \ldots & \ldots \\ \boldsymbol{b}_r^{\mathrm{T}} \boldsymbol{s} \, \boldsymbol{b}_1 & \boldsymbol{b}_r^{\mathrm{T}} \boldsymbol{s} \, \boldsymbol{b}_2 & \ldots & \boldsymbol{b}_r^{\mathrm{T}} \boldsymbol{s} \, \boldsymbol{b}_r \end{pmatrix} = \boldsymbol{B}^{\mathrm{T}} \boldsymbol{s} \, \boldsymbol{B} \, . \tag{8.16}$$

For a moment, let the input data themselves be independent. Thus, for any fixed i, the order of the x_{il} is indeterminate so that individuals may be interchanged. Clearly, permutations do not alter the $\bar{\beta}_k$, they alter, however, the $\bar{\beta}_{kl}$. At the same time, the diagonal elements of the matrix (8.13) are not affected, the non-diagonal elements, however, are. Moreover, the diagonal as well as the off-diagonal elements of the matrix (8.16) will alter. Again, this complies perfectly with the properties of the underlying statistical model and in particular with our aim to introduce confidence intervals according to Student.

Each of the quantities $\bar{\beta}_{kl}$, as defined in (8.8), implies a set of input data $x_{1l}, x_{2l}, \ldots, x_{ml}$, the elements of which may or may not be dependent. But let us remind that the consecutive sets $l = 1, \ldots, n$ are considered independent. From there, the $\bar{\beta}_{kl}$; $l = 1, 2, \ldots, n$ are normally distributed and independent. Hence, we expect the intervals

$$\bar{\beta}_k \pm \frac{t_P(n-1)}{\sqrt{n}} \sqrt{\sum_{i,j}^{m} b_{ik} b_{jk}\, s_{ij}}\,; \quad k = 1, \ldots, r \tag{8.17}$$

to localize the components μ_k; $k = 1, \ldots, r$ of the vector

$$\boldsymbol{\mu}_{\bar{\beta}} = (\mu_1 \quad \mu_2 \quad \ldots \quad \mu_r)^{\mathrm{T}} = \boldsymbol{\beta}_0 + \boldsymbol{B}^{\mathrm{T}} \boldsymbol{f} \tag{8.18}$$

with probability P as given by Student's $t_P(n-1)$.

8.2 Propagation of Systematic Errors

From (8.4) we read the vector

$$\boldsymbol{f}_{\bar{\beta}} = \boldsymbol{B}^{\mathrm{T}} \boldsymbol{f} \tag{8.19}$$

of the propagated systematic errors, written in components

$$f_{\bar{\beta}_k} = \sum_{i=1}^{m} b_{ik} f_i\,; \quad k = 1, \ldots, r\,. \tag{8.20}$$

The worst-case estimations are

$$f_{s,\bar{\beta}_k} = \sum_{i=1}^{m} |\, b_{ik} \,|\, f_{s,i}\,. \tag{8.21}$$

For equal systematic errors, $f_i = f$, $f_{s,i} = f_s\,;\ i = 1, \ldots, m$ the estimations turn out more favorable,

$$f_{s,\bar{\beta}_k} = f_s \left| \sum_{i=1}^{m} b_{ik} \right|\,. \tag{8.22}$$

According to (8.21), the formalism seems to respond to a rising number m of input data with a possibly inappropriate growth in $f_{s,\bar{\beta}_k}$, thus blotting out the gain of input information. This however does not apply, since the method of least squares brings about as much as a weighting of errors thus preventing pile ups. An example was given in (6.25).

8.3 Uncertainties of the Estimators

Combining the confidence intervals (8.17) with the worst-case estimations (8.21) issues the results

$$\bar{\beta}_k - u_{\bar{\beta}_k} \le \beta_{0,\,k} \le \bar{\beta}_k + u_{\bar{\beta}_k}\,; \quad k = 1, \ldots, r$$

$$u_{\bar{\beta}_k} = \frac{t_P\,(n-1)}{\sqrt{n}} \sqrt{\sum_{i,j}^{m} b_{ik} b_{jk} s_{ij}} + \sum_{i=1}^{m} |\; b_{ik} \;|\; f_{s,i} \tag{8.23}$$

or, if instead of (8.21) reference is taken to (8.22),

$$\bar{\beta}_k - u_{\bar{\beta}_k} \le \beta_{0,\,k} \le \bar{\beta}_k + u_{\bar{\beta}_k}\,; \quad k = 1, \ldots, r$$

$$u_{\bar{\beta}_k} = \frac{t_P\,(n-1)}{\sqrt{n}} \sqrt{\sum_{i,j}^{m} b_{ik} b_{jk} s_{ij}} + f_s \left| \sum_{i=1}^{m} b_{ik} \right| . \tag{8.24}$$

As illustrated in Fig. 8.1, we should not be exclusively fixated on the numerical values of the least squares estimates. Rather, considering one and the same physical quantity, say, $\beta_{0,k}$, there may be a set of varying uncertainty intervals coming from different adjustments. As each adjustment will certainly imply differing physical and metrological details, the actual positions and length of the respective uncertainty intervals cannot be expected to be identical. Nevertheless, they should collectively localize the common true value.

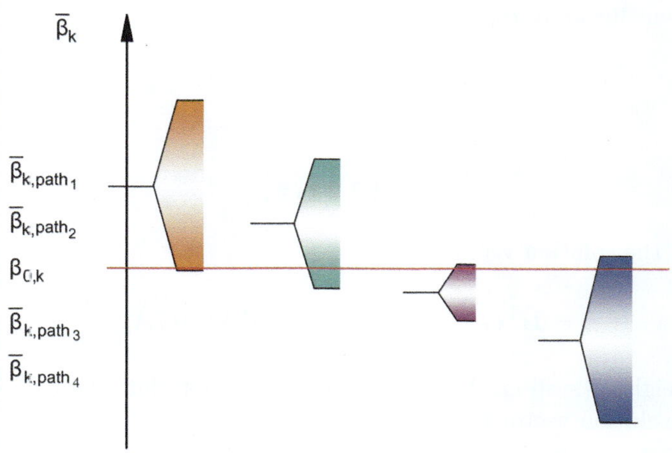

Fig. 8.1. Uncertainty intervals coming from different experiments and aiming at one and the same physical quantity, say $\beta_{0,k}$, should mutually overlap

8.4 Weighting Factors

To boost the influence of the more accurate input data and to weaken the influence of the less accurate ones, so-called weight factors may be introduced. For convenience, weight factors are gathered within a weight matrix. The classical Gaussian error calculus, which dismisses unknown systematic errors, defines the weight matrix of an inconsistent linear system via the Gauss–Markov theorem. In a sense, this weight matrix attempts to issue an objective weighting procedure. The uncertainties of the adjusted estimators are considered minimal and thus called "optimal." Unfortunately, the theorem has never been operative. Firstly, the so-defined weight matrix is based on the unknown theoretical variances and the unknown covariances[1] of the input data and thus does not exist. Secondly, the theorem is known to fail in the presence of biases. Thirdly, at any rate, the uncertainties still depend on the choice of the design matrix $\boldsymbol{A}$.

Nevertheless, weight factors prove useful, given the altered results continue to localize the true values. However, as there is no theorem producing a weight matrix, be it an "optimal one" or not, it is left to the experimenters to decide on which of the input data to boost and which to weaken.

Let us consider a diagonal matrix of weights

$$\boldsymbol{G} = \mathrm{diag}\{g_1, g_2, \ldots, g_m\} \tag{8.25}$$

where the g_i; $i = 1, \ldots, m$ are taken to be the reciprocals of the uncertainties of the input data

$$g_i = 1/u_{\bar{x}_i}\,; i = 1, \ldots, m\,. \tag{8.26}$$

The smaller the uncertainty $u_{\bar{x}_i}$, the larger the weight. Left-multiplying the inconsistent linear system

$$\boldsymbol{A}\boldsymbol{\beta} \approx \bar{\boldsymbol{x}}$$

by $\boldsymbol{G}$,

$$\boldsymbol{G}\boldsymbol{A}\boldsymbol{\beta} \approx \boldsymbol{G}\bar{\boldsymbol{x}}\,, \tag{8.27}$$

produces the solution vector

$$\bar{\boldsymbol{\beta}} = \tilde{\boldsymbol{B}}^{\mathrm{T}}\boldsymbol{G}\bar{\boldsymbol{x}}\,, \quad \tilde{\boldsymbol{B}} = (\boldsymbol{G}\boldsymbol{A})\left[(\boldsymbol{G}\boldsymbol{A})^{\mathrm{T}}(\boldsymbol{G}\boldsymbol{A})\right]^{-1} \tag{8.28}$$

of the weighted system. A weight matrix does not shift the components of the true solution vector $\boldsymbol{\beta}_0$ as

[1] To recall, theoretical covariances have a fixed sign; empirical covariances may be either positive or negative.

$$\boldsymbol{G A \beta}_0 = \boldsymbol{G x}_0$$

re-establishes

$$\boldsymbol{\beta}_0 = \left[(\boldsymbol{GA})^{\mathrm{T}} (\boldsymbol{GA}) \right]^{-1} (\boldsymbol{GA})^{\mathrm{T}} \boldsymbol{G x}_0$$

regardless of the weights chosen. This property per se proves self-evident; nevertheless, it turns out crucial in regard to the localization of the estimators' true values.

To assign uncertainties to the solution vector of the weighted system we start from

$$\begin{aligned} \bar{\boldsymbol{\beta}} &= \tilde{\boldsymbol{B}}^{\mathrm{T}} \boldsymbol{G} \left[\boldsymbol{x}_0 + (\bar{\boldsymbol{x}} - \boldsymbol{\mu}) + \boldsymbol{f} \right] \\ &= \boldsymbol{\beta}_0 + \tilde{\boldsymbol{B}}^{\mathrm{T}} \boldsymbol{G} \left(\bar{\boldsymbol{x}} - \boldsymbol{\mu} \right) + \tilde{\boldsymbol{B}}^{\mathrm{T}} \boldsymbol{G f} \, . \end{aligned} \tag{8.29}$$

The empirical variance–covariance matrix of the solution vector takes the form

$$\boldsymbol{s}_{\bar{\boldsymbol{\beta}}} = \tilde{\boldsymbol{B}}^{\mathrm{T}} \left(\boldsymbol{G}^{\mathrm{T}} \boldsymbol{s} \, \boldsymbol{G} \right) \tilde{\boldsymbol{B}} \tag{8.30}$$

with $\boldsymbol{s}$ as given in (8.13). Let

$$\tilde{\boldsymbol{\mu}}_{\bar{\boldsymbol{\beta}}} = \boldsymbol{\beta}_0 + \tilde{\boldsymbol{B}}^{\mathrm{T}} \boldsymbol{G f} \tag{8.31}$$

denote the expectation of the solution vector (8.29). Hence, there are confidence intervals

$$\bar{\beta}_k \pm \frac{t_P(n-1)}{\sqrt{n}} \sqrt{\sum_{i,j}^{m} \tilde{b}_{ik} \tilde{b}_{jk} g_i g_j \; s_{ij}} \; ; \quad k = 1, \ldots, r \tag{8.32}$$

localizing the components of the vector $\tilde{\boldsymbol{\mu}}_{\bar{\boldsymbol{\beta}}}$ with probability P as given by Student's $t_P(n-1)$.

Finally, we consider the vector of the propagated systematic errors

$$\tilde{\boldsymbol{f}}_{\bar{\boldsymbol{\beta}}} = \tilde{\boldsymbol{B}}^{\mathrm{T}} \boldsymbol{G f} \tag{8.33}$$

with components

$$\tilde{f}_{\bar{\beta}_k} = \sum_{i=1}^{m} \tilde{b}_{ik} g_i f_i \, ; \quad k = 1, \ldots, r \tag{8.34}$$

the worst-case estimates of which are given by

$$\tilde{f}_{s,\bar{\beta}_k} = \sum_{i=1}^{m} \mid \tilde{b}_{ik} \mid g_i f_{s,i} \, ; \quad k = 1, \ldots, r \, . \tag{8.35}$$

In case of equal systematic errors this turns into

$$\tilde{f}_{s,\bar{\beta}_k} = f_s \left| \sum_{i=1}^{m} \tilde{b}_{ik} g_i \right| \; ; \quad k = 1, \ldots, r \, . \tag{8.36}$$

After all, the result of the weighted adjustment is given by

$$\bar{\beta}_k - u_{\bar{\beta}_k} \le \beta_{0,k} \le \bar{\beta}_k + u_{\bar{\beta}_k} \; ; \quad k = 1, \ldots, r$$

$$u_{\bar{\beta}_k} = \frac{t_P(n-1)}{\sqrt{n}} \sqrt{\sum_{i,j}^{m} \tilde{b}_{ik} \tilde{b}_{jk} g_i g_j s_{ij}} + \sum_{i=1}^{m} | \, \tilde{b}_{ik} \, | \, g_i f_{s,i} \tag{8.37}$$

or, alternatively, by

$$\bar{\beta}_k - u_{\bar{\beta}_k} \le \beta_{0,k} \le \bar{\beta}_k + u_{\bar{\beta}_k} \; ; \quad k = 1, \ldots, r$$

$$u_{\bar{\beta}_k} = \frac{t_P(n-1)}{\sqrt{n}} \sqrt{\sum_{i,j}^{m} \tilde{b}_{ik} \tilde{b}_{jk} g_i g_j s_{ij}} + f_s \left| \sum_{i=1}^{m} \tilde{b}_{ik} g_i \right| \, . \tag{8.38}$$

Remarkably enough, whether or not there is a weight matrix and in case there is one independent of the weights chosen, the inequalities given in (8.23), (8.24), (8.37), and (8.38) encase the true values $\beta_{0,k}\, ; k = 1, . . ., r$. – In a sense, this observation may be seen to re-establish the order of least squares.

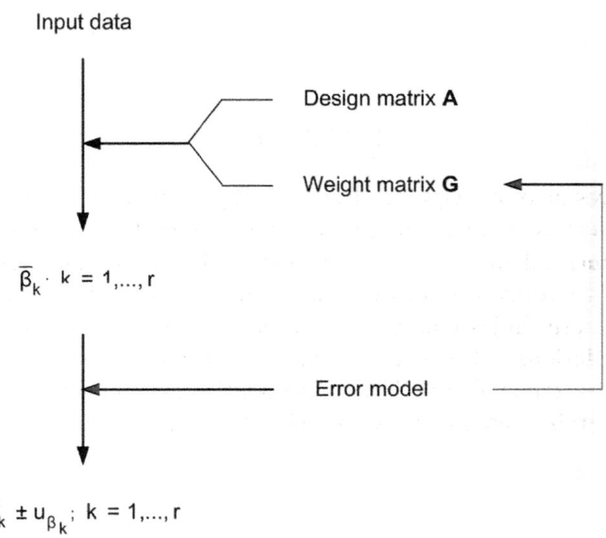

Fig. 8.2. The method of least squares is tied down to the input data, the design matrix $\boldsymbol{A}$, the weight matrix $\boldsymbol{G}$, and the error model

As Fig. 8.2 illustrates, least squares adjustments are tied down to

- the input data,
- the design matrix $\boldsymbol{A}$,
- the weight matrix $\boldsymbol{G}$ and, in particular, to
- the error model.

8.5 Example

Under the error model discussed, the method of least squares provides robust and reliable estimators. Nevertheless, it might be of interest to unravel the method's localization properties in detail and to explore where its safety might expire. To this end, we resort to a numerical example.

Data Simulation

Let us consider a linear system

$$\boldsymbol{A}\boldsymbol{\beta}_0 = \boldsymbol{x}_0$$

with design matrix $\boldsymbol{A}$, true solution vector $\boldsymbol{\beta}_0$, and true vector of input data $\boldsymbol{x}_0$ as given by

$$\boldsymbol{A} = \begin{pmatrix} -1 & -3 & 1 & 0 & 2 \\ 0 & -1 & 2 & -3 & 1 \\ 3 & 1 & -2 & -1 & 4 \\ -2 & 1 & -1 & 3 & 2 \\ 3 & -1 & 2 & 1 & 0 \\ -2 & 2 & 0 & -1 & 3 \\ 0 & 2 & -2 & 3 & -1 \end{pmatrix} ; \quad \boldsymbol{x}_0 = \begin{pmatrix} 6 \\ -3 \\ 15 \\ 19 \\ 11 \\ 13 \\ 5 \end{pmatrix} ; \quad \boldsymbol{\beta}_0 = \begin{pmatrix} 1 \\ 2 \\ 3 \\ 4 \\ 5 \end{pmatrix} .$$

We superpose random and systematic errors onto the components of the true vector $\boldsymbol{x}_0$. To visualize the influence of the design matrix, we assume the random errors to follow identical normal densities with theoretical variances

$$\sigma_i = 2 * 10^{-4} ; \quad i = 1, \ldots, 5 .$$

To begin with, Fig. 8.3 considers random errors alone. We observe that the difference between $\bar{\beta}_3$ and $\beta_{0,3}$ nearly exhausts the confidence interval.

Further, Fig. 8.4 explores the influence of systematic errors alone. Here, the interplay between the signs of the elements of the matrix $\boldsymbol{B}$ and the signs of the actual systematic errors f_i; $i = 1, \ldots, 5$ are crucial. We put

$$f_{s,i} = \sqrt{3} * \sigma_i ; \quad i = 1, \ldots, 5$$

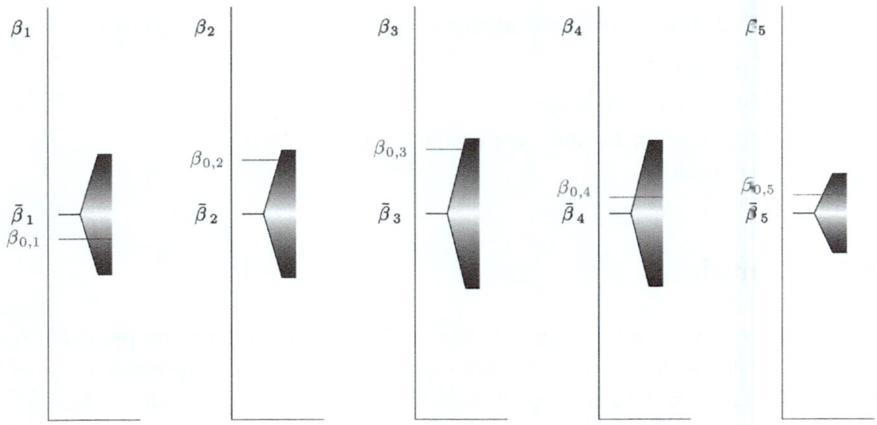

Fig. 8.3. Data simulation, random errors alone

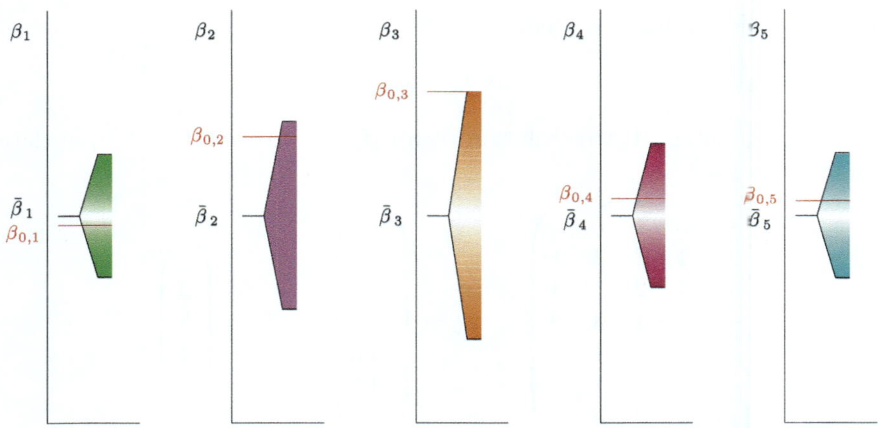

Fig. 8.4. Data simulation, systematic errors alone; $f_i = \text{sign}\,(b_{i,3}) f_{s,i}$; $i = 1, \ldots, 5$

and, in order to fathom the worst possiblity,

$$f_i = \text{sign}(b_{i,3}) * f_{s,i}\,; \quad i = 1, \ldots, 5\,.$$

Clearly, this choice will exhaust the systematic uncertainty of $\bar{\beta}_3$. It appears meaningless to speculate about the chances of a coincidence of the two events yanked out with respect to $\bar{\beta}_3$. We may, however, take the example to underscore the safety of the approach.

Part IV

Essence of Metrology

9 Dissemination of Units

Any system of physical units suffers from the distinction between defined and realized units. Defined units are theoretically conceived, faultless quantities. Realized units, by contrast, are to be implemented on the part of measuring devices and therefore affected by imperfections. This is why realized units are charged by uncertainties. On behalf of the *Bureau International des Poids et Mesures* (BIPM) [1], the universally accepted system of units is the SI.

9.1 Working Standards

So-called working standards are intended for everyday use. They are to be derived from the realized SI units or primary standards. Let us postpone the design of multiples and sub-multiples of units and firstly tackle the linking of standards of equal nominal values.

To meet the demands of practice, hierarchies of standards of falling accuracies have been established. Let $H_k;\ k = 0, 1, 2, \ldots$ designate "the ranks" of the standards and assume increasing k-s to express decreasing accuracies. Let H_0 mark the primary standard. Figure 9.1 sketches the flow of the true values within a hierarchy of SI standards of equal nominal value. The uncertainty of each standard is meant to localize the pertaining true value $\beta_{0,k}$.

Suppose the comparator operates according to

$$\beta_k = \beta_{k-1} + x_k\,; \quad k = 1, 2, \ldots \tag{9.1}$$

where β_k designates a standard of rank H_k, which is to be calibrated, β_{k-1} a standard of rank H_{k-1}, which accomplishes the link-up and, finally, x_k the difference displayed by the comparator. The link-up implies a loss of accuracy as the uncertainty of the primary standard adds to the uncertainty of the linking procedure. Let us consider n repeated measurements

$$x_{kl} = x_{0,k} + (x_{kl} - \mu_k) + f_k\,; \quad l = 1, \ldots, n\,. \tag{9.2}$$

As usual, $x_{0,k}$ designates the true value of the displayed difference x_{kl}, $x_{kl} - \mu_k$ a random error, and f_k a systematic error. The n repeated measurements issue an arithmetic mean and an empirical variance

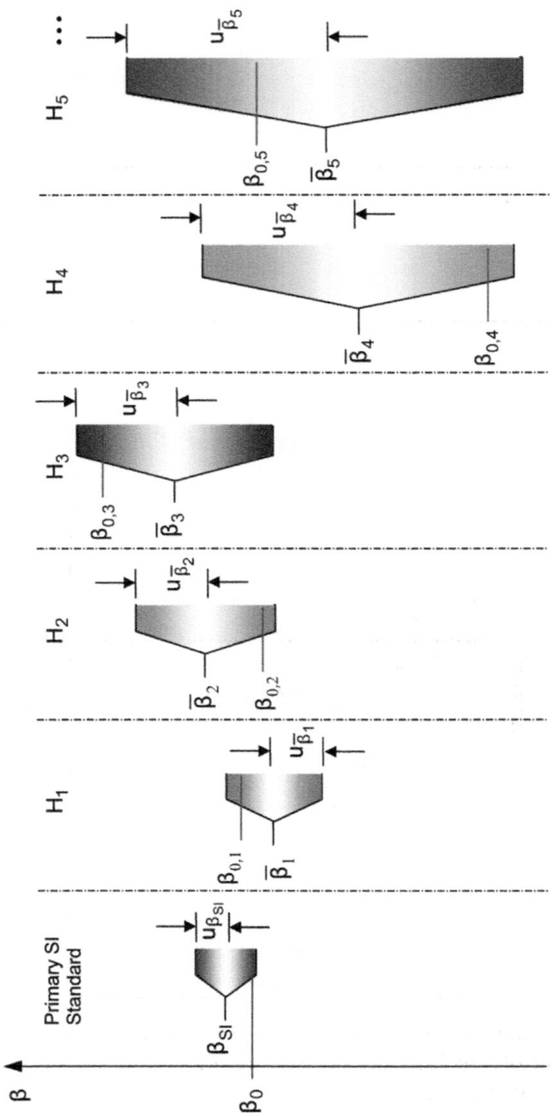

Fig. 9.1. A hierarchy of SI standards of equal nominal value and ranks H_1, H_2, ... implying growing uncertainties. The uncertainty of each standard should localize the pertaining true value $\beta_{0,k}$

$$\bar{x}_k = \frac{1}{n}\sum_{l=1}^{n} x_{kl}\,, \qquad s_{x_k}^2 = \frac{1}{n-1}\sum_{l=1}^{n}\left(x_{kl}-\bar{x}_k\right)^2\,. \tag{9.3}$$

Let the systematic error f_k be confined to

$$-f_{s,k} \le f_k \le f_{s,k}\,. \tag{9.4}$$

Thus, we consider the interval

$$\bar{x}_k - u_{\bar{x}_k} \leq x_{0,k} \leq \bar{x}_k + u_{\bar{x}_k}\,; \quad u_{\bar{x}_k} = \frac{t_P(n-1)}{\sqrt{n}}\, s_{x_k} + f_{s,k} \tag{9.5}$$

to localize the true value $x_{0,k}$ of the displayed differences x_{kl}; $l = 1, \ldots, n$.

Let β_0 denote the unknown true value of a primary standard of nominal value $N(\beta)$, say, $N(\beta) = 1\,\mathrm{kg}$ so that

$$N(\beta) = \beta_0 + f_{N(\beta)}\,, \qquad -u_{N(\beta)} \leq f_{N(\beta)} \leq u_{N(\beta)}\,. \tag{9.6}$$

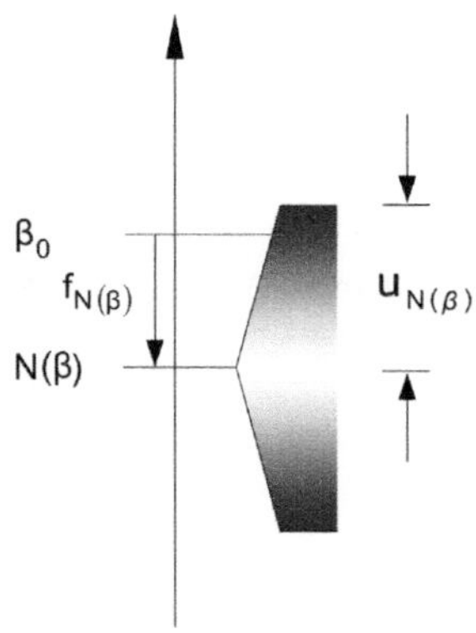

Fig. 9.2. Nominal value $N(\beta)$ and true value β_0 of a primary standard

Figure 9.2 illustrates the difference between β_0 and $N(\beta)$. Whenever the primary standard is used, β_0 is effective. Let us link a working standard β_1 of true value $\beta_{0,1}$ to the primary standard. The physical error equation reads

$$\beta_{1l} = \beta_0 + x_{1l} = \beta_0 + x_{0,1} + (x_{1l} - \mu_1) + f_1\,; \quad l = 1, \ldots, n\,. \tag{9.7}$$

The sum of β_0 and $x_{0,1}$ on the right-hand side expresses the traceability of $\beta_{0,1}$ to β_0 via the true indication $x_{0,1}$ of the comparator

$$\beta_{0,1} = \beta_0 + x_{0,1}\,. \tag{9.8}$$

Averaging the n values β_{1l}; $l = 1, \ldots, n$ as given in (9.7) produces

$$\bar{\beta}_1 = \beta_0 + \bar{x}_1\,. \tag{9.9}$$

Still, $\bar{\beta}_1$ is metrologically undefined. In order to arrive at a serviceable statement, we have to substitute $N(\beta)$ for β_0

$$\bar{\beta}_1 = N(\beta) + \bar{x}_1 \tag{9.10}$$

and to add the associated systematic error $f_{N(\beta)}$ to the uncertainty of the modified mean

$$u_{\bar{\beta}_1} = \frac{t_P(n-1)}{\sqrt{n}} s_{x_1} + f_{s,1} + u_{N(\beta)} \,. \tag{9.11}$$

According to

$$\bar{\beta}_1 - u_{\bar{\beta}_1} \le \beta_{0,1} \le \bar{\beta}_1 + u_{\bar{\beta}_1} \tag{9.12}$$

we consider the interval $\bar{\beta}_1 \pm u_{\bar{\beta}_1}$ to localize the true value $\beta_{0,1}$ of the lower ranked standard $\bar{\beta}_1$. In general, the true value $\beta_{0,1}$ of $\bar{\beta}_1$ will differ from the true value β_0 of the primary standard $N(\beta)$. Let us repeat the procedure for $k = 2$

$$\beta_{2l} = \beta_{0,1} + x_{2l} = \beta_{0,1} + x_{0,2} + (x_{2l} - \mu_2) + f_2 \,; \quad l = 1, \ldots, n \,.$$

Again, the right-hand side expresses the traceability of $\beta_{0,2}$ to $\beta_{0,1}$ via $x_{0,2}$, the true indication of the comparator,

$$\beta_{0,2} = \beta_{0,1} + x_{0,2} = \beta_0 + x_{0,1} + x_{0,2} \,. \tag{9.13}$$

The average of the n values β_{2l}; $l = 1, \ldots, n$ yields

$$\bar{\beta}_2 = \beta_{0,1} + \bar{x}_2 \,.$$

As $\beta_{0,1}$ is unknown, we have to substitute $\bar{\beta}_1$ for $\beta_{0,1}$

$$\bar{\beta}_2 = \bar{\beta}_1 + \bar{x}_2 \tag{9.14}$$

and to add the uncertainty of $\bar{\beta}_1$ to the uncertainty of the modified mean $\bar{\beta}_2$

$$u_{\bar{\beta}_2} = \frac{t_P(n-1)}{\sqrt{n}} s_{x_1} + f_{s,1} + \frac{t_P(n-1)}{\sqrt{n}} s_{x_2} + f_{s,2} + u_{N(\beta)} \,. \tag{9.15}$$

After all, we consider the interval $\bar{\beta}_2 \pm u_{\bar{\beta}_2}$ to localize the true value $\beta_{0,2}$ of the lower ranked standard $\bar{\beta}_2$, written explicitly

$$\bar{\beta}_2 - u_{\bar{\beta}_2} \le \beta_{0,2} \le \bar{\beta}_2 + u_{\bar{\beta}_2} \,. \tag{9.16}$$

Ultimately, a standard of rank k reveals traceability according to

$$\beta_{0,k} = \beta_0 + \sum_{i=1}^{k} x_{0,i} \,. \tag{9.17}$$

9.2 Key Comparisons

So-called key comparisons ensure the national metrology institutes to rely on equivalent SI standards. Let $\bar{\beta}^{(1)}, \bar{\beta}^{(2)}, \ldots, \bar{\beta}^{(m)}$ denote a selected group of standards of the same nominal value, of the same hierarchical rank, and with true values $\beta_0^{(i)}$; $i = 1, \ldots, m$, Fig. 9.3. The true values of the standards need not coincide. However, it is indispensable that the standards' uncertainties localize the respective true values. Key comparisons are implemented via round robins. Here, a suitable transfer standard, say, T is passed on from one participant to the next. Each time the transfer standard gets calibrated, Fig. 9.4. During the portage the physical properties of the transfer standard are expected to remain constant. If this happens to apply, each link-up relates to the same true value. Afterward, the mutual consistency of the results is to be scrutinized. The i-th participant, $i \in 1, \ldots, m$, links the transfer standard

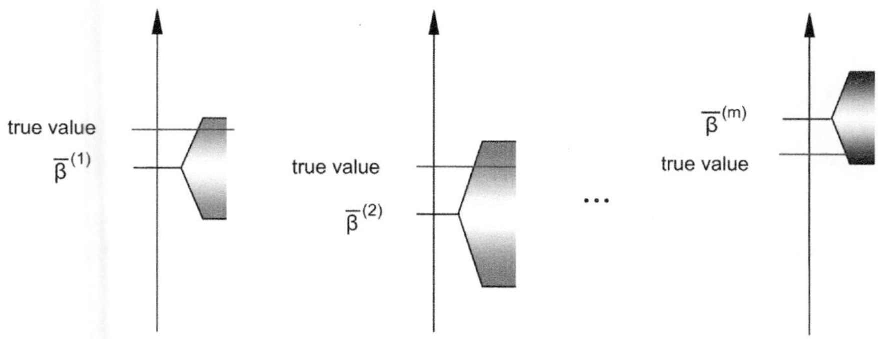

Fig. 9.3. The true values of standards holding the same metrological rank need not coincide nor must their uncertainties overlap. However, the respective uncertainties should localize the pertaining true values

T to his laboratory standard $\bar{\beta}^{(i)}$ where the uncertainties of the laboratory standards are considered to localize the belonging true values $\beta_0^{(i)}$,

$$\bar{\beta}^{(i)} - u_{\bar{\beta}^{(i)}} \le \beta_0^{(i)} \le \bar{\beta}^{(i)} + u_{\bar{\beta}^{(i)}}\,; \quad i = 1, \ldots, m\,.$$

For convenience, let us rewrite the inequalities as a set of equations

$$\bar{\beta}^{(i)} = \beta_0^{(i)} + f_{\bar{\beta}^{(i)}}\,, \quad -u_{\bar{\beta}^{(i)}} \le f_{\bar{\beta}^{(i)}} \le u_{\bar{\beta}^{(i)}}\,; \quad i = 1, \ldots, m\,. \tag{9.18}$$

The physical error equations of the m link-ups read

$$\begin{aligned} T_l^{(i)} &= \beta_0^{(i)} + x_l^{(i)}\,; \quad l = 1, \ldots, n \\ &= \beta_0^{(i)} + x_0^{(i)} + \left(x_l^{(i)} - \mu_{x^{(i)}}\right) + f_{\bar{x}^{(i)}}\,; \quad i = 1, \ldots, m \end{aligned}$$

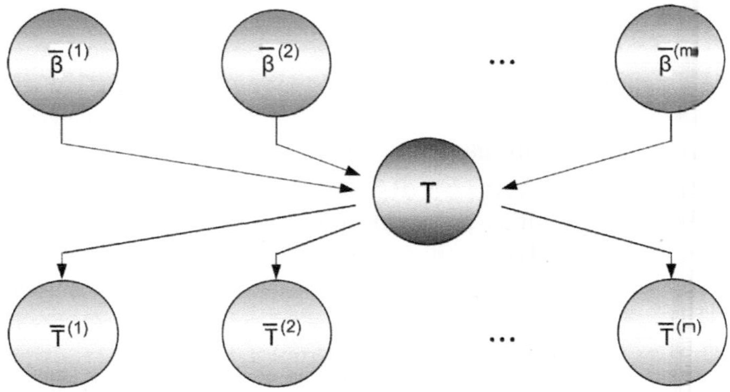

Fig. 9.4. Round robin for a group of m standards $\bar{\beta}^{(i)}$; $i = 1, \dots, m$ implemented via a transfer standard T

and

$$\begin{aligned}\bar{T}^{(i)} &= \beta_0^{(i)} + \bar{x}^{(i)} ; \quad i = 1, \dots, m \\ &= \beta_0^{(i)} + x_0^{(i)} + \left(\bar{x}^{(i)} - \mu_{x^{(i)}}\right) + f_{x^{(i)}} .\end{aligned}$$

For practical reasons, we have to substitute estimators $\bar{\beta}^{(i)}$ for the inaccessible true values $\beta_0^{(i)}$ and to include the imported uncertainty in the uncertainty $u_{\bar{T}^{(i)}}$ of the respective mean $\bar{T}^{(i)}$. Thus we have

$$\bar{T}^{(i)} = \bar{\beta}^{(i)} + \bar{x}^{(i)} ; \quad u_{\bar{T}^{(i)}} = u_{\bar{\beta}^{(i)}} + u_{\bar{x}^{(i)}} ; \quad i = 1, \dots, m . \tag{9.19}$$

Let T_0 designate the true value of the transfer standard. Then, as traceability suggests to put

$$T_0 = \beta_0^{(i)} + x_0^{(i)} ; \quad i = 1, \dots, m \tag{9.20}$$

we have

$$\left| \bar{T}^{(i)} - T_0 \right| \le u_{\bar{T}^{(i)}} ; \quad i = 1, \dots, m . \tag{9.21}$$

Figure 9.5 depicts the result of a consistent round robin. We might wish to quantify differences of the kind $\bar{T}^{(i)} - \bar{T}^{(j)}$. Here, we should observe

$$\left| \bar{T}^{(i)} - \bar{T}^{(j)} \right| \le u_{\bar{\beta}^{(i)}} + u_{\bar{x}^{(i)}} + u_{\bar{\beta}^{(j)}} + u_{\bar{x}^{(j)}} ; \quad i = 1, \dots, m .$$

For completeness, we finally discuss what is called the key comparison reference value, abbreviated KCRV. The KCRV is nothing else but the weighted grand mean[1] as defined in (6.23)

[1] As addressed in Sect. 6.2, a group of means with unequal true values should not be averaged.

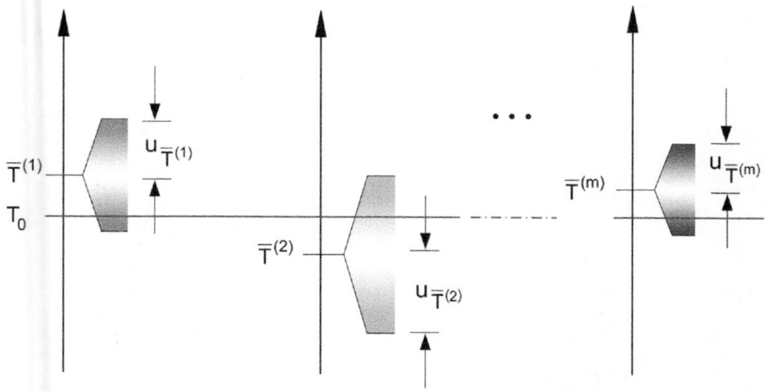

Fig. 9.5. In case of consistency one and the same horizontal line intersects each of the uncertainties $u_{\bar{T}^{(i)}}$

$$\bar{\beta} = \sum_{i=1}^{m} w_i \bar{T}^{(i)} \,, \quad w_i = \frac{g_i^2}{\sum\limits_{i=1}^{m} g_i^2} \,, \quad g_i = \frac{1}{u_{\bar{T}^{(i)}}} \,. \tag{9.22}$$

We derive the uncertainties $u_{\bar{d}_i}$ of the m differences

$$\bar{d}_i = \bar{T}^{(i)} - \bar{\beta} \,; \quad i = 1, \ldots, m \,. \tag{9.23}$$

To this end, we insert (9.19) into (9.23), so that

$$\bar{d}_i = \bar{\beta}^{(i)} + \bar{x}^{(i)} - \sum_{j=1}^{m} w_j \left(\bar{\beta}^{(j)} + \bar{x}^{(j)} \right) .$$

Due to (9.18) we have

$$\bar{d}_i = \beta_0^{(i)} + f_{\bar{\beta}^{(i)}} + \bar{x}^{(i)} - \sum_{j=1}^{m} w_j \left(\beta_0^{(j)} + f_{\bar{\beta}^{(j)}} + \bar{x}^{(j)} \right) .$$

Inserting the error equations of the $\bar{x}^{(i)}$; $i = 1, \ldots, m$ yields

$$\begin{aligned} \bar{d}_i &= \beta_0^{(i)} + f_{\bar{\beta}^{(i)}} + x_0^{(i)} + \left(\bar{x}^{(i)} - \mu_{x^{(i)}} \right) + f_{x^{(i)}} \\ &\quad - \sum_{j=1}^{m} w_j \left(\beta_0^{(j)} + f_{\bar{\beta}^{(j)}} + x_0^{(j)} + \left(\bar{x}^{(j)} - \mu_{x^{(j)}} \right) + f_{x^{(j)}} \right) . \end{aligned}$$

Putting

$$f_i = f_{x^{(i)}} + f_{\bar{\beta}^{(i)}} \tag{9.24}$$

and considering (9.20), we arrive at

$$
\begin{aligned}
\bar{d}_i &= T_0 + \left(\bar{x}^{(i)} - \mu_{x^{(i)}}\right) + f_i - \sum_{j=1}^{m} w_j \left[T_0 + \left(\bar{x}^{(j)} - \mu_{x^{(j)}}\right) + f_j\right] \\
&= \left(\bar{x}^{(i)} - \mu_{x^{(i)}}\right) - \sum_{j=1}^{m} w_j \left(\bar{x}^{(j)} - \mu_{x^{(j)}}\right) + f_i - \sum_{j=1}^{m} w_j f_j \, .
\end{aligned}
\tag{9.25}
$$

Reverting to individual measurements $x_l^{(i)}$ we find

$$
\begin{aligned}
\bar{d}_i &= \frac{1}{n}\sum_{l=1}^{n}\left(x_l^{(i)} - \mu_{x^{(i)}}\right) - \sum_{j=1}^{m} w_j \left[\frac{1}{n}\sum_{l=1}^{n}\left(x_l^{(j)} - \mu_{x^{(j)}}\right)\right] + f_i - \sum_{j=1}^{m} w_j f_j \\
&= \frac{1}{n}\sum_{l=1}^{n}\left[\left(x_l^{(i)} - \mu_{x^{(i)}}\right) - \sum_{j=1}^{m} w_j \left(x_l^{(j)} - \mu_{x^{(j)}}\right)\right] + f_i - \sum_{j=1}^{m} w_j f_j \, ,
\end{aligned}
\tag{9.26}
$$

i.e.

$$
\begin{gathered}
\bar{d}_{i,l} = \left(x_l^{(i)} - \mu_{x^{(i)}}\right) - \sum_{j=1}^{m} w_j \left(x_l^{(j)} - \mu_{x^{(j)}}\right) + f_i - \sum_{j=1}^{m} w_j f_j \, ; \\
i = 1, \dots, m \, ; \quad l = 1, \dots, n \, .
\end{gathered}
\tag{9.27}
$$

Hence, we are in a position to define the n differences

$$
\bar{d}_{i,l} - \bar{d}_i = \left(x_l^{(i)} - \bar{x}^{(i)}\right) - \sum_{j=1}^{m} w_j \left(x_l^{(j)} - \bar{x}^{(j)}\right) . \tag{9.28}
$$

Let the empirical variances and covariances of the participants

$$
s_{x^{(i)}x^{(j)}} \equiv s_{ij} = \frac{1}{n-1}\sum_{l=1}^{n}\left(x_l^{(i)} - \bar{x}^{(i)}\right)\left(x_l^{(j)} - \bar{x}^{(j)}\right) ; \quad i, j = 1, \dots, m
$$

define an empirical variance–covariance matrix

$$
\boldsymbol{s} = \begin{pmatrix} s_{11} & s_{12} & \dots & s_{1m} \\ s_{21} & s_{22} & \dots & s_{2m} \\ \dots & \dots & \dots & \dots \\ s_{m1} & s_{m2} & \dots & s_{mm} \end{pmatrix} , \qquad s_{ii} \equiv s_i^2
$$

and the weights w_i an auxiliary vector

$$
\boldsymbol{w} = \begin{pmatrix} w_1 & w_2 & \dots & w_m \end{pmatrix}^{\mathrm{T}} .
$$

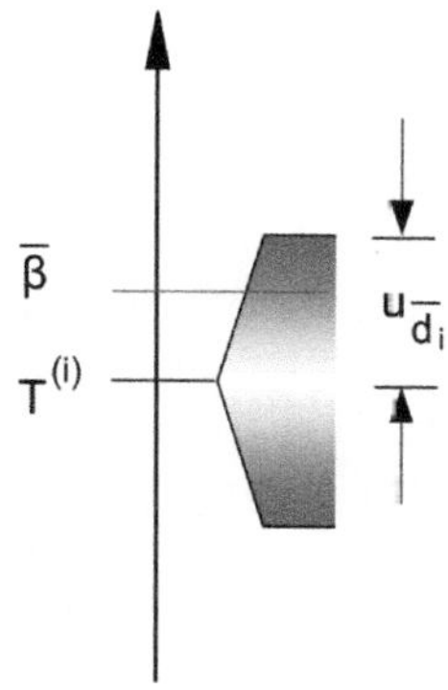

Fig. 9.6. None of the quantities $|\bar{T}^{(i)} - \bar{\beta}|$ should exceed the respective uncertainty $u_{\bar{d}_i}$

After all, (9.28) produces

$$s_{\bar{d}_i}^2 = \frac{1}{n-1}\sum_{l=1}^{n}\left(\bar{d}_{i,l} - \bar{d}_i\right)^2 = s_i^2 - 2\sum_{j=1}^{m} w_j\, s_{ij} + \boldsymbol{w}^{\mathrm{T}}\boldsymbol{s}\,\boldsymbol{w}\,. \tag{9.29}$$

The worst case estimation of (9.24) yields

$$f_{s,i} = f_{s,x^{(i)}} + u_{\bar{\beta}^{(i)}}\,; \quad i = 1,\dots,m\,.$$

Hence, the uncertainties of the differences (9.23) take the form

$$\begin{aligned} u_{\bar{d}_i} &= \frac{t_P(n-1)}{\sqrt{n}}\sqrt{s_i^2 - 2\sum_{j=1}^{m} w_j\, s_{ij} + \boldsymbol{w}^{\mathrm{T}}\boldsymbol{s}\,\boldsymbol{w}} \\ &\quad + (1 - w_i)\, f_{s,i} + \sum_{j=1 j\neq i}^{m} w_j f_{s,j}\,; \quad i = 1,\dots,m\,. \end{aligned} \tag{9.30}$$

We are free to add to the second term on the right $\pm w_i f_{s,i}$,

$$(1 - w_i)\, f_{s,i} + \sum_{j=1 j\neq i}^{m} w_j f_{s,j} \pm w_i f_{s,i} = f_{s,i} - 2w_i f_{s,i} + \sum_{j=1}^{m} w_j f_{s,j}$$

so that

$$\begin{aligned} u_{\bar{d}_i} &= \frac{t_P(n-1)}{\sqrt{n}}\sqrt{s_i^2 - 2\sum_{j=1}^{m} w_j\, s_{ij} + \boldsymbol{w}^{\mathrm{T}}\boldsymbol{s}\,\boldsymbol{w}} \\ &\quad + f_{s,i} - 2w_i f_{s,i} + \sum_{j=1}^{m} w_j f_{s,j}\,; \quad i = 1,\dots,m\,. \end{aligned} \tag{9.31}$$

As the $d_{0,i}$ vanish, the quotations $\mid \bar{d}_i - d_{0,i} \mid \le u_{\bar{d}_i}$ turn into

$$\left| \bar{T}^{(i)} - \bar{\beta} \right| \le u_{\bar{d}_i} \, ; \quad i = 1, \ldots, m \, . \tag{9.32}$$

Given everything has gone well, Fig. 9.6 would illustrate the result. In case of equal weights, we set $w_i = 1/m$. Whether or not (9.32) proves practical appears somewhat questionable. Let us recall, the grand mean $\bar{\beta}$ as defined in (9.22) is established through the calibrations $\bar{T}^{(i)}$ themselves. Consequently, we cannot rule out that just by chance, correct calibrations come out incorrect and incorrect ones correct. After all, unrecognized effects might either compensate or boost each other.

On the other hand, it appears simpler and might possibly be even more beneficial to check whether the uncertainties of the $\bar{T}^{(i)}$; $i = 1, \ldots, m$ mutually overlap. If so, this at least announces compatibility of the $\bar{\beta}^{(1)}$, $\bar{\beta}^{(2)}$, $\ldots, \bar{\beta}^{(m)}$.

10 Multiples and Sub-multiples

To establish a group of standards of varying nominal values, either multiples or sub-multiples, requires sophisticated linking procedures. As an example, we shall consider the down-scaling of the base unit of the mass, the kilogram.

10.1 Calibration Chains

Via regulation, the sub-multiples and multiples of the kilogram are

$$\ldots, 10\,\mathrm{g}\,,\ 20\,\mathrm{g}\,,\ 50\,\mathrm{g}\,,\ 100\,\mathrm{g}\,,\ 200\,\mathrm{g}\,,\ 500\,\mathrm{g}\,,\ 1\,\mathrm{kg}\,,\ 2\,\mathrm{kg}\,,\ 5\,\mathrm{kg}\,,\ 10\,\mathrm{kg}\,,\ \ldots \tag{10.1}$$

Let us confine ourselves to down-scaling and to consider weights of nominal values

$$\begin{gathered} N(m_1) = 500\,\mathrm{g}\,, \quad N(m_2) = 200\,\mathrm{g}\,, \quad N(m_3) = 200\,\mathrm{g} \\ N(m_4) = 100\,\mathrm{g}\,, \quad N(m_5) = 100\,\mathrm{g}\,. \end{gathered} \tag{10.2}$$

For convenience, we admit differences

$$\beta_k = m_k - N(m_k)\,; \quad k = 1, 2, \ldots \tag{10.3}$$

to exist between the physical masses m_k and their nominal values $N(m_k)$.

To control the link-up procedures, we additionally include a so-called *check-weight*, say,

$$N(m_6) = 300\,\mathrm{g}$$

of known mass m_6 and known uncertainty u_{m_6} and expect the two intervals

$$(m_6 \pm u_{m_6})_{\text{before calibration}} \qquad \text{and} \qquad (m_6 \pm u_{m_6})_{\text{after calibration}}$$

to overlap.

We assume the link-up to be established through a given primary standard of nominal value $N = 1\,\mathrm{kg}$ and physically true value N_0. We let $\bar{N}$ designate the associate mean with uncertainty $u_{\bar{N}}$, Fig. 10.1

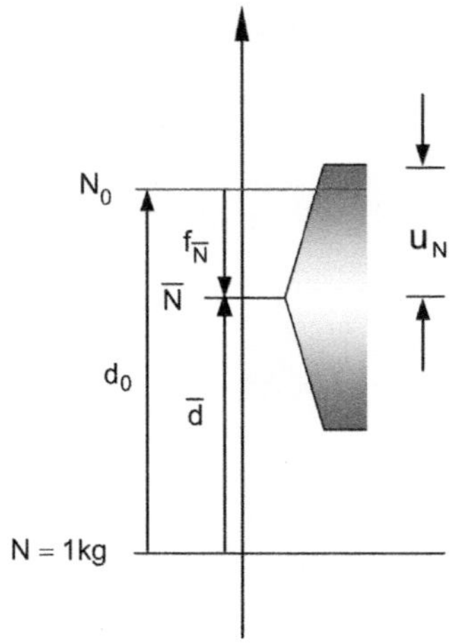

Fig. 10.1. Link-up standard with nominal value $N = 1\,\mathrm{kg}$, true value N_0, and mean value $\bar{N}$

$$\begin{gathered}\bar{N} - u_{\bar{N}} \le N_0 \le \bar{N} + u_{\bar{N}} \\ \bar{N} = N_0 + f_{\bar{N}}\,, \qquad -u_{\bar{N}} \le f_{\bar{N}} \le u_{\bar{N}}\,.\end{gathered} \tag{10.4}$$

We shall always liken groups of weights of equal nominal values, e.g.

$$\left.\begin{array}{lll}\{m_1 + m_2 + m_3 + m_6\} & \text{nominal value} & 1.2\,\mathrm{kg} \\ \{m_4 + m_5 + N_0\} & \text{nominal value} & 1.2\,\mathrm{kg}\end{array}\right\} \quad \text{1st comparison}$$

and assume $l = 1, \ldots, n$ repeated measurements. The observational equation reads

$$\{m_1 + m_2 + m_3 + m_6\} - \{m_4 + m_5 + N_0\} \approx \bar{x}_1\,; \quad \bar{x}_1 = \frac{1}{n}\sum_{=1}^{n} x_{1l}\,.$$

In physical terms, the true value of the primary standard enters. However, as N_0 is unknown, we have to substitute $\bar{N}$ for N_0 and to increase the uncertainty $u_{\bar{x}_1}$ of the right-hand side by $u_{\bar{N}}$ up to $u_{\bar{N}} + u_{\bar{x}_1}$

$$\begin{gathered}\{m_1 + m_2 + m_3 + m_6\} - \left\{m_4 + m_5 + \bar{N}\right\} \approx \bar{x}_1 \\ u_{\text{right-hand side}} = u_{\bar{N}} + u_{\bar{x}_1}\,.\end{gathered}$$

Referring to the deviations as defined in (10.3), we have

$$\beta_1 + \beta_2 + \beta_3 - \beta_4 - \beta_5 + \beta_6 \approx \bar{x}_1 + \bar{d} \tag{10.5}$$

where

$$\bar{d} = \bar{N} - N \tag{10.6}$$

denotes the difference between $\bar{N}$ and the nominal value N. Obviously, the true value of $\bar{d}$ is

$$d_0 = N_0 - N \tag{10.7}$$

and the associated error equation

$$\bar{d} = d_0 + f_{\bar{N}}\,; \quad -u_{\bar{N}} \le f_{\bar{N}} \le u_{\bar{N}}\,. \tag{10.8}$$

As we take from (10.5), the first row of the design matrix $\boldsymbol{A}$ is constituted by numbers $+1$ or -1, depending on whether the weight is a member of the first or of the second group of weights. Since there are other comparisons which do not concern all weights of the set (10.2), as e.g.

$$\left.\begin{array}{lll} \{m_6\} & \text{nominal value} & 300\,\text{g} \\ \{m_3 + m_5\} & \text{nominal value} & 300\,\text{g} \end{array}\right\} \quad \text{2nd comparison}$$

implying

$$-\beta_3 - \beta_5 + \beta_6 \approx \bar{x}_2$$

the design matrix also exhibit zeros. After all, the first two rows of the design matrix $\boldsymbol{A}$ read

$$\begin{array}{cccccc} 1 & 1 & 1 & -1 & -1 & 1 \\ 0 & 0 & -1 & 0 & -1 & 1 \end{array} \tag{10.9}$$

As there are $r = 6$ unknowns, $m > 6$ observational equations are needed. The rank of $\boldsymbol{A}$ is to be chosen to be 6.

The exemplary weighing scheme presented below is made up of $m = 16$ comparisons. The first five comparisons include the link-up standard $\bar{N}$, which is indicated by a (+) sign to the right of the matrix. For convenience we introduce an auxiliary vector

$$\bar{\boldsymbol{d}} = (\underbrace{\bar{d} \quad \ldots \quad \bar{d}}_{p} \mid \underbrace{0 \quad \ldots \quad 0}_{m-p})^{\mathrm{T}} \tag{10.10}$$

where $p = 5$. Furthermore, the scheme assumes 14 comparisons to include the check-weight. Setting

$$\boldsymbol{\beta} = (\beta_1 \quad \beta_2 \quad \ldots \quad \beta_r)^{\mathrm{T}}\,, \quad \bar{\boldsymbol{x}} = (\bar{x}_1 \quad \bar{x}_2 \quad \ldots \quad \bar{x}_m)^{\mathrm{T}}\,; \quad r = 6\,, \ m = 16$$

and

$$
\boldsymbol{A} = \begin{pmatrix}
1 & 1 & 1 & -1 & -1 & 1 \\
1 & -1 & 1 & 1 & 1 & 1 \\
1 & 1 & -1 & 1 & 1 & 1 \\
1 & 1 & 0 & 1 & -1 & 1 \\
1 & 0 & 1 & -1 & 1 & 1 \\
1 & -1 & 1 & -1 & -1 & -1 \\
1 & -1 & -1 & 1 & 1 & -1 \\
1 & 1 & -1 & -1 & -1 & -1 \\
1 & -1 & 0 & 1 & -1 & -1 \\
1 & 0 & -1 & -1 & 1 & -1 \\
0 & -1 & 0 & -1 & 0 & 1 \\
0 & -1 & 0 & -1 & 0 & 1 \\
0 & 0 & -1 & 0 & -1 & 1 \\
0 & 0 & -1 & 0 & -1 & 1 \\
0 & 0 & 0 & 1 & -1 & 0 \\
0 & 0 & 0 & 1 & -1 & 0
\end{pmatrix}
\begin{matrix} (+) \\ (+) \\ (+) \\ (+) \\ (+) \\ \\ \\ \\ \\ \\ \\ \\ \\ \\ \\ \\ \end{matrix}
$$

the inconsistent linear system takes the form

$$\boldsymbol{A\beta} \approx \bar{\boldsymbol{x}} + \bar{\boldsymbol{d}}\,. \tag{10.11}$$

The least squares solution vector is

$$
\begin{aligned}
\bar{\boldsymbol{\beta}} &= \left(\boldsymbol{A}^{\mathrm{T}}\boldsymbol{A}\right)^{-1}\boldsymbol{A}^{\mathrm{T}}\left(\bar{\boldsymbol{x}} + \bar{\boldsymbol{d}}\right) = \boldsymbol{B}^{\mathrm{T}}\left(\bar{\boldsymbol{x}} + \bar{\boldsymbol{d}}\right) \\
\boldsymbol{B} &= \boldsymbol{A}\left(\boldsymbol{A}^{\mathrm{T}}\boldsymbol{A}\right)^{-1} = (b_{ik})\;; \quad i = 1,\ldots,m\,; \quad k = 1,\ldots,r\,,
\end{aligned} \tag{10.12}
$$

in components

$$\bar{\beta}_k = \sum_{i=1}^{m} b_{ik}\bar{x}_i + \bar{d}\sum_{i=1}^{p} b_{ik}\,; \quad k = 1,\ldots,r\,. \tag{10.13}$$

Inserting the error equations

$$\bar{x}_i = x_{0,i} + (\bar{x}_i - \mu_i) + f_i\,, \quad \bar{d} = d_0 + f_{\bar{N}}$$

we find

$$\bar{\beta}_k = \sum_{i=1}^{m} b_{ik}\left[x_{0,i} + (\bar{x}_i - \mu_i) + f_i\right] + (d_0 + f_{\bar{N}})\sum_{i=1}^{p} b_{ik}\,; \quad k = 1,\ldots,r\,.$$

Obviously, the

$$\beta_{0,k} = \sum_{i=1}^{m} b_{ik}x_{0,i} + d_0\sum_{i=1}^{p} b_{ik}\,; \quad k = 1,\ldots,r \tag{10.14}$$

designate the flow of the true values $\beta_{0,k}$ of the estimators $\bar{\beta}_k$. The propagated systematic errors turn out as

$$f_{s,\bar{\beta}_k} = \sum_{i=1}^{m} | b_{ik} | f_{s,i} + u_{\bar{N}} \left| \sum_{i=1}^{p} b_{ik} \right| ; \quad k = 1, \ldots, r . \tag{10.15}$$

Returning to individual measurements, (10.13) yields

$$\bar{\beta}_{kl} = \sum_{i=1}^{m} b_{ik} x_{il} + \bar{d} \sum_{i=1}^{p} b_{ik} ; \quad k = 1, \ldots, r . \tag{10.16}$$

Hence, the differences

$$\bar{\beta}_{kl} - \bar{\beta}_k = \sum_{i=1}^{m} b_{ik} \left(x_{il} - \bar{x}_i \right) ; \quad k = 1, \ldots, r ; \quad l = 1, \ldots, n$$

produce the empirical variances and covariances of the components of the solution vector

$$\begin{aligned} s_{\beta_k \beta_{k'}} &= \frac{1}{n-1} \sum_{l=1}^{n} \left(\bar{\beta}_{kl} - \bar{\beta}_k \right) \left(\bar{\beta}_{k'l} - \bar{\beta}_{k'} \right) \\ &= \frac{1}{n-1} \sum_{l=1}^{n} \left[\sum_{i=1}^{m} b_{ik} \left(x_{il} - \bar{x}_i \right) \right] \left[\sum_{j=1}^{m} b_{jk'} \left(x_{jl} - \bar{x}_j \right) \right] \\ &= \sum_{i,j} b_{ik} b_{jk'} \, s_{ij} , \quad s_{\beta_k \beta_k} \equiv s_{\beta_k}^2 ; \quad k, k' = 1, \ldots, r . \end{aligned} \tag{10.17}$$

Here, the

$$s_{ij} = \frac{1}{n-1} \sum_{l=1}^{n} \left(x_{il} - \bar{x}_i \right) \left(x_{jl} - \bar{x}_j \right) ; \quad i, j = 1, \ldots, m$$

denote the empirical variances and covariances of the input data which we gather within a matrix $\boldsymbol{s}$, as done in (8.13). Thus, we have

$$\boldsymbol{s}_{\bar{\boldsymbol{\beta}}} = \begin{pmatrix} s_{\bar{\beta}_1 \bar{\beta}_1} & s_{\bar{\beta}_1 \bar{\beta}_2} & \ldots & s_{\bar{\beta}_1 \bar{\beta}_r} \\ s_{\bar{\beta}_2 \bar{\beta}_1} & s_{\bar{\beta}_2 \bar{\beta}_2} & \ldots & s_{\bar{\beta}_2 \bar{\beta}_r} \\ \ldots & \ldots & \ldots & \ldots \\ s_{\bar{\beta}_r \bar{\beta}_1} & s_{\bar{\beta}_r \bar{\beta}_2} & \ldots & s_{\bar{\beta}_r \bar{\beta}_r} \end{pmatrix} = \boldsymbol{B}^{\mathrm{T}} \boldsymbol{s} \, \boldsymbol{B} . \tag{10.18}$$

After all, the uncertainties of the adjusted weights $\bar{\beta}_k$; $k = 1, \ldots, r$ take the form

$$\begin{aligned} u_{\bar{\beta}_k} &= \frac{t_P(n-1)}{\sqrt{n}} s_{\bar{\beta}_k} + f_{s,\bar{\beta}_k} ; \quad k = 1, \ldots, r \\ &= \frac{t_P(n-1)}{\sqrt{n}} \sqrt{\sum_{i,j} b_{ik} b_{jk} s_{ij}} + \sum_{i=1}^{m} | b_{ik} | f_{s,i} + u_{\bar{N}} \left| \sum_{i=1}^{p} b_{ik} \right| . \end{aligned} \tag{10.19}$$

Let us finally introduce weighting factors as discussed in Sect. 8.4. Left-multiplying the inconsistent linear system (10.11) by a weighting matrix

$$\boldsymbol{G} = \mathrm{diag}\{g_1, g_2, \ldots, g_m\}$$

produces

$$\boldsymbol{GA\beta} \approx \boldsymbol{G}(\bar{\boldsymbol{x}} + \bar{\boldsymbol{d}}) \tag{10.20}$$

with $\bar{d}$ as given in (10.10). Referring to

$$\tilde{\boldsymbol{B}} = (\boldsymbol{GA})\left[(\boldsymbol{GA})^{\mathrm{T}}(\boldsymbol{GA})\right]^{-1} = \left(\tilde{b}_{ik}\right); \qquad i = 1, \ldots, m; \quad k = 1, \ldots, r$$

the components of the solution vector read

$$\bar{\beta}_k = \sum_{i=1}^{m} \tilde{b}_{ik}\, g_i\, \bar{x}_i + \bar{d} \sum_{i=1}^{p} \tilde{b}_{ik}\, g_i\,; \quad k = 1, \ldots, r\,. \tag{10.21}$$

The error equations

$$\bar{x}_i = x_{0,i} + (\bar{x}_i - \mu_i) + f_i\,, \quad \bar{d} = d_0 + f_{\bar{N}}$$

yield

$$\bar{\beta}_k = \sum_{i=1}^{m} \tilde{b}_{ik}\, g_i\, [x_{0,i} + (\bar{x}_i - \mu_i) + f_i] + (d_0 + f_{\bar{N}}) \sum_{i=1}^{p} \tilde{b}_{ik}\, g_i\,; \quad k = 1, \ldots, r\,.$$

As a weight matrix does not shift the components of the true solution vector $\boldsymbol{\beta}_0$, we have

$$\beta_{0,k} = \sum_{i=1}^{m} \tilde{b}_{ik}\, g_i\, x_{0,i} + d_0 \sum_{i=1}^{p} \tilde{b}_{ik}\, g_i\,; \quad k = 1, \ldots, r \tag{10.22}$$

Furthermore, we observe propagated systematic errors

$$f_{s,\bar{\beta}_k} = \sum_{i=1}^{m} |\, \tilde{b}_{ik} \,|\, g_i\, f_{s,i} + u_{\bar{N}} \left| \sum_{i=1}^{p} \tilde{b}_{ik}\, g_i \right|; \quad k = 1, \ldots, r\,. \tag{10.23}$$

Splitting (10.21) up into individual measurements

$$\bar{\beta}_{kl} = \sum_{i=1}^{m} \tilde{b}_{ik}\, g_i\, x_{il} + \bar{d} \sum_{i=1}^{p} \tilde{b}_{ik}\, g_i\,; \quad k = 1, \ldots, r\,. \tag{10.24}$$

we obtain the differences

$$\bar{\beta}_{kl} - \bar{\beta}_k = \sum_{i=1}^{m} \tilde{b}_{ik}\, g_i\, (x_{il} - \bar{x}_i)\,; \quad k = 1, \ldots, r; \quad l = 1, \ldots, n$$

Thus, the empirical variances and covariances of the components of the solution vector take the form

$$
\begin{aligned}
s_{\beta_k\beta_{k'}} &= \frac{1}{n-1}\sum_{l=1}^{n}\left(\bar\beta_{kl}-\bar\beta_k\right)\left(\bar\beta_{k'l}-\bar\beta_{k'}\right) \\
&= \frac{1}{n-1}\sum_{l=1}^{n}\left[\sum_{i=1}^{m}\tilde b_{ik}\,g_i\,(x_{il}-\bar x_i)\right]\left[\sum_{j=1}^{m}\tilde b_{jk'}\,g_j\,(x_{jl}-\bar x_j)\right] \qquad (10.25)\\
&= \sum_{i,j}^{m}\tilde b_{ik}\tilde b_{jk'}\,g_ig_j\,s_{ij}\,, \quad s_{\beta_k\beta_k}\equiv s^2_{\beta_k}\,; \quad k,k'=1,\dots,r
\end{aligned}
$$

where

$$
s_{ij}=\frac{1}{n-1}\sum_{l=1}^{n}(x_{il}-\bar x_i)(x_{jl}-\bar x_j)\,; \quad i,j=1,\dots,m
$$

denote the elements of empirical variance–covariance matrix $\boldsymbol{s}$ of the input data. Hence,

$$
\boldsymbol{s}_{\bar\beta}=\begin{pmatrix} s_{\bar\beta_1\bar\beta_1} & s_{\bar\beta_1\bar\beta_2} & \dots & s_{\bar\beta_1\bar\beta_r}\\ s_{\bar\beta_2\bar\beta_1} & s_{\bar\beta_2\bar\beta_2} & \dots & s_{\bar\beta_2\bar\beta_r}\\ \dots & \dots & \dots & \dots\\ s_{\bar\beta_r\bar\beta_1} & s_{\bar\beta_r\bar\beta_2} & \dots & s_{\bar\beta_r\bar\beta_r}\end{pmatrix}=\tilde{\boldsymbol{B}}^{\mathrm T}\left(\boldsymbol{G}^{\mathrm T}\boldsymbol{s}\,\boldsymbol{G}\right)\tilde{\boldsymbol{B}}\,. \qquad (10.26)
$$

After all, the weighted adjustment produces uncertainties

$$
u_{\bar\beta_k}=\frac{t_P(n-1)}{\sqrt n}\,s_{\bar\beta_k}+f_{s,\bar\beta_k}\,; \quad k=1,\dots,r
$$

$$
=\frac{t_P(n-1)}{\sqrt n}\sqrt{\sum_{i,j}^{m}\tilde b_{ik}\tilde b_{jk}\,g_ig_j\,s_{ij}}+\sum_{i=1}^{m}\mid\tilde b_{ik}\mid g_i\,f_{s,i}+u_{\bar N}\left|\sum_{i=1}^{p}\tilde b_{ik}\,g_i\right|\,. \qquad (10.27)
$$

Data Simulation

Deploying the preceding weighing scheme we link the set of masses

$$
N(m_1)=500\,\mathrm g\,,\quad N(m_2)=200\,\mathrm g\,,\quad N(m_3)=200\,\mathrm g
$$

$$
N(m_4)=100\,\mathrm g\,,\quad N(m_5)=100\,\mathrm g
$$

and a check-weight

$$N(m_6) = 300\,\mathrm{g}\,,$$

holding a formal position, to a standard of nominal value

$$N = 1\,\mathrm{kg}\,.$$

Let us assume

$$\begin{aligned} N_0 &= 1000\,,\,000\,080\,\mathrm{g} \\ \bar{N} &= 1000\,,\,000\,100\,\mathrm{g} \\ u_{\bar{N}} &= 0\,,\,000\,050\,\mathrm{g} \\ d_0 &= 0\,,\,000\,080\,\mathrm{g} \\ \bar{d} &= 0\,,\,000\,100\,\mathrm{g} \end{aligned}$$

$$\begin{aligned} m_{0,1} &= 500\,,\,010\,000\,\mathrm{g} \\ m_{0,2} &= 200\,,\,020\,000\,\mathrm{g} \\ m_{0,3} &= 199\,,\,970\,000\,\mathrm{g} \\ m_{0,4} &= 100\,,\,005\,000\,\mathrm{g} \\ m_{0,5} &= 99\,,\,993\,000\,\mathrm{g} \\ m_{0,6} &= 300\,,\,050\,000\,\mathrm{g}\,. \end{aligned}$$

and further that the first five comparisons, labeled by a (+) sign, are carried out on a balance with a standard deviation of $\sigma_1 = 0,\ 000\ 010$ g and the remaining eleven on a balance with $\sigma_2 = 0,\ 000\ 0002$ g. As the link-up mass $\bar{N}$ occurs likewise in each of the first five comparisons, the error propagation asks us to feed in the pertaining systematic error flatly five times. We suppose

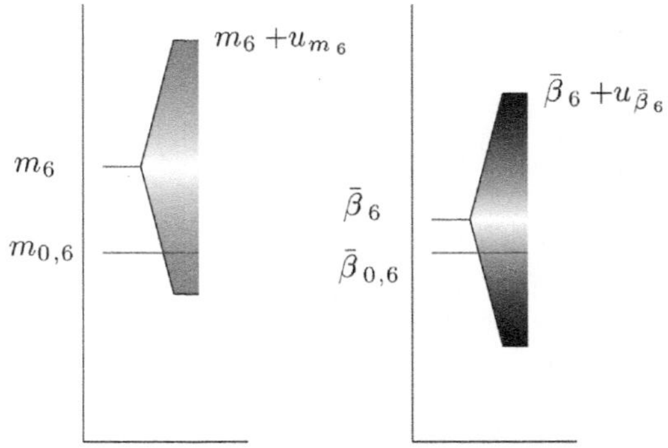

Fig. 10.2. The a priori and a posteriori intervals $m_6 \pm u_{m_6}$ and $\bar{\beta}_6 \pm u_{\bar{\beta}_6}$, allocated to the check weight, should localize the respective true values $m_{0,6}$ and $\beta_{0,6}$

the remaining eleven comparisons being free of systematic errors. Table 10.1 renders the simulated input data $\bar{x}_i\,; i = 1, \ldots, 16$, the underlying true values $x_{0,i}$ and the uncertainties $u_{\bar{x}_i}$.

Figure 10.2 suggests, here of course in a more formal sense, that the a priori interval $m_6 \pm u_{m_6}$ and the a posteriori interval $\bar{\beta}_6 \pm u_{\bar{\beta}_6}$, as allocated to the check-weight on the part of the adjustment, are expected to overlap. Ultimately, Table 10.2 summarizes the results.

Table 10.1. Data Simulation: Input Data, True Values, and Uncertainties

i	$\bar{x}_i$	$x_{0,i}$	$u_{\bar{x}_i}$
1	0.051 915 0	0.051 920 0	0.000 014 0
2	0.007 914 0	0.007 920 0	0.000 015 0
3	0.107 909 0	0.107 920 0	0.000 010 0
4	0.091 923 0	0.091 920 0	0.000 015 0
5	0.017 917 0	0.017 920 0	0.000 016 0
6	−0.087 999 9	−0.088 000 0	0.000 001 3
7	−0.032 000 0	−0.032 000 0	0.000 001 7
8	0.011 998 4	0.012 000 0	0.000 001 2
9	−0.047 999 6	−0.048 000 0	0.000 001 2
10	−0.021 998 5	−0.022 000 0	0.000 001 6
11	0.024 999 6	0.025 000 0	0.000 002 1
12	0.025 001 2	0.025 000 0	0.000 002 2
13	0.086 999 5	0.087 000 0	0.000 001 9
14	0.086 999 7	0.087 000 0	0.000 001 1
15	0.011 999 9	0.012 000 0	0.000 002 3
16	0.011 999 0	0.012 000 0	0.000 001 1

Table 10.2. Link-up of a Mass Decade: Estimates of the Mass Differences, True Values, and Uncertainties

k	$\bar{\beta}_k$	$\beta_{0\,k}$	$u_{\bar{\beta}_k}$
1	0.010007	0.0100	0.000029
2	0.020002	0.0200	0.000012
3	−0.029997	−0.0300	0.000012
4	0.005001	0.0050	0.000006
5	−0.006998	−0.0070	0.000006
6	0.050004	0.0500	0.000017

10.2 Pairwise Comparisons

Occasionally, within a set of physical quantities of the same quality (masses, lengths, voltages, etc.) only differences between two individuals are measurable. Consider, e.g., $r = 4$ quantities $\beta_k;\ k = 1, \ldots, r$. Then, there are either $r(r-1)/2$ or $r(r-1)$, i.e. six or twelve measurable differences, respectively, depending on whether not only the $(\beta_i - \beta_j)$ but also their inversions $(\beta_j - \beta_i)$ matter.

Let us refer to the first, simpler situation

$$\begin{aligned}
\beta_1 - \beta_2 &\approx \bar{x}_1 \\
\beta_1 - \beta_3 &\approx \bar{x}_2 \\
\beta_1 - \beta_4 &\approx \bar{x}_3 \\
\beta_2 - \beta_3 &\approx \bar{x}_4 \\
\beta_2 - \beta_4 &\approx \bar{x}_5 \\
\beta_3 - \beta_4 &\approx \bar{x}_6 \,.
\end{aligned}$$

Obviously, the first and the second relation establish the fourth one, the first and the third the fifth one and, finally, the second and the third the sixth one. Consequently, the design matrix

$$\boldsymbol{A} = \begin{pmatrix} 1 & -1 & 0 & 0 \\ 1 & 0 & -1 & 0 \\ 1 & 0 & 0 & -1 \\ 0 & 1 & -1 & 0 \\ 0 & 1 & 0 & -1 \\ 0 & 0 & 1 & -1 \end{pmatrix}$$

has rank 3. As has been outlined in [28], Sect. 7.3 constraints are needed. To have an example, let us consider

$$h_1\beta_1 + h_2\beta_2 + h_3\beta_3 + h_4\beta_4 = y \,. \tag{10.28}$$

Introducing an auxiliary vector $\boldsymbol{H} = (h_1 \quad h_2 \quad h_3 \quad h_4)$ we may write

$$\boldsymbol{H}\boldsymbol{\beta} = y \,.$$

As has been shown [28], the solution vector

$$\bar{\boldsymbol{\beta}} = (\boldsymbol{A}^{\mathrm{T}}\boldsymbol{A} + \boldsymbol{H}^{\mathrm{T}}\boldsymbol{H})^{-1}(\boldsymbol{A}^{\mathrm{T}}\boldsymbol{x} + \boldsymbol{H}^{\mathrm{T}}y) \tag{10.29}$$

satisfies

$$\boldsymbol{H}\bar{\boldsymbol{\beta}} = y \,.$$

To abbreviate the notation, we resort to auxiliary matrices

$$\boldsymbol{\Gamma} = \boldsymbol{A}(\boldsymbol{A}^{\mathrm{T}}\boldsymbol{A} + \boldsymbol{H}^{\mathrm{T}}\boldsymbol{H})^{-1} \quad \text{and} \quad \boldsymbol{\Lambda} = \boldsymbol{H}(\boldsymbol{A}^{\mathrm{T}}\boldsymbol{A} + \boldsymbol{H}^{\mathrm{T}}\boldsymbol{H})^{-1}$$

so that

$$\bar{\boldsymbol{\beta}} = \boldsymbol{\Gamma}^{\mathrm{T}}\boldsymbol{x} + \boldsymbol{\Lambda}^{\mathrm{T}}y$$

Putting $m = r(r-1)/2$ and

$$\boldsymbol{\Gamma} = (\gamma_{ik}), \quad \boldsymbol{\Lambda} = (\lambda_{1k}); \quad i = 1, \ldots, m; \quad k = 1, \ldots, r$$

the components of the solution vector turn out as

$$\bar{\beta}_k = \sum_{i=1}^{m} \gamma_{ik}\, \bar{x}_i + \lambda_{1k}\, y; \quad k = 1, \ldots, r\,. \tag{10.30}$$

Inserting the error equations

$$\bar{x}_i = x_{0,i} + (\bar{x}_i - \mu_i) + f_i\,; \quad i = 1, \, \ldots \, , m$$

in (10.30)

$$\bar{\beta}_k = \sum_{i=1}^{m} \gamma_{ik}[x_{0,i} + (\bar{x}_i - \mu_i) + f_i] + \lambda_{1k}\, y\,; \quad k = 1, \, \ldots \, , r \tag{10.31}$$

issues the flow of true values $\beta_{0,k}$ of the components $\bar{\beta}_k$

$$\beta_{0,k} = \sum_{i=1}^{m} \gamma_{ik} x_{0,i} + \lambda_{1k}\, y\,; \quad k = 1, \, \ldots \, , r \tag{10.32}$$

and, beyond, the propagated systematic errors

$$f_{\bar{\beta}_k} = \sum_{i=1}^{m} \gamma_{ik} f_i\,; \quad k = 1, \ldots, r \tag{10.33}$$

Subtracting (10.30) from

$$\bar{\beta}_{kl} = \sum_{i=1}^{m} \gamma_{ik} x_{il} + \lambda_{1k}\, y\,; \quad l = 1, \ldots, n; \quad k = 1, \, \ldots \, , r$$

produces the differences

$$\bar{\beta}_{kl} - \bar{\beta}_k = \sum_{i=1}^{m} \gamma_{ik}(x_{il} - \bar{x}_i)\,; \quad l = 1, \, \ldots \, , n \quad k = 1, \, \ldots \, , r\,. \tag{10.34}$$

Thus, the solution vector's empirical variances and covariances are

$$
\begin{aligned}
s_{\bar{\beta}_k \bar{\beta}_{k'}} &= \frac{1}{n-1} \sum_{l=1}^{n} \left(\bar{\beta}_{kl} - \bar{\beta}_k\right) \left(\bar{\beta}_{k'l} - \bar{\beta}_{k'}\right) \\
&= \frac{1}{n-1} \sum_{l=1}^{n} \left[\sum_{i=1}^{m} \gamma_{ik} \left(x_{il} - \bar{x}_i\right) \right] \left[\sum_{j=1}^{m} \gamma_{jk'} \left(x_{jl} - \bar{x}_j\right) \right] \qquad (10.35) \\
&= \sum_{i,j}^{m} \gamma_{ik}\, \gamma_{jk'}\, s_{ij}\,, \qquad s_{\bar{\beta}_k \bar{\beta}_k} \equiv s^2_{\bar{\beta}_k}\,; \quad k, k' = 1, \ldots, r
\end{aligned}
$$

where, again, the

$$
s_{ij} = \frac{1}{n-1} \sum_{l=1}^{n} \left(x_{il} - \bar{x}_i\right) \left(x_{jl} - \bar{x}_j\right)\,; \quad i, j = 1, \ldots, m
$$

denote the elements of the empirical variance–covariance matrix $\boldsymbol{s}$ of the input data as given in (8.13).

For convenience, we gather the $s_{\bar{\beta}_k \bar{\beta}_{k'}}$ within a matrix

$$
\boldsymbol{s}_{\bar{\boldsymbol{\beta}}} = \begin{pmatrix} s_{\bar{\beta}_1 \bar{\beta}_1} & s_{\bar{\beta}_1 \bar{\beta}_2} & \ldots & s_{\bar{\beta}_1 \bar{\beta}_r} \\ s_{\bar{\beta}_2 \bar{\beta}_1} & s_{\bar{\beta}_2 \bar{\beta}_2} & \ldots & s_{\bar{\beta}_2 \bar{\beta}_r} \\ \ldots & \ldots & \ldots & \ldots \\ s_{\bar{\beta}_r \bar{\beta}_1} & s_{\bar{\beta}_r \bar{\beta}_2} & \ldots & s_{\bar{\beta}_r \bar{\beta}_r} \end{pmatrix} = \boldsymbol{\Gamma}^{\mathrm{T}} \boldsymbol{s}\, \boldsymbol{\Gamma}\,. \qquad (10.36)
$$

After all, the overall uncertainties

$$
\begin{aligned}
u_{\bar{\beta}_k} &= \frac{t_P(n-1)}{\sqrt{n}}\, s_{\bar{\beta}_k} + f_{s,\bar{\beta}_k}\,; \qquad k = 1, \ldots, r \\
&= \frac{t_P(n-1)}{\sqrt{n}} \sqrt{\sum_{i,j}^{m} \gamma_{ik}\, \gamma_{jk}\, s_{ij}} + \sum_{i=1}^{m} |\, \gamma_{ik} \,|\, f_{s,i}
\end{aligned} \qquad (10.37)
$$

produce intervals $\bar{\beta}_k \pm u_{\bar{\beta}_k}$; $k = 1, \ldots, r$ which are meant to localize the true values $\bar{\beta}_{0,k}$ as given in (10.32).

11 Founding Pillars

The consistency of physical units and physical constants for one and the traceability of measuring processes for another constitute the founding pillars of metrology.

11.1 Consistency

Since the days of James Clerk Maxwell, metrologists strive to link physical units to atomic constants. Should this ever be accomplished, the realizations of the individual primary standards were still charged by imperfections rendering the system of physical units and physical constants inconsistent. To clear discrepancies and contradictions a least squares adjustment procuring consistent units and constants proves beneficial.

The idea is to submit an appropriate set of fundamental physical constants, as determined experimentally, to a least squares adjustment. To emphasize: The least squares procedure is *commissioned* to numerically shift those constants belatedly, so that any inconsistencies due to experimental imperfections vanish. Clearly, this is a gimmick, however perfectly admissible – as long as the intervals

$$\bar{\beta}_k - u_{\bar{\beta}_k} \le \beta_{0,k} \le \bar{\beta}_k + u_{\bar{\beta}_k}\,; \quad k = 1, \ldots, r\,,$$

as issued by least squares, continue to localize the respective true values.

Let us once more allude to the role of weighting factors which we know to shift estimators and shrink belonging uncertainties, Sect. 2.5. Nevertheless, with respect to the error model pursued here, the setting of weighting factors is in order as the intervals spanned by the shifted estimators and shrunken uncertainties maintain the localization of true values, which has been discussed extensively.

Hence, a least squares adjustment of fundamental physical constants is in a position to provide physics with a consistent system of physical units and constants.

11.2 Traceability

The everlasting pursuit of true values carries into effect what metrology is per se. Table 11.1 resumes the varying appearances of traceability met so far.

Table 11.1. Traceability in Error Propagation and Basic Link Ups

Case	Transfer of true values	Reference
Definition	$\beta_{0,k} - \beta_{0,i} = x_0$	(1.1)
Functions	$x_{0,1}, x_{0,2}, \ldots \quad \rightarrow \quad \Phi(x_{0,1}, x_{0,2}, \ldots)$	Fig. 7.1
Working standards	$\beta_{0,k} = \beta_0 + \sum_{i=1}^{k} x_{0,i}\,; \quad k = 1, 2 \ldots$	(9.17)
Key comparisons	$T_0 = \beta_0^{(i)} + x_0^{(i)}\,; \quad i = 1, \ldots, m$	(9.20)
Least squares	$\boldsymbol{\beta}_0 = \boldsymbol{B}^{\mathrm{T}} \boldsymbol{x}_0$	(2.15)
Calibration chains	$\beta_{0,k} = \sum_{i=1}^{m} b_{ik} x_{0,i} + d_0 \sum_{i=1}^{p} b_{ik}\,; \quad k = 1, \ldots, r$	(10.14)
Pairwise comparisons	$\beta_{0,k} = \sum_{i=1}^{m} \gamma_{ik} x_{0,i} + \lambda_{1k}\, y\,; \quad k = 1, \ldots, r$	(11.32)

As metrology heavily rests on calibration chains, the keeping of traceability proves essential for the consistency of units, measures, and physical constants.

Part V

Fitting of Straight Lines

12 Preliminaries

The fitting of straight lines is based on data pairs

$$(x_1, y_1), \quad (x_2, y_2), \quad \ldots \quad , (x_m, y_m)\,; \qquad m > 2\,.$$

12.1 Distinction of Cases

We shall see about three fitting situations, Table 12.1:

Table 12.1. Fitting of Straight Lines, Three Cases

Case	Abscissas	Ordinates
(i)	Error-free	Individual measurements
(ii)	Error-free	Repeated measurements
(iii)	Repeated measurements	Repeated measurements

Case (i) considers the metrological standard situation. While the abscissas are assumed correct, the ordinates are taken to be individual measurements. Each ordinate is supposed to hold a particular random error, stemming from one and the same normal distribution, and a common unknown systematic error. Case (ii), likewise relying on exact abscissas, accounts for repeated measurements of the ordinates. Here, the scattering of the random errors as well as the actual values of the unknown systematic errors may vary from ordinate to ordinate. Case (iii), finally, considers erroneous abscissas and erroneous ordinates with varying theoretical variances and varying unknown systematic errors.

There are differing paths to fit a straight line. We may consider distances parallel to the y-axis, parallel to the x-axis, or perpendicular to the straight line itself. However, regardless of the approach pursued, we expect the error model to issue uncertainties which localize the straight line's true parameters—though the uncertainties themselves may differ.

For an illustration, we consider the y-intercept of a straight line fit $y(x) = \beta_1 + \beta_2 x$. Estimators $\bar{\beta}_1$ and uncertainties $u_{\bar{\beta}_1}$ of the paths may differ; nevertheless, the results, say,

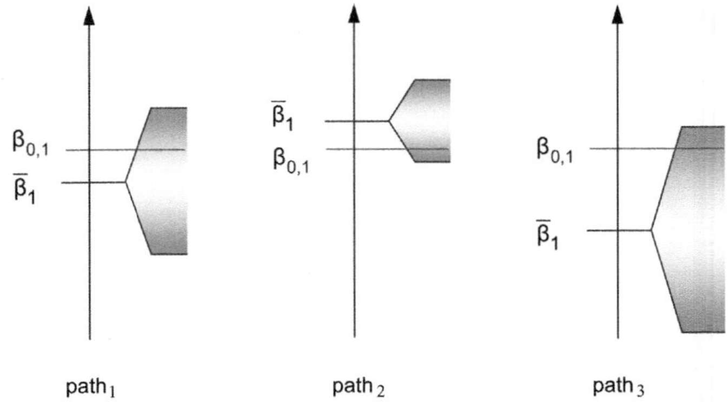

Fig. 12.1. Localization property of results $\bar{\beta}_1 \pm u_{\bar{\beta}_1}$ of varying paths; $\beta_{0,1}$ marking the true value

$$\bar{\beta}_1 \pm u_{\bar{\beta}_1}\,|_{path\,1}; \quad \bar{\beta}_1 \pm u_{\bar{\beta}_1}\,|_{path\,2}; \quad \bar{\beta}_1 \pm u_{\bar{\beta}_1}\,|_{path\,3}$$

are likewise expected to localize the true value $\beta_{0,1}$, Fig. 12.1.

12.2 True Straight Line

Let

$$y(x) = \beta_{0,1} + \beta_{0,2}\,x \tag{12.1}$$

be the equation of a straight line with parameters $\beta_{0,1}$ and $\beta_{0,2}$ designating the y-intercept and slope, respectively. Let there be m data pairs

$$(x_{0,1}, y_{0,1}), \quad (x_{0,2}, y_{0,2}), \quad \ldots \quad , (x_{0,m}, y_{0,m}) \tag{12.2}$$

fulfilling (12.1), in matrix form

$$\boldsymbol{A}\boldsymbol{\beta}_0 = \boldsymbol{y}_0 \tag{12.3}$$

with

$$\boldsymbol{A} = \begin{pmatrix} 1 & x_{0,1} \\ 1 & x_{0,2} \\ \ldots & \ldots \\ 1 & x_{0,m} \end{pmatrix}, \quad \boldsymbol{\beta}_0 = \begin{pmatrix} \beta_{0,1} \\ \beta_{0,2} \end{pmatrix}, \quad \boldsymbol{y}_0 = \begin{pmatrix} y_{0,1} \\ y_{0,2} \\ \ldots \\ y_{0,m} \end{pmatrix}. \tag{12.4}$$

Given the design matrix $\boldsymbol{A}$ has rank 2, the linear system (12.3) reproduces

$$\boldsymbol{\beta}_0 = \boldsymbol{B}^{\mathrm{T}}\boldsymbol{y}_0\,, \qquad \boldsymbol{B} = \boldsymbol{A}\left(\boldsymbol{A}^{\mathrm{T}}\boldsymbol{A}\right)^{-1}. \tag{12.5}$$

We address (12.1) as the true straight line, (12.2) as true input data and $\boldsymbol{\beta}_0$ as true solution vector. On substituting erroneous input data for the true ones, (12.3) breaks down. As measured data suffer from measuring errors, the true straight line proves inaccessible. Nevertheless, we may fit to the erroneous input data a least squares line and assess the true parameters $\beta_{0,1}$, $\beta_{0,2}$ via estimators $\bar{\beta}_1$, $\bar{\beta}_2$ and associated uncertainties $u_{\bar{\beta}_1}$, $u_{\bar{\beta}_2}$.

Should no set of true data exist, the adjustment would be ill defined right from the outset. This, in fact, breeds a problem: The least squares solution vector, which is due to an orthogonal projection, is quite insensitive to the metrological properties of the input data. Thus, whenever the design matrix has full rank there is a solution vector, be the input data meaningful or not. This observation however does not abate the usefulness of the method of least squares per se, rather it puts the responsibility of the experimenter in demand: The method of least squares is limited to accomplish a smart reallocation of measurement errors but is in no way apt to cure inappropriate physical conditions.

After all, the existence of a true straight line is a must.

13 Straight Lines: Case (i)

Case (i), Table 12.1 assumes correct abscissas and individually measured ordinates.

13.1 Fitting Conditions

Given $m > 2$ data pairs

$$(x_{0,1}, y_1), \quad (x_{0,2}, y_2), \quad \dots \quad (x_{0,m}, y_m) \tag{13.1}$$

with error-free abscissas $x_{0,i}$ and measured, i.e. erroneous, ordinates

$$y_i = y_{0,i} + (y_i - \mu_{y_i}) + f_{y_i}\,; \quad i = 1, \dots, m\,. \tag{13.2}$$

As usual, we formally assign random variables Y_i to the measured ordinates y_i so that the μ_{y_i} denote the expectations $E\{Y_i\}$ of the Y_i.

We suppose the random errors $(y_i - \mu_{y_i})\,;\ i = 1, \dots, m$ to stem from one and the same normal distribution and thus to have a common theoretical variance, say, σ_y^2. At present, we let each of the ordinates be charged by the same unknown systematic error

$$\begin{gathered} f_{y_i} = f_y\,; \quad i = 1, \dots, m \\ -f_{s,y} \le f_y \le f_{s,y}\,. \end{gathered} \tag{13.3}$$

As yet the uncertainties u_{y_i} of the ordinates are unknown. Thus we are not in a position to localize the true values $y_{0,i}$ of the y_i via

$$y_i \pm u_{y_i}\,; \quad i = 1, \dots, m\,.$$

We shall, however, be in a position to deliver the u_{y_i} belatedly.

13.2 Orthogonal Projection

The inconsistent, over-determined, linear system to be submitted to least squares reads

$$\beta_1 + \beta_2 \, x_{0,i} \approx y_i \, ; \quad i = 1, \ldots, m > 2 \, . \tag{13.4}$$

We are looking for a least squares line

$$y(x) = \bar{\beta}_1 + \bar{\beta}_2 \, x \tag{13.5}$$

fitting the data (13.1). With

$$\boldsymbol{A} = \begin{pmatrix} 1 & x_{0,1} \\ 1 & x_{0,2} \\ \cdots & \cdots \\ 1 & x_{0,m} \end{pmatrix}, \qquad \boldsymbol{\beta} = \begin{pmatrix} \beta_1 \\ \beta_2 \end{pmatrix}, \qquad \boldsymbol{y} = \begin{pmatrix} y_1 \\ y_2 \\ \cdots \\ y_m \end{pmatrix}$$

the matrix notation of (13.4) is

$$\boldsymbol{A}\boldsymbol{\beta} \approx \boldsymbol{y} \, . \tag{13.6}$$

The orthogonal projection of the vector $\boldsymbol{y}$ of observations onto the column space of the matrix $\boldsymbol{A}$ produces the solution vector

$$\bar{\boldsymbol{\beta}} = \boldsymbol{B}^{\mathrm{T}} \boldsymbol{y} \, , \qquad \bar{\beta}_k = \sum_{i=1}^{m} b_{ik} \, y_i \, ; \qquad k = 1, 2 \tag{13.7}$$

where

$$\boldsymbol{B} = \boldsymbol{A} \left(\boldsymbol{A}^{\mathrm{T}} \boldsymbol{A} \right)^{-1} = \begin{pmatrix} b_{11} & b_{12} \\ b_{21} & b_{22} \\ \cdots & \cdots \\ b_{m1} & b_{m2} \end{pmatrix}$$

the matrix elements being

$$\boldsymbol{B} = \frac{1}{D} \begin{bmatrix} \sum\limits_{j=1}^{m} x_{0,j}^2 - x_{0,1} \sum\limits_{j=1}^{m} x_{0,j} & -\sum\limits_{j=1}^{m} x_{0,j} + m \, x_{0,1} \\ \sum\limits_{j=1}^{m} x_{0,j}^2 - x_{0,2} \sum\limits_{j=1}^{m} x_{0,j} & -\sum\limits_{j=1}^{m} x_{0,j} + m \, x_{0,2} \\ \cdots & \cdots \\ \sum\limits_{j=1}^{m} x_{0,j}^2 - x_{0,m} \sum\limits_{j=1}^{m} x_{0,j} & -\sum\limits_{j=1}^{m} x_{0,j} + m \, x_{0,m} \end{bmatrix} \tag{13.8}$$

with

$$D = | \, \boldsymbol{A}^{\mathrm{T}} \boldsymbol{A} \, | = m \sum_{j=1}^{m} x_{0,j}^2 - \left[\sum_{j=1}^{m} x_{0,j} \right]^2 .$$

13.3 Uncertainties of the Input Data

The presence of unknown systematic errors, as specified in (13.3), does not suspend the minimized sum of squared residuals

$$\bar{Q} = \left(\boldsymbol{y} - \boldsymbol{A}\bar{\boldsymbol{\beta}}\right)^{\mathrm{T}} \left(\boldsymbol{y} - \boldsymbol{A}\bar{\boldsymbol{\beta}}\right) \tag{13.9}$$

to issues an estimate

$$s_y^2 = \frac{\bar{Q}}{m-2} \tag{13.10}$$

of the theoretical variance of the input data,

$$E\{S_y^2\} = \sigma_y^2 \,.$$

The empirical variance s_y^2 has degrees of freedom $\nu = m - 2$. Let

$$\boldsymbol{f} = f_y \begin{pmatrix} 1 \\ \cdots \\ 1 \end{pmatrix}$$

designates an $m \times 1$–vector. Then, for (13.10) to be valid, the condition

$$\boldsymbol{f} = \boldsymbol{P}\boldsymbol{f} \,, \qquad \boldsymbol{P} = \boldsymbol{A}(\boldsymbol{A}^{\mathrm{T}}\boldsymbol{A})^{-1}\boldsymbol{A}^{\mathrm{T}} \tag{13.11}$$

has to be fulfilled [28].

Disregarding random errors for the moment, we have

$$\boldsymbol{A} \begin{pmatrix} \beta_{0,1} + f_y \\ \beta_{0,2} \end{pmatrix} = \begin{pmatrix} y_{0,1} \\ \cdots \\ y_{0,m} \end{pmatrix} + f_y \begin{pmatrix} 1 \\ \cdots \\ 1 \end{pmatrix}$$

and thus

$$\boldsymbol{A} \begin{pmatrix} f_y \\ 0 \end{pmatrix} = f_y \begin{pmatrix} 1 \\ \cdots \\ 1 \end{pmatrix} = \boldsymbol{f} \,.$$

Left-multiplying by $\boldsymbol{P}$ yields

$$\boldsymbol{A} \begin{pmatrix} f_y \\ 0 \end{pmatrix} = \boldsymbol{P}\boldsymbol{f}$$

so that $\boldsymbol{f} = \boldsymbol{P}\boldsymbol{f}$.

Hence, we may narrow down the positions of the expectations

$$E\{Y_i\} = \mu_{y_i}\,; \quad i = 1, \ldots, m$$

by confidence intervals

$$y_i - t_P\,(m-2)\,s_y \le \mu_{y_i} \le y_i + t_P\,(m-2)\,s_y\,; \quad i = 1, \ldots, m \tag{13.12}$$

of probability P. But then, the overall uncertainties of the measured ordinates $y_i\,;\ i = 1, \ldots, m$ turn out to be

$$u_{y_i} = t_P\,(m-2)\,s_y + f_{s,y}\,; \quad i = 1, \ldots, m\,. \tag{13.13}$$

After all, the intervals

$$y_i - u_{y_i} \le y_{0,i} \le y_i + u_{y_i}\,; \quad i = 1, \ldots, m\,. \tag{13.14}$$

are intended to localize the true values $y_{0,i}$. Albeit our initial knowledge was meager and implied no hints to suchlike statements, our requesting the y_i to establish a straight line obviously revealed this information.

To emphasize, the proceeding applies only, given the y_i; $i = 1, \ldots, m$ are charged by one and the same systematic error.

13.4 Uncertainties of the Components of the Solution Vector

Random Errors

Though we have got no repeated measurements, we may still endue $\bar{\boldsymbol{\beta}}$ with an empirical variance–covariance matrix. Let us first state the expectations of the components $\bar{\beta}_1$ and $\bar{\beta}_2$,

$$E\{\bar{\beta}_1\} = \mu_{\bar{\beta}_1} = \sum_{i=1}^{m} b_{i1}\,\mu_{y_i}\,, \qquad E\{\bar{\beta}_2\} = \mu_{\bar{\beta}_2} = \sum_{i=1}^{m} b_{i2}\,\mu_{y_i} \tag{13.15}$$

where

$$\mu_{y_i} = E\,\{Y_i\} = y_{0,i} + f_y\,. \tag{13.16}$$

This brings forth the theoretical variances and the theoretical covariance

$$\sigma_{\bar{\beta}_1}^2 = E\left\{\left(\bar{\beta}_1 - \mu_{\bar{\beta}_1}\right)^2\right\} = E\left\{\left[\sum_{i=1}^{m} b_{i1}\,(Y_i - \mu_{y_i})\right]^2\right\} = \sigma_y^2 \sum_{i=1}^{m} b_{i1}^2$$

$$\sigma_{\bar{\beta}_1\bar{\beta}_2} = \sigma_y^2 \sum_{i=1}^{m} b_{i1} b_{i2}\,, \qquad \sigma_{\bar{\beta}_2}^2 = \sigma_y^2 \sum_{i=1}^{m} b_{i2}^2\,.$$

Obviously, the empirical counterparts are

$$s_{\bar{\beta}_1}^2 = s_y^2 \sum_{i=1}^{m} b_{i1}^2 \,, \quad s_{\bar{\beta}_1 \bar{\beta}_2} = s_y^2 \sum_{i=1}^{m} b_{i1} b_{i2} \,, \quad s_{\bar{\beta}_2}^2 = s_y^2 \sum_{i=1}^{m} b_{i2}^2 \,. \tag{13.17}$$

After all, the empirical variance–covariance matrix of the solution vector turns out to be

$$\boldsymbol{s}_{\bar{\beta}} = \begin{pmatrix} s_{\bar{\beta}_1 \bar{\beta}_1} & s_{\bar{\beta}_1 \bar{\beta}_2} \\ & \\ s_{\bar{\beta}_2 \bar{\beta}_1} & s_{\bar{\beta}_2 \bar{\beta}_2} \end{pmatrix} ; \qquad s_{\bar{\beta}_1 \bar{\beta}_1} \equiv s_{\bar{\beta}_1}^2 \,, \qquad s_{\bar{\beta}_2 \bar{\beta}_2} \equiv s_{\bar{\beta}_2}^2$$

the elements of which having degrees of freedom $\nu = m - 2$.

Systematic Errors

The formal decomposition

$$\bar{\beta}_k = \sum_{i=1}^{m} b_{ik} \left[y_{0,i} + (y_i - \mu_{y_i}) + f_y \right] ; \qquad k = 1, 2 \tag{13.18}$$

issues the propagated systematic errors

$$f_{\bar{\beta}_k} = f_y \sum_{i=1}^{m} b_{ik} \, ; \quad k = 1, 2 \, .$$

As

$$\sum_{i=1}^{m} b_{i1} = 1 \, , \qquad \sum_{i=1}^{m} b_{i2} = 0 \, , \tag{13.19}$$

we observe

$$f_{\bar{\beta}_1} = f_y$$
$$f_{\bar{\beta}_2} = 0 \, . \tag{13.20}$$

Overall Uncertainties

We localize the parameters $\mu_{\bar{\beta}_1} = \beta_{0,1} + f_y$ and $\mu_{\bar{\beta}_2} = \beta_{0,2}$ as given in (13.15) via confidence intervals

$$\bar{\beta}_1 - t_P(m-2) \, s_{\bar{\beta}_1} \le \mu_{\bar{\beta}_1} \le \bar{\beta}_1 + t_P(m-2) \, s_{\bar{\beta}_1}$$

$$\bar{\beta}_2 - t_P(m-2) \, s_{\bar{\beta}_2} \le \mu_{\bar{\beta}_2} \le \bar{\beta}_2 + t_P(m-2) \, s_{\bar{\beta}_2}$$

of probability P. Hence, the uncertainties of the components $\bar{\beta}_1$ and $\bar{\beta}_2$ take the form

$$u_{\bar{\beta}_1} = t_P(m-2) \, s_{\bar{\beta}_1} + f_{s,y}$$
$$u_{\bar{\beta}_2} = t_P(m-2) \, s_{\bar{\beta}_2} \, . \tag{13.21}$$

The unknown systematic error shifts the straight line parallel to itself. As this does not affect the slope, f_y does not enter $u_{\bar{\beta}_2}$.

13.5 Uncertainty Band

Random Errors

Inserting the formal decompositions

$$\bar{\beta}_k = \sum_{i=1}^{m} b_{ik} y_i = \beta_{k,0} + \sum_{i=1}^{m} b_{ik}(y_i - \mu_{y_i}) + f_y \sum_{i=1}^{m} b_{ik}\,; \quad k = 1, 2 \tag{13.22}$$

into the fitted straight line

$$y(x) = \bar{\beta}_1 + \bar{\beta}_2\, x$$

produces

$$y(x) = \beta_{0,1} + \beta_{0,2}\, x + \sum_{i=1}^{m} (b_{i1} + b_{i2} x)(y_i - \mu_{y_i}) + f_y\,. \tag{13.23}$$

The expectation

$$E\{Y(x)\} = \mu_{y(x)} = \beta_{0,1} + \beta_{0,2}\, x + f_y$$

induces us to introduce the x-dependent theoretical variance

$$\sigma^2_{y(x)} = E\{(Y(x) - \mu_{\bar{y}(x)})^2\} = \sigma_y^2 \sum_{i=1}^{m} (b_{i1} + b_{i2} x)^2\,; \quad \sigma_y^2 = E\left\{(Y_i - \mu_{y_i})^2\right\}\,.$$

Thus

$$t_P(m-2)\, s_y \sqrt{\sum_{i=1}^{m} (b_{i1} + b_{i2}\, x)^2} \tag{13.24}$$

addresses the uncertainty due to random errors.

Systematic Errors

The formal decomposition (13.23) issues

$$f_{y(x)} = f_y\,. \tag{13.25}$$

Overall Uncertainty

Combining (13.24) with the worst-case estimation of (13.25) puts forth the uncertainty band

$$y(x) \pm u_{y(x)}$$

$$u_{y(x)} = t_P(m-2)\, s_y \sqrt{\sum_{i=1}^{m} (b_{i1} + b_{i2}\, x)^2} + f_{s,y}\,. \tag{13.26}$$

We expect the bordering lines, set up in symmetry to the least squares line (13.5), to localize the true straight line (12.1).

13.6 *EP*-Region

The *EP*-region localizes the pair of true values $(\beta_{0,1}, \beta_{0,2})$ with respect to the pair of estimators $(\bar{\beta}_1, \bar{\beta}_2)$.

Confidence Ellipse

Combining

$$\mu_{\bar{\beta}_1} = \beta_{0,1} + f_y\,, \quad \mu_{\bar{\beta}_2} = \beta_{0,2}$$

with

$$\bar{\beta}_k = \sum_{i=1}^{m} b_{ik}\left[y_{0,i} + (y_i - \mu_{y_i}) + f_y\right]; \qquad k = 1, 2$$

yields

$$\bar{\beta}_k = \mu_{\bar{\beta}_k} + \sum_{i=1}^{m} b_{ik}\,(y_i - \mu_{y_i})\,; \qquad k = 1, 2\,.$$

As discussed in Appendix C, Hotellings's ellipse is given by

$$s_{\bar{\beta}_2\bar{\beta}_2}(\bar{\beta}_1 - \mu_{\bar{\beta}_1})^2 - 2s_{\bar{\beta}_1\bar{\beta}_2}(\bar{\beta}_1 - \mu_{\bar{\beta}_1})(\bar{\beta}_2 - \mu_{\bar{\beta}_2}) + s_{\bar{\beta}_1\bar{\beta}_1}(\bar{\beta}_2 - \mu_{\bar{\beta}_2})^2$$
$$= t^2(2, m-2)\,|\boldsymbol{s}_{\bar{\beta}}|\,.$$

From this we draw the confidence ellipse

$$\begin{aligned} s_{\bar{\beta}_2\bar{\beta}_2}(\beta_1 - \bar{\beta}_1)^2 - 2s_{\bar{\beta}_1\bar{\beta}_2}(\beta_1 - \bar{\beta}_1)(\beta_2 - \bar{\beta}_2) + s_{\bar{\beta}_1\bar{\beta}_1}(\beta_2 - \bar{\beta}_2)^2 \\ = t_P^2(2, m-2)|\boldsymbol{s}_{\bar{\beta}}| \end{aligned} \tag{13.27}$$

being centered in $(\bar{\beta}_1, \bar{\beta}_2)$. We expect the confidence ellipse to localize the point

$$\boldsymbol{\mu}_{\bar{\beta}} = \begin{pmatrix} \mu_{\bar{\beta}_1} \\ \mu_{\bar{\beta}_2} \end{pmatrix} \tag{13.28}$$

with probability P.

Security Polygon

Due to the measuring conditions of error-free abscissas and erroneous ordinates charged by one and the same unknown systematic error, the security polygon degenerates into an interval

$$\begin{aligned} -f_{s,y} \le f_{\bar{\beta}_1} \le f_{s,y} \\ f_{\bar{\beta}_2} = 0\,. \end{aligned} \tag{13.29}$$

For convenience, let us visualize this interval as a "stick" of length $2f_{s,y}$.

Mergence of Ellipse and Interval

The mergence of the confidence ellipse (13.27) and the degenerated security polygon (13.29) is discussed in Appendix G.

Figure 13.1 displays the true straight line (12.1), the fitted straight line (13.5), the uncertainties (13.21) of the estimators $\bar{\beta}_1, \bar{\beta}_2$, the uncertainty band (13.26) and, finally, the EP-region, meant to localize the tuple of true values $\beta_{0,1}$ and $\beta_{0,2}$.

The illustrations are based on simulated data implying known true values and, in particular, extensive graphical scale transformations, Appendix A.

Equivalence of Uncertainty and EP-Region

Intuitively, we expect the uncertainty band and the EP-region as depicted in Fig. 13.1 to be equivalent. Remarkably enough, it proves doable to transfer one into the other. The procedure is outlined in Chap. 16.

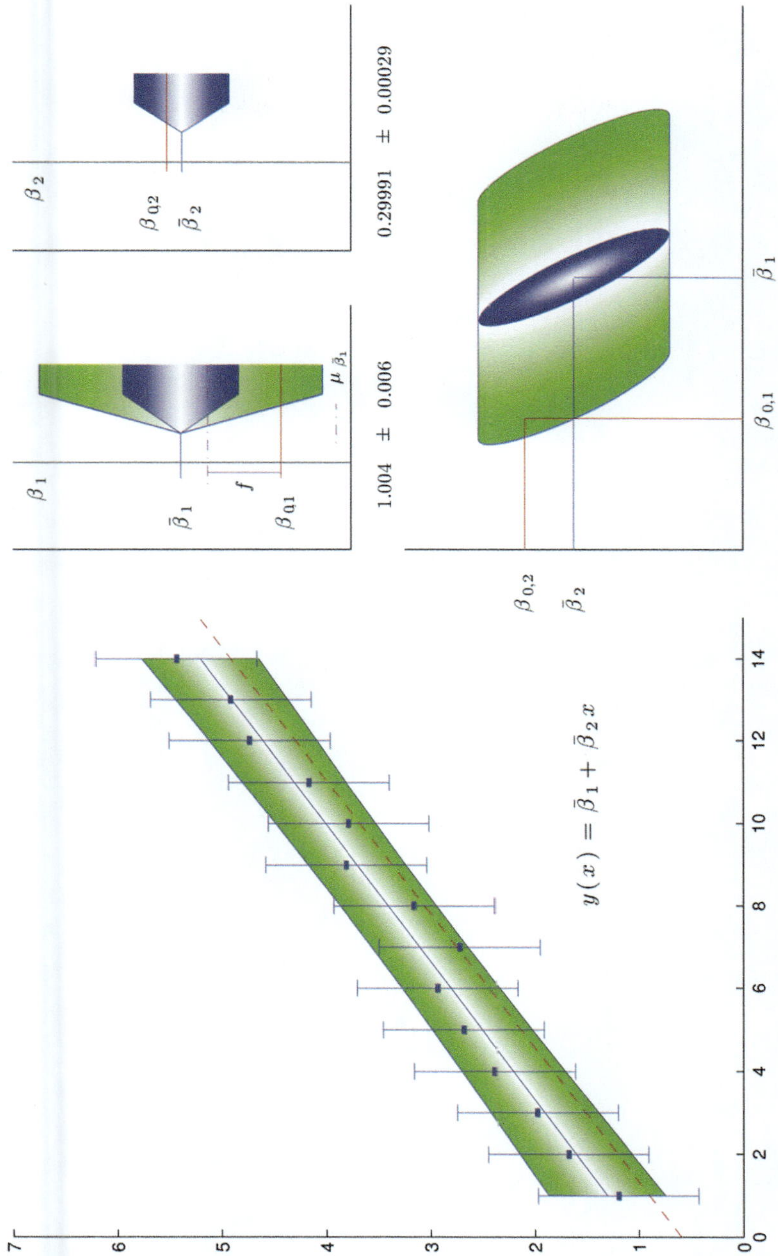

Fig. 13.1. Straight lines, case (i). *Left*: least squares line, uncertainty band, and true straight line (*dashed*). *Top right*: uncertainty intervals localizing the straight line's true parameters $\beta_{0,1}$ and $\beta_{0,2}$. Equal systematic errors do not affect $\bar{\beta}_2$. *Bottom right*: EP-region, localizing the pair of true values $\beta_{0,1}$, $\beta_{0,2}$

14 Straight Lines: Case (ii)

Case (ii), Table 12.1 assumes correct abscissas and repeatedly measured ordinates. As now the empirical variances of the ordinates are directly accessible the scattering of the random errors may vary from ordinate to ordinate. Beyond, varying unknown systematic errors are admitted. As a matter of course, the minimized sum of squared residuals still enters the construction of the solution vector, it proves however no longer serviceable with respect to assessing uncertainties.

14.1 Fitting Conditions

We refer to $m > 2$ data pairs

$$(x_{0,1}, \bar{y}_1), \quad (x_{0,2}, \bar{y}_2), \quad \dots \quad (x_{0,m}, \bar{y}_m) \tag{14.1}$$

with error-free abscissas and ordinates being arithmetic means

$$\bar{y}_i = \frac{1}{n}\sum_{l=1}^{n} y_{il} = y_{0,i} + (\bar{y}_i - \mu_{\bar{y}_i}) + f_{\bar{y}_i}; \quad i = 1, \dots, m$$

$$\mu_{\bar{y}_i} = E\{\bar{Y}_i\}; \qquad -f_{s,\bar{y}_i} \le f_{\bar{y}_i} \le f_{s,\bar{y}_i}\,.$$

According to the idea to stick to well-defined measuring conditions, each of the $\bar{y}_i$ implies the same number n of repeated measurements. The i-specific empirical variances

$$s_{\bar{y}_i}^2 = \frac{1}{n-1}\sum_{l=1}^{n}(y_{il} - \bar{y}_i)^2\,; \quad i = 1, \dots, m$$

and the limits $\pm f_{s,\bar{y}_i}$ of the unknown systematic errors issue the uncertainties $u_{\bar{y}_i}$ of the input data. We have

$$\bar{y}_i = u_{\bar{y}_i}\,, \quad u_{\bar{y}_i} = \frac{t_P(n-1)}{\sqrt{n}}\, s_{\bar{y}_i} + f_{s,\bar{y}_i}\,; \quad i = 1, \dots, m\,. \tag{14.2}$$

We expect the intervals

$$\bar{y}_i - u_{\bar{y}_i} \le y_{0,i} \le \bar{y}_i + u_{\bar{y}_i}\,; \quad i = 1, \dots, m\,.$$

to localize the true values $y_{0,i}$ of the ordinates.

14.2 Orthogonal Projection

Let us transfer the inconsistent, over-determined, linear system

$$\beta_1 + \beta_2 x_{0,i} \approx \bar{y}_i \,; \quad i = 1, \ldots, m > 2 \tag{14.3}$$

to

$$\boldsymbol{A\beta} \approx \bar{\boldsymbol{y}}$$

where

$$\boldsymbol{A} = \begin{pmatrix} 1 & x_{0,1} \\ 1 & x_{0,2} \\ \ldots & \ldots \\ 1 & x_{0,m} \end{pmatrix}, \quad \boldsymbol{\beta} = \begin{pmatrix} \beta_1 \\ \beta_2 \end{pmatrix}, \quad \bar{\boldsymbol{y}} = \begin{pmatrix} \bar{y}_1 \\ \bar{y}_2 \\ \ldots \\ \bar{y}_m \end{pmatrix}.$$

The orthogonal projection yields

$$\bar{\boldsymbol{\beta}} = \boldsymbol{B}^{\mathrm{T}} \boldsymbol{y} \,, \qquad \bar{\beta}_k = \sum_{i=1}^{m} b_{ik} \bar{y}_i \,; \qquad k = 1, 2 \tag{14.4}$$

with

$$\boldsymbol{B} = \boldsymbol{A} \left(\boldsymbol{A}^{\mathrm{T}} \boldsymbol{A} \right)^{-1} = (b_{ik}) \,; \quad i = 1, \ldots, m; \quad k = 1, 2$$

The components $\bar{\beta}_k$; $k = 1, 2$ of the solution vector $\bar{\boldsymbol{\beta}}$ appoint the least squares line

$$y(x) = \bar{\beta}_1 + \bar{\beta}_2 \, x \,. \tag{14.5}$$

14.3 Uncertainties of the Components of the Solution Vector

Random Errors

In view of Sect. 7.3, we notionally introduce an ensemble of least squares lines based on the data sets $(y_{1l}, y_{2l}, \ldots, y_{ml})$; $l = 1, \ldots, n$ issuing estimators

$$\bar{\beta}_{kl} = \sum_{i=1}^{m} b_{ik} \, y_{il} \,, \qquad l = 1, \ldots, n \,; \quad k = 1, 2 \,. \tag{14.6}$$

Inserting

$$\bar{y}_i = \frac{1}{n} \sum_{l=1}^{n} y_{il} \,; \quad i = 1, \ldots, m$$

into (14.4) yields

$$\bar{\beta}_k = \sum_{i=1}^{m} b_{ik}\left[\frac{1}{n}\sum_{l=1}^{n} y_{il}\right] = \frac{1}{n}\sum_{l=1}^{n}\left[\sum_{i=1}^{m} b_{ik}y_{il}\right]$$

$$= \frac{1}{n}\sum_{l=1}^{n}\bar{\beta}_{kl}\,;\; k = 1,2\,.$$

The differences

$$\bar{\beta}_{kl} - \bar{\beta}_k = \sum_{i=1}^{m} b_{ik}\left(y_{il} - \bar{y}_i\right)$$

produce the elements

$$s_{\bar{\beta}_k\bar{\beta}_{k'}} = \frac{1}{n-1}\sum_{l=1}^{n}\left(\bar{\beta}_{kl} - \bar{\beta}_k\right)\left(\bar{\beta}_{k'l} - \bar{\beta}_{k'}\right)$$

$$= \frac{1}{n-1}\sum_{l=1}^{n}\left[\sum_{i=1}^{m} b_{ik}\left(y_{il} - \bar{y}_i\right)\right]\left[\sum_{j=1}^{m} b_{jk'}\left(y_{jl} - \bar{y}_j\right)\right]$$

$$= \sum_{i,j=1}^{m} b_{ik}\,b_{jk'}\,s_{ij}\,;\quad k, k' = 1,2$$

of the empirical variance–covariance matrix of the solution vector in which the

$$s_{ij} = \frac{1}{n-1}\sum_{l=1}^{n}(y_{il} - \bar{y}_i)(y_{jl} - \bar{y}_j)\,;\quad i,j = 1,\dots,m$$

designate the elements

$$\boldsymbol{s} = (s_{ij})\,;\quad i,j = 1,\dots,m \tag{14.7}$$

of the empirical variance–covariance matrix of the input data, each having degrees of freedom $\nu = n - 1$. After all, we have

$$\boldsymbol{s}_{\bar{\beta}} = \begin{pmatrix} s_{\bar{\beta}_1\bar{\beta}_1} & s_{\bar{\beta}_1\bar{\beta}_2} \\ & \\ s_{\bar{\beta}_2\bar{\beta}_1} & s_{\bar{\beta}_2\bar{\beta}_2} \end{pmatrix} = \boldsymbol{B}^T\boldsymbol{s}\,\boldsymbol{B}, \qquad s_{\bar{\beta}_k\bar{\beta}_k} \equiv s^2_{\bar{\beta}_k}\,. \tag{14.8}$$

Systematic Errors

The formal decompositions

$$\bar{\beta}_k = \sum_{i=1}^{m} b_{ik}\left[y_{0,i} + (\bar{y}_i - \mu_{\bar{y}_i}) + f_{\bar{y}_i}\right];\quad k = 1,2 \tag{14.9}$$

issue the propagated systematic errors

$$f_{\bar{\beta}_k} = \sum_{i=1}^{m} b_{ik} f_{\bar{y}_i} \; ; \quad k = 1, 2 \tag{14.10}$$

their worst-case estimations being

$$f_{s,\bar{\beta}_k} = \sum_{i=1}^{m} \mid b_{ik} \mid f_{s,\bar{y}_i} \; ; \quad k = 1, 2 \, . \tag{14.11}$$

In case of equal $f_{\bar{y}_i}$ and $f_{s,\bar{y}_i}$,

$$\begin{gathered} f_{\bar{y}_i} = f_y \, , \qquad f_{s,\bar{y}_i} = f_{s,y} \, ; \qquad i = 1, \ldots m \\ -f_{s,y} \le f_y \le f_{s,y} \, , \end{gathered} \tag{14.12}$$

due to (13.19) we have

$$\begin{aligned} f_{\bar{\beta}_1} &= f_y \\ f_{\bar{\beta}_2} &= 0 \, . \end{aligned} \tag{14.13}$$

Confidence Intervals and Overall Uncertainties

We localize the expectations

$$\begin{aligned} E\{\bar{\beta}_1\} &= \mu_{\bar{\beta}_1} = \beta_{0,1} + f_{\bar{\beta}_1} \\ E\{\bar{\beta}_2\} &= \mu_{\bar{\beta}_2} = \beta_{0,2} + f_{\bar{\beta}_2} \end{aligned} \tag{14.14}$$

through confidence intervals

$$\bar{\beta}_1 - \frac{t_P(n-1)}{\sqrt{n}} \, s_{\bar{\beta}_1} \le \mu_{\bar{\beta}_1} \le \bar{\beta}_1 + \frac{t_P(n-1)}{\sqrt{n}} \, s_{\bar{\beta}_1}$$

$$\bar{\beta}_2 - \frac{t_P(n-1)}{\sqrt{n}} \, s_{\bar{\beta}_2} \le \mu_{\bar{\beta}_2} \le \bar{\beta}_2 + \frac{t_P(n-1)}{\sqrt{n}} \, s_{\bar{\beta}_2}$$

of probability P. Thus, the overall uncertainties of the $\bar{\beta}_k$ take the form

$$\begin{aligned} u_{\bar{\beta}_1} &= \frac{t_P(n-1)}{\sqrt{n}} \, s_{\bar{\beta}_1} + f_{s,\bar{\beta}_1} \\ u_{\bar{\beta}_2} &= \frac{t_P(n-1)}{\sqrt{n}} \, s_{\bar{\beta}_2} + f_{s,\bar{\beta}_2} \, . \end{aligned} \tag{14.15}$$

Equal systematic errors give way to

$$\begin{aligned} u_{\bar{\beta}_1} &= \frac{t_P(n-1)}{\sqrt{n}}\, s_{\bar{\beta}_1} + f_{s,y} \\ u_{\bar{\beta}_2} &= \frac{t_P(n-1)}{\sqrt{n}}\, s_{\bar{\beta}_2} \end{aligned} \qquad (14.16)$$

as $f_{\bar{\beta}_2} = 0$. At any rate, the result reads

$$\bar{\beta}_k \pm u_{\bar{\beta}_k}\,; \quad k = 1, 2\,.$$

14.4 Uncertainty Band

Random Errors

The $\bar{\beta}_{kl}$ as defined in (14.6) suggest an ensemble of n straight lines

$$y_l(x) = \bar{\beta}_{1l} + \bar{\beta}_{2l}\, x\,; \qquad l = 1, \dots, n\,.$$

Summed over l reestablishes the least squares line

$$y(x) = \frac{1}{n} \sum_{l=1}^{n} y_l(x) = \bar{\beta}_1 + \bar{\beta}_2\, x\,.$$

Due to

$$\bar{\beta}_{kl} - \bar{\beta}_k = \sum_{i=1}^{m} b_{ik}\,(y_{il} - \bar{y}_i)$$

we have

$$\begin{aligned} y_l(x) - y(x) &= (\bar{\beta}_{1l} - \bar{\beta}_1) + (\bar{\beta}_{2l} - \bar{\beta}_2)\, x \\ &= \sum_{i=1}^{m} (b_{i1} + b_{i2}\, x)(y_{il} - \bar{y}_i)\,. \end{aligned}$$

Hence, for any fixed x the empirical variance is given by

$$s_{y(x)}^2 = \frac{1}{n-1} \sum_{l=1}^{n} (y_l(x) - y(x))^2 = \boldsymbol{b}^T \boldsymbol{s}\, \boldsymbol{b}\,. \qquad (14.17)$$

Here

$$\boldsymbol{b} = [\ (b_{11} + b_{12}x) \qquad (b_{21} + b_{22}x) \quad \dots \quad (b_{m1} + b_{m2}x)\ ]^T$$

denotes an auxiliary vector and $\boldsymbol{s}$ the matrix (14.7).

After all, the term

$$\frac{t_P(n-1)}{\sqrt{n}}\, s_{y(x)} \qquad (14.18)$$

devises the uncertainty due to random errors.

Systematic Errors

Inserting

$$\bar{\beta}_k = \beta_{0,k} + \sum_{i=1}^{m} b_{ik}(\bar{y}_i - \mu_{\bar{y}_i}) + \sum_{i=1}^{m} b_{ik} f_{\bar{y}_i}\,; \quad k = 1,2 \tag{14.19}$$

into

$$y(x) = \bar{\beta}_1 + \bar{\beta}_2\, x$$

yields

$$\begin{aligned} y(x) &= \beta_{0,1} + \beta_{0,2}\, x \\ &\quad + \sum_{i=1}^{m}(b_{i1} + b_{i2}x)(\bar{y}_i - \mu_{\bar{y}_i}) \\ &\quad + \sum_{i=1}^{m}(b_{i1} + b_{i2}x) f_{\bar{y}_i}\,. \end{aligned} \tag{14.20}$$

Hence, for any fixed x the propagated systematic error is issued as

$$f_{y(x)} = \sum_{i=1}^{m}(b_{i1} + b_{i2}x) f_{\bar{y}_i}$$

its worst-case estimation being

$$f_{s,y(x)} = \sum_{i=1}^{m} \mid b_{i1} + b_{i2}x \mid f_{s,\bar{y}_i}\,. \tag{14.21}$$

Ultimately, due to (13.19), equal systematic errors as defined in (14.12) issue

$$f_{y(x)} = f_{s,y}\,. \tag{14.22}$$

Overall Uncertainty

From (14.18) and (14.21) we draw the uncertainty band

$$\begin{aligned} & y(x) \pm u_{y(x)} \\ & u_{y(x)} = \frac{t_P(n-1)}{\sqrt{n}}\, s_{y(x)} + \sum_{i=1}^{m} \mid b_{i1} + b_{i2}x \mid f_{s,\bar{y}_i}\,. \end{aligned} \tag{14.23}$$

In case of equal systematic errors this turns into

$$y(x) \pm u_{y(x)}$$

$$u_{y(x)} = \frac{t_P(n-1)}{\sqrt{n}}\, s_{y(x)} + f_{s,y}\,. \tag{14.24}$$

At any rate, the uncertainty band is meant to localize the true straight line (12.1).

14.5 *EP*-Region

The *EP*-region localizes the pair of true values $(\beta_{0,1}, \beta_{0,2})$ with respect to the pair of estimators $(\bar{\beta}_1, \bar{\beta}_2)$.

Confidence Ellipsoid

Considering (14.9),

$$\bar{\beta}_k = \mu_{\bar{\beta}_k} + \sum_{i=1}^{m} b_{ik}(\bar{y}_i - \mu_{\bar{y}_i})\,, \quad \mu_{\bar{\beta}_k} = \beta_{0,k} + \sum_{i=1}^{m} b_{ik} f_{\bar{y}_i}\,; \quad k = 1, 2$$

and $\boldsymbol{s}_{\bar{\boldsymbol{\beta}}}$ as given in (14.8), Hotelling's ellipse takes the form

$$(\bar{\boldsymbol{\beta}} - \boldsymbol{\mu}_{\bar{\boldsymbol{\beta}}})^{\mathrm{T}} \boldsymbol{s}_{\bar{\boldsymbol{\beta}}}^{-1} (\bar{\boldsymbol{\beta}} - \boldsymbol{\mu}_{\bar{\boldsymbol{\beta}}}) = \frac{t_P^2(2, n-1)}{n}\,.$$

The associated confidence ellipse reads

$$(\boldsymbol{\beta} - \bar{\boldsymbol{\beta}})^{\mathrm{T}} \boldsymbol{s}_{\bar{\boldsymbol{\beta}}}^{-1} (\boldsymbol{\beta} - \bar{\boldsymbol{\beta}}) = \frac{t_P^2(2, n-1)}{n}\,. \tag{14.25}$$

Inserting

$$\boldsymbol{s}_{\bar{\boldsymbol{\beta}}}^{-1} = \left(\boldsymbol{B}^{\mathrm{T}} \boldsymbol{s}\, \boldsymbol{B}\right)^{-1} = \frac{1}{|\, \boldsymbol{s}_{\bar{\boldsymbol{\beta}}} \,|} \begin{pmatrix} s_{\bar{\beta}_2\bar{\beta}_2} & -s_{\bar{\beta}_1\bar{\beta}_2} \\ -s_{\bar{\beta}_2\bar{\beta}_1} & s_{\bar{\beta}_1\bar{\beta}_1} \end{pmatrix}$$

$$|\, \boldsymbol{s}_{\bar{\boldsymbol{\beta}}} \,| = s_{\bar{\beta}_1\bar{\beta}_1} s_{\bar{\beta}_2\bar{\beta}_2} - s_{\bar{\beta}_1\bar{\beta}_2}^2$$

we obtain

$$s_{\bar{\beta}_2\bar{\beta}_2}\left(\beta_1 - \bar{\beta}_1\right)^2 - 2 s_{\bar{\beta}_1\bar{\beta}_2}\left(\beta_1 - \bar{\beta}_1\right)\left(\beta_2 - \bar{\beta}_2\right) + s_{\bar{\beta}_1\bar{\beta}_1}\left(\beta_2 - \bar{\beta}_2\right)^2$$

$$= |\, \boldsymbol{s}_{\bar{\boldsymbol{\beta}}} \,|\, \frac{t_P^2(2, n-1)}{n}\,. \tag{14.26}$$

The confidence ellipse is expected to localize the point

$$\boldsymbol{\mu}_{\bar{\boldsymbol{\beta}}} = \begin{pmatrix} \mu_{\bar{\beta}_1} \\ \mu_{\bar{\beta}_2} \end{pmatrix} \tag{14.27}$$

with probability P.

Security Polygon

While the $f_{\bar{y}_1}, f_{\bar{y}_1}, \ldots, f_{\bar{y}_m}$ are "scanning" the set of points lying within or on the faces of the m-dimensional hypercuboid

$$-f_{s,\bar{y}_i} \le f_{\bar{y}_i} \le f_{s,\bar{y}_i} \; ; \quad i = 1, \ldots, m$$

the propagated systematic errors (14.10),

$$f_{\bar{\beta}_k} = \sum_{i=1}^{m} b_{ik} f_{\bar{y}_i} ; \quad k = 1, 2 \, ,$$

span a polygon, Appendix F. Putting, however, $f_{\bar{y}_i} = f_y \, ; \; f_{s,\bar{y}_i} = f_{s,y};$ $i = 1, \ldots, m$ the polygon degenerates into an interval

$$\begin{aligned} -f_{s,y} \le f_{\bar{\beta}_1} &\le f_{s,y} \\ f_{\bar{\beta}_2} &= 0 \, . \end{aligned} \tag{14.28}$$

For convenience, we take it as a "stick" of length $2f_{s,y}$, Appendix G.

Mergence of the Confidence Ellipse Either with an Interval or with a Polygon

The mergence of a confidence ellipse either with a non-degenerated or with a degenerated security polygon is addressed in Appendix G.

Figure 14.1 displays the true straight line (12.1), the fitted straight line (14.5), the uncertainties (14.15) of the estimators $\bar{\beta}_1, \bar{\beta}_2$, the uncertainty band (14.23) and, finally, the EP-region, meant to localize the tuple of true values $\beta_{0,1}$ and $\beta_{0,2}$.

Figure 14.2 considers equal systematic errors and renders the true straight line (12.1), the fitted straight line (14.5), the uncertainties (14.16) of the estimators $\bar{\beta}_1, \bar{\beta}_2$, the uncertainty band (14.24) and, ultimately, the EP-region, intended to localize the tuple of true values $\beta_{0,1}$ and $\beta_{0,2}$.

The illustrations are based on simulated data implying known true values and, in particular, extensive graphical scale transformations, Appendix A.

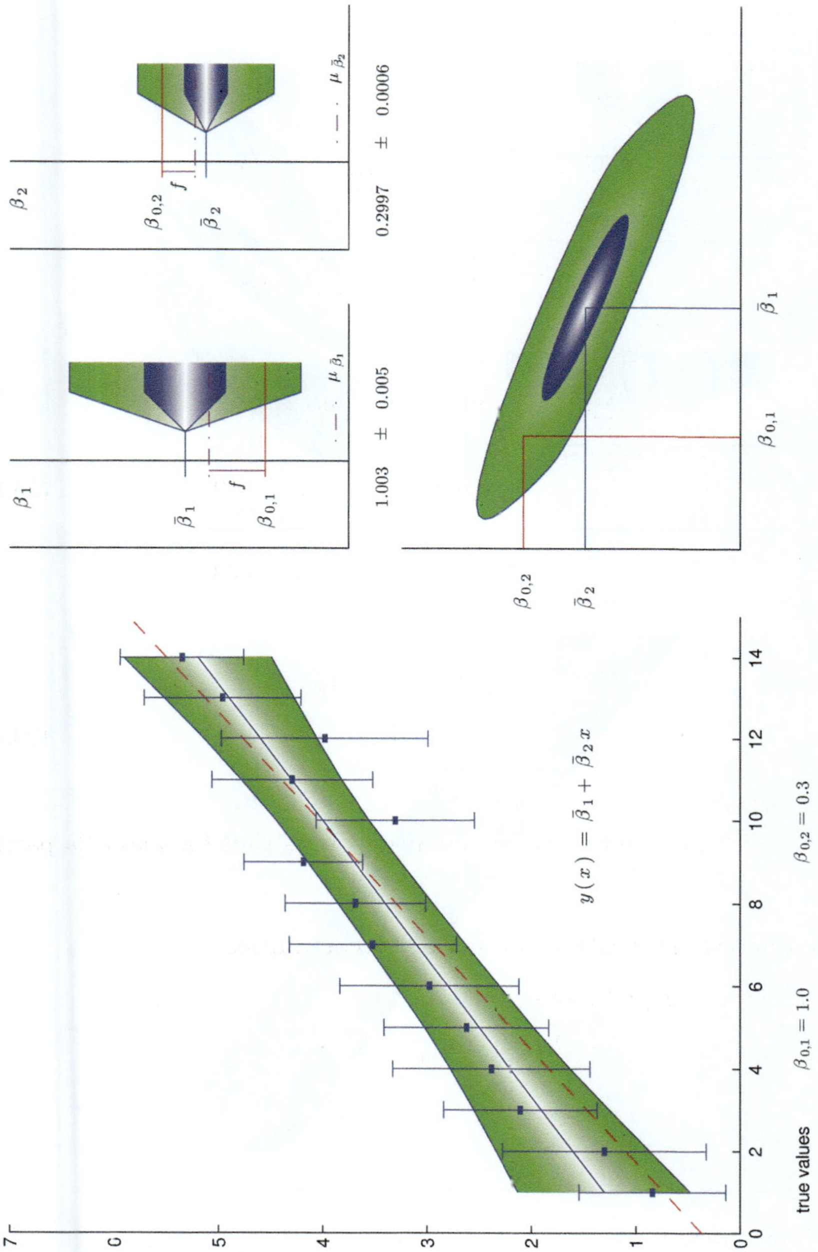

Fig. 14.1. Straight line, case (ii). *Left*: least squares line, uncertainty band, and true straight line (*dashed*). *Top right*: uncertainty intervals localizing the straight line's true parameters $\beta_{0,1}$ and $\beta_{0,2}$. *Bottom right*: EP-region, localizing the tuple of true values $\beta_{0,1}$, $\beta_{0,2}$

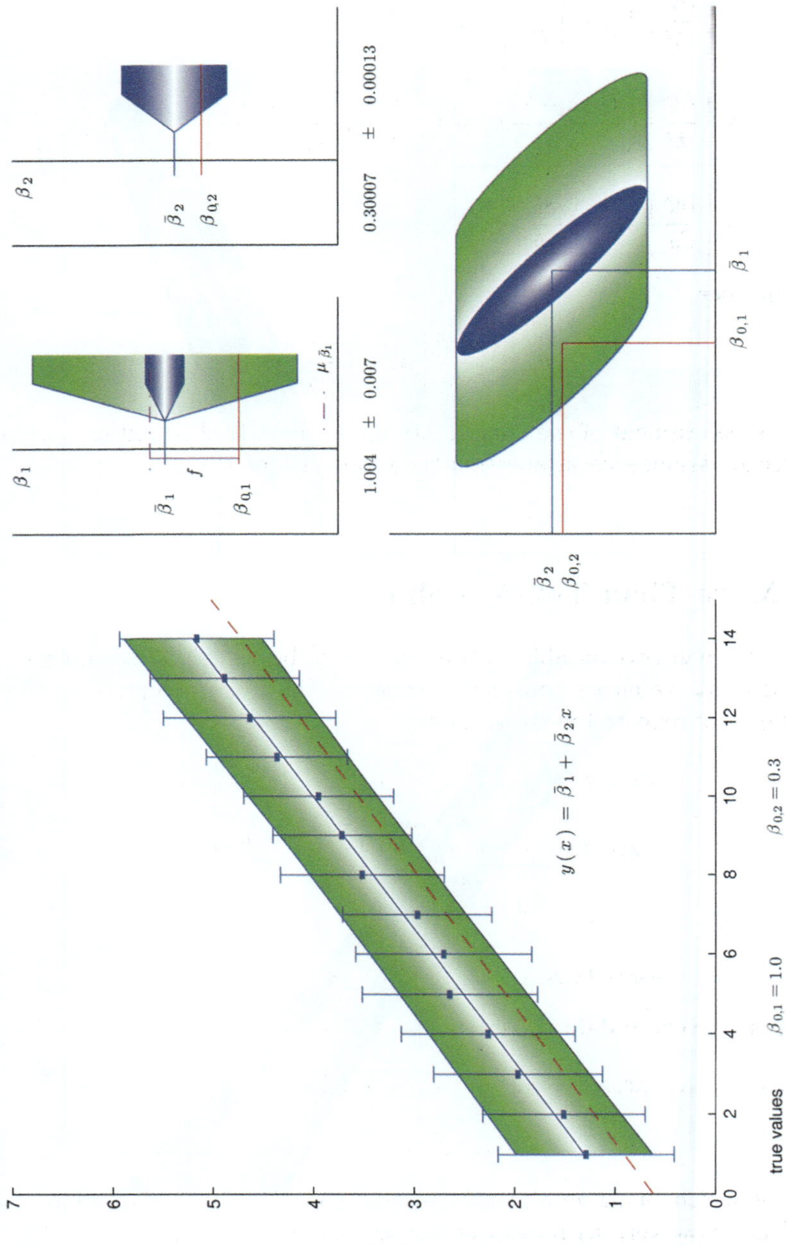

Fig. 14.2. Straight lines, case (ii), equal systematic errors. *Left*: least squares line, uncertainty band, and true straight line (*dashed*). *Top right*: uncertainty intervals localizing the straight line's true parameters $\beta_{0,1}$ and $\beta_{0,2}$. Equal systematic errors do not affect $\bar{\beta}_2$. *Bottom right*: EP-region, localizing the tuple of true values $\beta_{0,1}$, $\beta_{0,2}$

15 Straight Lines: Case (iii)

Case (iii), Table 12.1 considers erroneous abscissas and erroneous ordinates. Presumably, this is the most realistic situation causing the assessment of uncertainties, however, to be somewhat intricate.

15.1 Fitting Conditions

Let there be $m > 2$ pairs of arithmetic means

$$(\bar{x}_1, \bar{y}_1), \quad (\bar{x}_2, \bar{y}_2), \quad \ldots \quad (\bar{x}_m, \bar{y}_m) \tag{15.1}$$

each covering n repeated measurements

$$\bar{x}_i = \frac{1}{n}\sum_{l=1}^{n} x_{il} = x_{0,i} + (\bar{x}_i - \mu_{\bar{x}_i}) + f_{\bar{x}_i}\,; \quad i = 1, \ldots, m$$

$$E\{\bar{X}_i\} = \mu_{\bar{x}_i}\,; \qquad -f_{s,\bar{x}_i} \le f_{\bar{x}_i} \le f_{s,\bar{x}_i}$$

and

$$\bar{y}_i = \frac{1}{n}\sum_{l=1}^{n} y_{il} = y_{0,i} + (\bar{y}_i - \mu_{\bar{y}_i}) + f_{\bar{y}_i}\,; \quad i = 1, \ldots, m$$

$$E\{\bar{Y}_i\} = \mu_{\bar{y}_i}\,; \qquad -f_{s,\bar{y}_i} \le f_{\bar{y}_i} \le f_{s,\bar{y}_i}\,.$$

The empirical variances

$$s_{\bar{x}_i}^2 = \frac{1}{n-1}\sum_{l=1}^{n}(x_{il} - \bar{x}_i)^2\,, \quad s_{\bar{y}_i}^2 = \frac{1}{n-1}\sum_{l=1}^{n}(y_{il} - \bar{y}_i)^2\,; \quad i = 1, \ldots, m$$

and the error limits $\pm f_{s,\bar{x}_1}$ and $\pm f_{s,\bar{y}_i}$ issue the uncertainties of the input data

$$\begin{aligned} &\bar{x}_i \pm u_{\bar{x}_i}\,, \quad u_{\bar{x}_i} = \frac{t_P(n-1)}{\sqrt{n}}\, s_{\bar{x}_i} + f_{s,\bar{x}_i} \\ & \qquad\qquad\qquad\qquad\qquad\qquad\qquad i = 1, \ldots, m\,. \\ &\bar{y}_i \pm u_{\bar{y}_i}\,, \quad u_{\bar{y}_i} = \frac{t_P(n-1)}{\sqrt{n}}\, s_{\bar{y}_i} + f_{s,\bar{y}_i}\,. \end{aligned} \tag{15.2}$$

The intervals $\bar{x}_i \pm u_{\bar{x}_i}$ and $\bar{y}_i \pm u_{\bar{y}_i}$ are intended to localize the true values $x_{0,i}$ and $y_{0,i}$, respectively.

15.2 Orthogonal Projection

The input data produce an inconsistent, over-determined, linear system

$$\beta_1 + \beta_2\,\bar{x}_i \approx \bar{y}_i\,; \qquad i = 1,\dots,m > 2, \tag{15.3}$$

in matrix form

$$\boldsymbol{A}\boldsymbol{\beta} \approx \bar{\boldsymbol{y}}$$

where

$$\boldsymbol{A} = \begin{pmatrix} 1 & \bar{x}_1 \\ 1 & \bar{x}_2 \\ \cdots & \cdots \\ 1 & \bar{x}_m \end{pmatrix}, \qquad \boldsymbol{\beta} = \begin{pmatrix} \beta_1 \\ \beta_2 \end{pmatrix}, \qquad \bar{\boldsymbol{y}} = \begin{pmatrix} \bar{y}_1 \\ \bar{y}_2 \\ \cdots \\ \bar{y}_m \end{pmatrix}.$$

The orthogonal projection yields

$$\bar{\boldsymbol{\beta}} = \boldsymbol{B}^{\mathrm{T}}\bar{\boldsymbol{y}}\,, \qquad \bar{\beta}_k = \sum_{i=1}^{m} b_{ik}\bar{y}_i\,; \qquad k = 1,2 \tag{15.4}$$

with

$$\boldsymbol{B} = \boldsymbol{A}\left(\boldsymbol{A}^{\mathrm{T}}\boldsymbol{A}\right)^{-1} = (b_{ik})\,; \quad i = 1,\dots,m; \quad k = 1,2\,.$$

The matrix $\boldsymbol{B}$ is given by

$$\boldsymbol{B} = \frac{1}{D}\begin{bmatrix} \sum\limits_{j=1}^{m}\bar{x}_j^2 - \bar{x}_1\sum\limits_{j=1}^{m}\bar{x}_j & -\sum\limits_{j=1}^{m}\bar{x}_j + m\bar{x}_1 \\ \sum\limits_{j=1}^{m}\bar{x}_j^2 - \bar{x}_2\sum\limits_{j=1}^{m}\bar{x}_j & -\sum\limits_{j=1}^{m}\bar{x}_j + m\bar{x}_2 \\ \cdots & \cdots \\ \sum\limits_{j=1}^{m}\bar{x}_j^2 - \bar{x}_m\sum\limits_{j=1}^{m}\bar{x}_j & -\sum\limits_{j=1}^{m}\bar{x}_j + m\bar{x}_m \end{bmatrix} \tag{15.5}$$

in which

$$D = |\,\boldsymbol{A}^{\mathrm{T}}\boldsymbol{A}\,|\,.$$

The $b_{i,k}$ bring to bear the errors of the abscissas. Clearly, these errors as well as the errors of the ordinates will have to be considered in the error propagation to come.

After all, the components $\bar{\beta}_k\,;\ k = 1,2$ of the solution vector $\bar{\boldsymbol{\beta}}$ devise the least squares line

$$y(x) = \bar{\beta}_1 + \bar{\beta}_2\,x\,. \tag{15.6}$$

15.3 Series Expansion of the Solution Vector

Let us submit the $\bar{\beta}_k$; $k = 1, 2$ to series expansions throughout a neighborhood of the point

$$(x_{0,1}\,, \dots\,, x_{0,m}\,;\; y_{0,1}\,, \dots\,, y_{0,m})\,,$$

firstly, with respect to the n points

$$(x_{1l}, x_{2l}, \dots, x_{ml}\,;\; y_{1l}, y_{2l}, \dots, y_{ml})\,; \quad l = 1, \dots, n$$

yielding

$$\begin{aligned}\bar{\beta}_{kl}(x_{1l}, \dots, x_{ml}\,;\; y_{1l}, \dots, y_{ml}) &= \bar{\beta}_k(x_{0,1}, \dots, x_{0,m};\; y_{0,1}, \dots, y_{0,m})\\ &+ \sum_{i=1}^{m} \frac{\partial \bar{\beta}_k}{\partial x_{0,i}}(x_{il} - \mu_{\bar{x}_i}) + \sum_{i=1}^{m} \frac{\partial \bar{\beta}_k}{\partial y_{0,i}}(y_{il} - \mu_{\bar{y}_i}) + \cdots \\ &+ \sum_{i=1}^{m} \frac{\partial \bar{\beta}_k}{\partial x_{0,i}} f_{\bar{x}_i} + \sum_{i=1}^{m} \frac{\partial \bar{\beta}_k}{\partial y_{0,i}} f_{\bar{y}_i} + \cdots\,; \quad k = 1, 2\end{aligned} \tag{15.7}$$

and, secondly, with respect to the sample-dependent means

$$(\bar{x}_1, \bar{x}_2, \dots, \bar{x}_m\,;\; \bar{y}_1, \bar{y}_2, \dots, \bar{y}_m)$$

issuing

$$\begin{aligned}\bar{\beta}_k(\bar{x}_1, \dots, \bar{x}_m;\; \bar{y}_1, \dots, \bar{y}_m) &= \bar{\beta}_k(x_{0,1}, \dots, x_{0,m};\; y_{0,1}, \dots, y_{0,m})\\ &+ \sum_{i=1}^{m} \frac{\partial \bar{\beta}_k}{\partial x_{0,i}}(\bar{x}_i - \mu_{\bar{x}_i}) + \sum_{i=1}^{m} \frac{\partial \bar{\beta}_k}{\partial y_{0,i}}(\bar{y}_i - \mu_{\bar{y}_i}) + \cdots \\ &+ \sum_{i=1}^{m} \frac{\partial \bar{\beta}_k}{\partial x_{0,i}} f_{\bar{x}_i} + \sum_{i=1}^{m} \frac{\partial \bar{\beta}_k}{\partial y_{0,i}} f_{\bar{y}_i} + \cdots\,; \quad k = 1, 2\,.\end{aligned} \tag{15.8}$$

As there is no other choice, we approximate the derivatives in $x_{0,1}\,, \dots, y_{0,m}$ through derivatives in $\bar{x}_1\,, \dots, \bar{y}_m$. Further, for convenience, we assign the coefficients

$$\begin{aligned} c_{i1} &= \frac{\partial \bar{\beta}_1}{\partial \bar{x}_i}\,, \qquad c_{i+m,1} = \frac{\partial \bar{\beta}_1}{\partial \bar{y}_i} \\ & \qquad\qquad\qquad\qquad\qquad\qquad i = 1,\, \dots\,, m \\ c_{i2} &= \frac{\partial \bar{\beta}_2}{\partial \bar{x}_i}\,, \qquad c_{i+m,2} = \frac{\partial \bar{\beta}_2}{\partial \bar{y}_i} \end{aligned}$$

to an auxiliary matrix

$$\boldsymbol{C}^{\mathrm{T}} = \begin{pmatrix} c_{11} & c_{21} & \cdots & c_{2m,1} \\ c_{12} & c_{22} & \cdots & c_{2m,2} \end{pmatrix} . \tag{15.9}$$

The coefficients c_{ik} are given in Appendix B. Beyond, we put

$$\begin{aligned} v_{il} &= x_{il} \,, & v_{i+m,l} &= y_{il} \\ \bar{v}_i &= \bar{x}_i \,, & \bar{v}_{i+m} &= \bar{y}_i \\ \mu_i &= \mu_{\bar{x}_i} \,, & \mu_{i+m} &= \mu_{\bar{y}_i} \\ f_i &= f_{\bar{x}_i} \,, & f_{i+m} &= f_{\bar{y}_i} \\ f_{s,i} &= f_{s,\bar{x}_i} \,, & f_{s,i+m} &= f_{s,\bar{y}_i} \,. \end{aligned}$$

After all, linearizing the expansions (15.7) and (15.8) produces

$$\begin{aligned} \bar{\beta}_{kl} &= \beta_{0,k} + \sum_{i=1}^{2m} c_{ik}(v_{il} - \mu_i) + \sum_{i=1}^{2m} c_{ik} f_i \,; \quad k = 1, 2 \,; \quad l = 1, \ldots, n \\ \bar{\beta}_k &= \beta_{0,k} + \sum_{i=1}^{2m} c_{ik}(\bar{v}_i - \mu_i) + \sum_{i=1}^{2m} c_{ik} f_i \end{aligned} \tag{15.10}$$

as $\bar{\beta}_k(x_{0,1}, \ldots, y_{0,m}) = \beta_{0,k}$. Subtraction yields

$$\bar{\beta}_{kl} - \bar{\beta}_k = \sum_{i=1}^{2m} c_{ik}(v_{il} - \bar{v}_i) \,; \quad k = 1, 2 \,. \tag{15.11}$$

Notionally, we consider an ensemble of least squares lines based on the respective l-th repeated measurements of the m measuring points $i = 1, \ldots, m$, i.e. on

$$(x_{1l}, \ldots, x_{ml} \,;\, y_{1l}, \ldots, y_{ml}) \,; \qquad l = 1, \ldots, n$$

so that

$$\bar{\beta}_k = \frac{1}{n} \sum_{l=1}^{n} \bar{\beta}_{kl} \,; \qquad k = 1, 2 \,. \tag{15.12}$$

While (15.11) devises the propagation of random errors, the propagated systematic errors are to be taken from

$$f_{\bar{\beta}_k} = \sum_{i=1}^{2m} c_{ik} f_i \,; \qquad k = 1, 2 \,. \tag{15.13}$$

15.4 Uncertainties of the Components of the Solution Vector

Random Errors

Invoking (15.11), the elements $s_{\bar{\beta}_k \bar{\beta}_{k'}}$ of the empirical variance–covariance matrix

$$\boldsymbol{s}_{\bar{\boldsymbol{\beta}}} = \begin{pmatrix} s_{\bar{\beta}_1 \bar{\beta}_1} & s_{\bar{\beta}_1 \bar{\beta}_2} \\ s_{\bar{\beta}_2 \bar{\beta}_1} & s_{\bar{\beta}_2 \bar{\beta}_2} \end{pmatrix}, \qquad s_{\bar{\beta}_k \bar{\beta}_k} \equiv s^2_{\bar{\beta}_k}$$

of the solution vector $\bar{\boldsymbol{\beta}}$ take the form

$$\begin{aligned} s_{\bar{\beta}_k \bar{\beta}_{k'}} &= \frac{1}{n-1} \sum_{l=1}^{n} \left(\bar{\beta}_{kl} - \bar{\beta}_k\right) \left(\bar{\beta}_{k'l} - \bar{\beta}_{k'}\right) \\ &= \frac{1}{n-1} \sum_{l=1}^{n} \left[\sum_{i=1}^{2m} c_{ik} \left(v_{il} - \bar{v}_i\right) \right] \left[\sum_{j=1}^{2m} c_{jk'} \left(v_{jl} - \bar{v}_j\right) \right] \\ &= \sum_{i,j=1}^{2m} c_{ik}\, c_{jk'}\, s_{ij} \, ; \quad k, k' = 1, 2 \end{aligned}$$

in which the

$$s_{ij} = \frac{1}{n-1} \sum_{l=1}^{n} (v_{il} - \bar{v}_i)(v_{jl} - \bar{v}_j) \, ; \quad i, j = 1, \dots, 2m$$

designate the elements of the empirical variance–covariance matrix

$$\boldsymbol{s} = (s_{ij}) \, ; \quad i, j = 1, \dots, 2m \tag{15.14}$$

of the input data, each having degrees of freedom $\nu = n - 1$. Eventually, deploying (15.9), the empirical variance–covariance matrix of the solution vector is issued by

$$\boldsymbol{s}_{\bar{\boldsymbol{\beta}}} = \boldsymbol{C}^{\mathrm{T}} \boldsymbol{s}\, \boldsymbol{C} \, . \tag{15.15}$$

Systematic Errors

In case of equal systematic errors and equal error bounds

$$\left.\begin{aligned} f_{\bar{x}_i} &= f_x \, , \quad f_{s,\bar{x}_i} = f_{s,x} \quad && - f_{s,x} \le f_x \le f_{s,x} \\ f_{\bar{y}_i} &= f_y \, , \quad f_{s,\bar{y}_i} = f_{s,y} \quad && - f_{s,y} \le f_y \le f_{s,y} \end{aligned}\right\} \quad i = 1, \dots, m \tag{15.16}$$

(15.13) passes into

$$f_{\bar{\beta}_k} = \sum_{i=1}^{m} c_{ik} f_{\bar{x}_i} + \sum_{i=1}^{m} c_{i+m,k} f_{\bar{y}_i}$$

$$= f_x \sum_{i=1}^{m} c_{ik} + f_y \sum_{i=1}^{m} c_{i+m,k} \,.$$

But as

$$\begin{aligned} &\sum_{i=1}^{m} c_{i1} = -f_{\bar{\beta}_2}\,, \qquad &&\sum_{i=1}^{m} c_{i+m,1} = 1 \\ &\sum_{i=1}^{m} c_{i,2} = 0\,, \qquad &&\sum_{i=1}^{m} c_{i+m,2} = 0 \end{aligned} \tag{15.17}$$

the $f_{\bar{\beta}_k}$ get reduced to

$$\begin{aligned} f_{\bar{\beta}_1} &= -f_x \bar{\beta}_2 + f_y \\ f_{\bar{\beta}_2} &= 0\,. \end{aligned} \tag{15.18}$$

Confidence Intervals and Overall Uncertainties

We localize the expectations

$$E\{\bar{\beta}_1\} = \mu_{\bar{\beta}_1} = \beta_{0,1} + \sum_{i=1}^{2m} c_{i1} f_i$$

$$E\{\bar{\beta}_2\} = \mu_{\bar{\beta}_2} = \beta_{0,2} + \sum_{i=1}^{2m} c_{i2} f_i$$

by confidence intervals

$$\bar{\beta}_1 - \frac{t_P(n-1)}{\sqrt{n}} s_{\bar{\beta}_1} \le \mu_{\bar{\beta}_1} \le \bar{\beta}_1 + \frac{t_P(n-1)}{\sqrt{n}} s_{\bar{\beta}_1}$$

$$\bar{\beta}_2 - \frac{t_P(n-1)}{\sqrt{n}} s_{\bar{\beta}_2} \le \mu_{\bar{\beta}_2} \le \bar{\beta}_2 + \frac{t_P(n-1)}{\sqrt{n}} s_{\bar{\beta}_2}$$

of probability P. Hence, the overall uncertainties of the components of the solution vector turn out to be

$$u_{\bar{\beta}_k} = \frac{t_P(n-1)}{\sqrt{n}} s_{\bar{\beta}_k} + \sum_{i=1}^{2m} |\, c_{ik} \,|\, f_{s,i}\,; \quad k = 1, 2\,. \tag{15.19}$$

Equal systematic errors, by contrast, suggest

$$
\begin{aligned}
u_{\bar{\beta}_1} &= \frac{t_P(n-1)}{\sqrt{n}}\, s_{\bar{\beta}_1} + f_{s,x} \mid \bar{\beta}_2 \mid + f_{s,y} \\
u_{\bar{\beta}_2} &= \frac{t_P(n-1)}{\sqrt{n}}\, s_{\bar{\beta}_2}\,.
\end{aligned}
\tag{15.20}
$$

Equal systematic errors do not affect the slope. They shift, however, the straight line parallel to itself. At any rate, the final result reads

$$\bar{\beta}_k \pm u_{\bar{\beta}_k}\,; \quad k = 1, 2\,.$$

15.5 Uncertainty Band

Random Errors

Let the $\bar{\beta}_{kl}$ as addressed in (15.10) induce an ensemble of n least square lines

$$y_l(x) = \bar{\beta}_{1l} + \bar{\beta}_{2l}\, x\,; \quad l = 1, \ldots, n$$

which, following (15.12), reestablishes (15.6)

$$y(x) = \frac{1}{n}\sum_{l=1}^{n} y_l(x)\,.$$

For any fixed x, the differences

$$
\begin{aligned}
y_l(x) - y(x) &= \left(\bar{\beta}_{1l} - \bar{\beta}_1\right) + \left(\bar{\beta}_{2l} - \bar{\beta}_2\right) x \\
&= \sum_{i=1}^{2m} (c_{i1} + c_{i2}x)(v_{i1} - \bar{v}_i)\,; \qquad l = 1, \ldots, n
\end{aligned}
$$

bring forth an empirical variance

$$s_{y(x)}^2 = \frac{1}{n-1}\sum_{l=1}^{n} \left(y_l(x) - y(x)\right)^2 = \boldsymbol{c}^T \boldsymbol{s}\, \boldsymbol{c}\,. \tag{15.21}$$

Here, $\boldsymbol{s}$ denotes the empirical variance–covariance matrix as quoted in (15.14) and $\boldsymbol{c}$ an auxiliary vector

$$\boldsymbol{c} = \left(c_{11} + c_{12}x \quad c_{21} + c_{22}x \quad \cdots \quad c_{2m,1} + c_{2m,2}x\right)^{\mathrm{T}}\,.$$

Systematic Errors

Inserting (15.10) into (15.6) produces

$$y(x) = \beta_{0,1} + \beta_{0,2}x + \sum_{i=1}^{2m}(c_{i1} + c_{i2}x)(\bar{v}_i - \mu_i) + \sum_{i=1}^{2m}(c_{i1} + c_{i2}x) f_i \,.$$

Thus, for a given x the systematic error is given by

$$f_{y(x)} = \sum_{i=1}^{2m}(c_{i1} + c_{i2}x) f_i \tag{15.22}$$

the worst-case estimation of which being

$$f_{s,y(x)} = \sum_{i=1}^{2m} \mid c_{i1} + c_{i2}\,x \mid f_{s,i} \,. \tag{15.23}$$

Considering equal systematic errors as introduced in (15.16), due to (15.17) we have

$$f_{s,y(x)} = \mid \bar{\beta}_2 \mid f_{s,x} + f_{s,y} \,. \tag{15.24}$$

Overall Uncertainty

After all, the uncertainty band takes the form

$$y(x) \pm u_{y(x)}$$

$$u_{y(x)} = \frac{t_P(n-1)}{\sqrt{n}}\, s_{y(x)} + \sum_{i=1}^{2m} \mid c_{i1} + c_{i2}\,x \mid f_{s,i} \,. \tag{15.25}$$

For equal systematic errors this passes into

$$y(x) \pm u_{y(x)}$$

$$u_{y(x)} = \frac{t_P(n-1)}{\sqrt{n}}\, s_{y(x)} + \mid \bar{\beta}_2 \mid f_{s,x} + f_{s,y} \,. \tag{15.26}$$

At any rate, we expect the uncertainty band to hold the true straight line (12.1).

15.6 *EP*-Region

The *EP*-region localizes the pair of true values $(\beta_{0,1}, \beta_{0,2})$ with respect to the pair of estimators $(\bar{\beta}_1, \bar{\beta}_2)$.

Confidence Ellipsoid

Let us rewrite (15.10) according to

$$\bar{\beta}_k = \mu_{\bar{\beta}_k} + \sum_{i=1}^{2m} c_{ik}\left(\bar{v}_i - \mu_i\right)\,; \quad \mu_{\bar{\beta}_k} = \beta_{0,k} + \sum_{i=1}^{2m} c_{ik} f_i\,; \quad k = 1, 2\,.$$

With $\boldsymbol{s}_{\bar{\boldsymbol{\beta}}}$ as given in (15.15), Hotelling's ellipse reads

$$(\bar{\boldsymbol{\beta}} - \mu_{\bar{\boldsymbol{\beta}}})^{\mathrm{T}} s_{\bar{\boldsymbol{\beta}}}^{-1} (\bar{\boldsymbol{\beta}} - \mu_{\bar{\boldsymbol{\beta}}}) = \frac{t_P^2(2, n-1)}{n}\,.$$

Hence, the confidence ellipse

$$(\boldsymbol{\beta} - \bar{\boldsymbol{\beta}})^{\mathrm{T}} s_{\bar{\boldsymbol{\beta}}}^{-1} (\boldsymbol{\beta} - \bar{\boldsymbol{\beta}}) = \frac{t_P^2(2, n-1)}{n} \tag{15.27}$$

localizes the point

$$\boldsymbol{\mu}_{\bar{\boldsymbol{\beta}}} = \begin{pmatrix} \mu_{\bar{\beta}_1} \\ \\ \mu_{\bar{\beta}_2} \end{pmatrix} \tag{15.28}$$

with probability P.

Security Polygon

Letting the f_i "scan" the set of points lying within or on the faces of the $2m$-dimensional hypercuboid

$$-f_{s,i} \le f_i \le f_{s,i} \quad i = 1, \ldots, 2m$$

the propagated systematic errors

$$f_{\bar{\beta}_k} = \sum_{i=1}^{2m} c_{ik} f_i\,; \quad k = 1, 2 \tag{15.29}$$

span a polygon. In case of equal systematic errors we have, considering (15.18),

$$\begin{aligned} f_{\bar{\beta}_1} &= -f_x\,\bar{\beta}_2 + f_y \\ f_{\bar{\beta}_2} &= 0\,. \end{aligned}$$

This causes the security polygon to degenerate into an interval

$$-\left(f_{s,x} \mid \bar{\beta}_2 \mid + f_{s,y}\right) \le f_{\bar{\beta}_1} \le f_{s,x} \mid \bar{\beta}_2 \mid + f_{s,y}\,. \tag{15.30}$$

Just for convenience, we notionally conceive (15.30) as a "stick" of length $2(f_{s,x} \mid \bar{\beta}_2 \mid + f_{s,y})$.

Mergence of Ellipse Either with Interval or with Polygon

The merging of a confidence ellipse either with a non-degenerated or with a degenerated security polygon is addressed in Appendix G.

Figure 15.1 displays the true straight line (12.1), the fitted straight line (15.6), the uncertainties (15.19) of the estimators $\bar{\beta}_1, \bar{\beta}_2$, the uncertainty band (15.25) and, finally, the EP-region, meant to localize the tuple of true values $\beta_{0,1}$ and $\beta_{0,2}$.

Figure 15.2 considers equal systematic errors and renders the true straight line (12.1), the fitted straight line (15.6), the uncertainties (15.20) of the estimators $\bar{\beta}_1, \bar{\beta}_2$, the uncertainty band (15.26) and, ultimately, the EP-region, intended to localize the tuple of true values $\beta_{0,1}$ and $\beta_{0,2}$.

Again, the illustrations are based on simulated data implying known true values and, in particular, extensive graphical scale transformations, Appendix A.

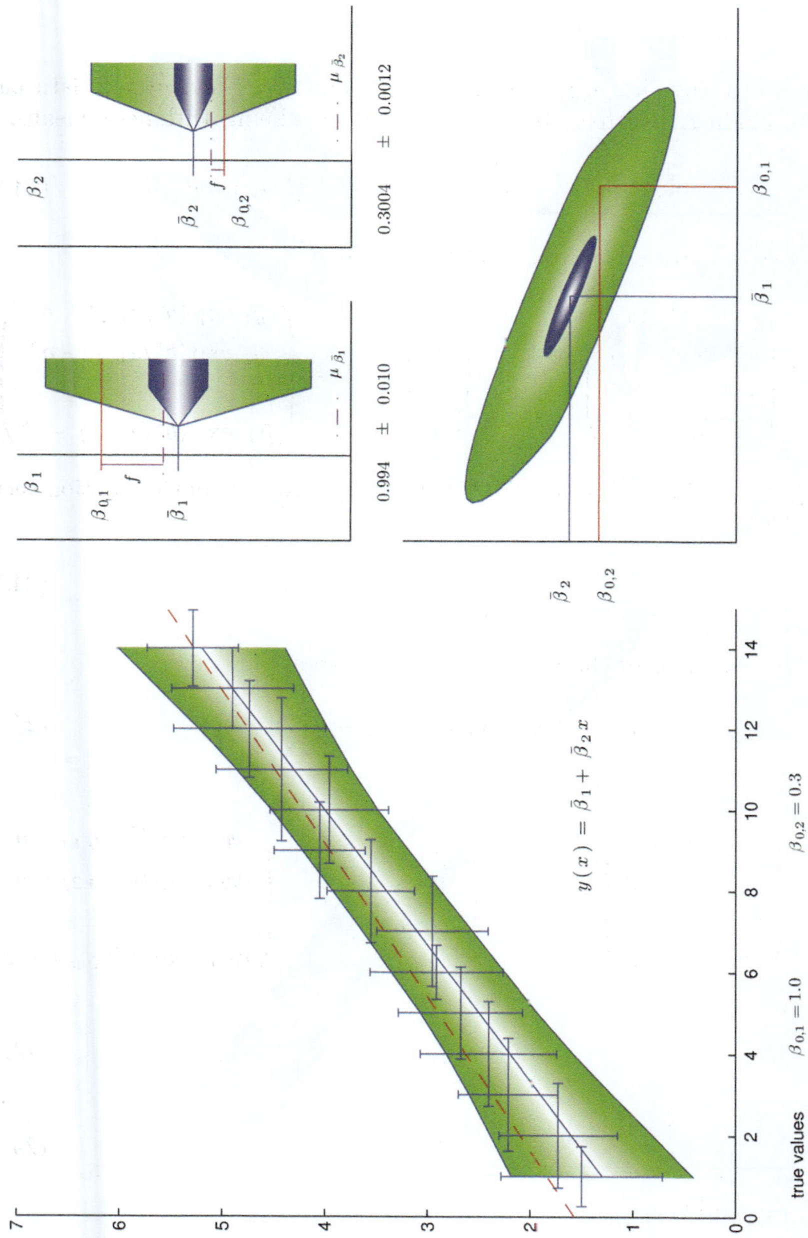

Fig. 15.1. Straight lines, case (iii). *Left*: least squares line, uncertainty band, and true straight line (*dashed*). *Top right*: uncertainty intervals localizing the straight line's true parameters $\beta_{0,1}$ and $\beta_{0,2}$. *Bottom right*: EP-region, localizing the tuple of true values $\beta_{0,1}$, $\beta_{0,2}$

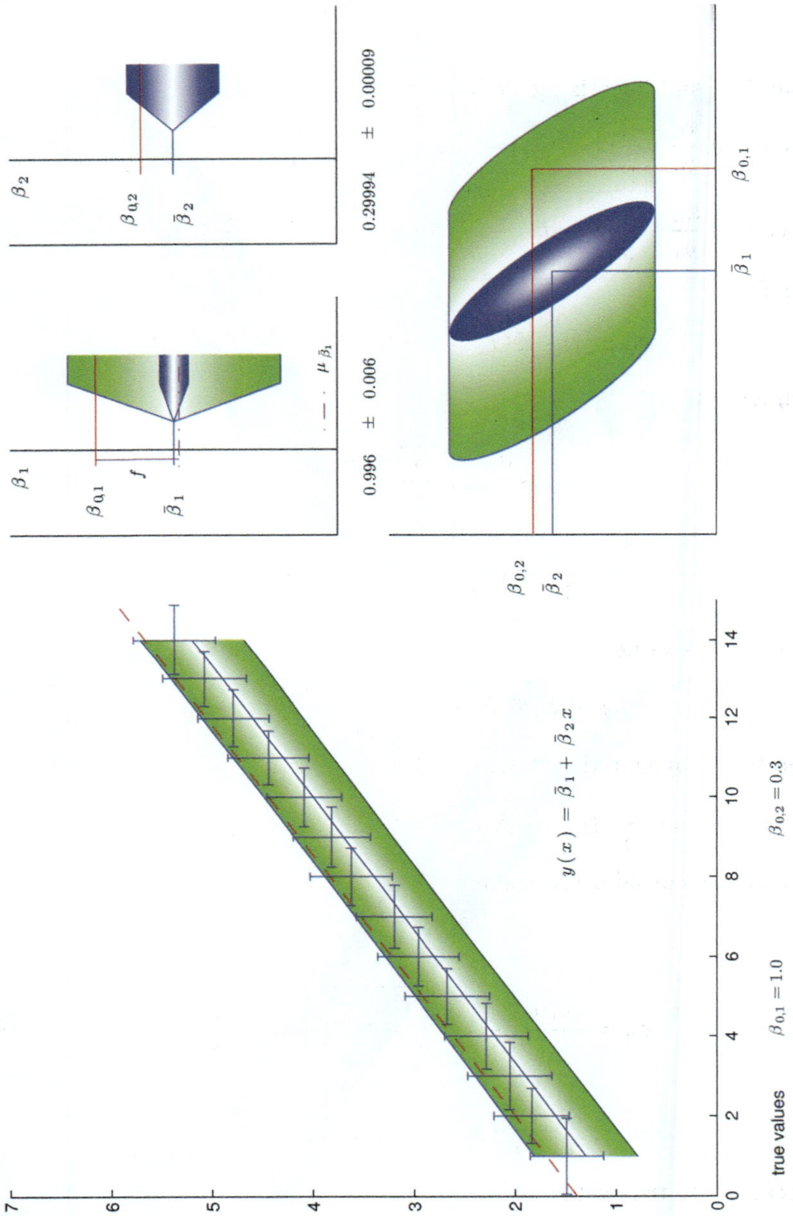

Fig. 15.2. Straight lines, case (iii), equal systematic errors. *Left*: least squares line, uncertainty band, and true straight line (*dashed*). *Top right*: uncertainty intervals localizing the straight line's true parameters $\beta_{0,1}$ and $\beta_{0,2}$. Equal systematic errors do not affect $\bar{\beta}_2$. *Bottom right*: EP-region, localizing the tuple of true values $\beta_{0,1}$, $\beta_{0,2}$

Part VI

Fitting of Planes

16 Preliminaries

The adjustment of planes calls for data triples

$$(x_1, y_1, z_1), \quad (x_2, y_2, z_2), \quad \ldots \quad , (x_m, y_m, z_m) \, ; \qquad m > 3 \, .$$

16.1 Distinction of Cases

We shall address three fitting situations, Table 16.1:

Table 16.1. Fitting of Planes, Three Cases

Case	x and y coordinates	z coordinates
(i)	Error-free	Individual measurements
(ii)	Error-free	Repeated measurements
(iii)	Repeated measurements	Repeated measurements

Case (i) assumes correct x, y-coordinates and erroneous, individually measured z-values. Each of the $z-$s is supposed to hold a particular random error, stemming from one and the same normal distribution, and a common unknown systematic error. Case (ii), likewise relying on exact x, y-coordinates, accounts for repeated measurements of the $z-$s. The scattering of the random errors as well as the actual values of the unknown systematic errors may vary from one measured z-value to the next. Case (iii), ultimately, admits measuring errors in all three coordinates with varying theoretical variances and varying unknown systematic errors.

16.2 True Plane

Let the equation of a plane

$$z(x, y) = \beta_{0,1} + \beta_{0,2} \, x + \beta_{0,3} \, y \tag{16.1}$$

be fulfilled by the m data triples

$$(x_{0,1}, y_{0,1}, z_{0,1}), \quad (x_{0,2}, y_{0,2}, z_{0,2}), \quad \dots \quad , (x_{0,m}, y_{0,m}, z_{0,m}) . \tag{16.2}$$

Written in matrices, this reads

$$\boldsymbol{A}\boldsymbol{\beta}_0 = \boldsymbol{z}_0 \tag{16.3}$$

where

$$\boldsymbol{A} = \begin{pmatrix} 1 & x_{0,1} & y_{0,1} \\ 1 & x_{0,2} & y_{0,2} \\ \dots & \dots & \dots \\ 1 & x_{0,m} & y_{0,m} \end{pmatrix}, \quad \boldsymbol{\beta}_0 = \begin{pmatrix} \beta_{0,1} \\ \beta_{0,2} \\ \beta_{0,3} \end{pmatrix}, \quad \boldsymbol{z}_0 = \begin{pmatrix} z_{0,1} \\ z_{0,2} \\ \dots \\ z_{0,m} \end{pmatrix}. \tag{16.4}$$

Given $\boldsymbol{A}$ has rank 3, (16.3) reproduces

$$\boldsymbol{\beta}_0 = \boldsymbol{B}^{\mathrm{T}}\boldsymbol{z}_0 , \qquad \boldsymbol{B} = \boldsymbol{A}\left(\boldsymbol{A}^{\mathrm{T}}\boldsymbol{A}\right)^{-1} . \tag{16.5}$$

We address (16.1) as the true plane, (16.2) as the true data triples and the vector $\boldsymbol{\beta}_0$ the true solution vector.

Feeding in empirical data clearly suspends (16.3) and thus conceals $\boldsymbol{\beta}_0$. Nevertheless, we may fit a least squares plane to the set of defective data and attempt to localize the components $\beta_{0,1}$, $\beta_{0,2}$, and $\beta_{0,3}$ of $\boldsymbol{\beta}_0$ via uncertainty intervals.

17 Planes: Case (i)

Case (i) of Table 16.1 assumes error-free x, y-coordinates and erroneous, individually measured z-coordinates.

17.1 Fitting Conditions

The adjustment relies on $m > 3$ data triples

$$(x_{0,1}, y_{0,1}, z_1), \quad (x_{0,2}, y_{0,2}, z_2), \quad \dots \quad (x_{0,m}, y_{0,m}, z_m). \tag{17.1}$$

The $x_{0,i}, y_{0,i}$ are considered correct and the z_i erroneous following

$$z_i = z_{0,i} + (z_i - \mu_{z_i}) + f_{z_i}\,; \quad i = 1, \dots, m \tag{17.2}$$

with expectations $\mu_{z_i} = E\{Z_i\}$. We presume the random errors $z_i - \mu_{z_i}$; $i = 1, \dots, m$ to be due to a common normal density and hence liable to a uniform theoretical variance, say, σ_z^2. At present, we let the z_i be charged by the same systematic error

$$\begin{gathered} f_{z_i} = f_z\,; \qquad i = 1, \dots, m \\ -f_{s,z} \le f_z \le f_{s,z}\,. \end{gathered} \tag{17.3}$$

For the time being, we are not in a position to specify the uncertainties of z_i-data.

17.2 Orthogonal Projection

The inconsistent, over-determined, linear system reads

$$\beta_1 + \beta_2\, x_{0,i} + \beta_3\, y_{0,i} \approx z_i\,; \qquad i = 1, \dots, m > 3\,. \tag{17.4}$$

We are in search of a least squares plane

$$z(x) = \bar{\beta}_1 + \bar{\beta}_2\, x + \bar{\beta}_3\, y \tag{17.5}$$

fitting the data set (17.1). Putting

$$\boldsymbol{A} = \begin{pmatrix} 1 & x_{0,1} & y_{0,1} \\ 1 & x_{0,2} & y_{0,2} \\ \cdots & \cdots & \cdots \\ 1 & x_{0,m} & y_{0,m} \end{pmatrix}, \qquad \boldsymbol{\beta} = \begin{pmatrix} \beta_1 \\ \beta_2 \\ \beta_3 \end{pmatrix}, \qquad \boldsymbol{z} = \begin{pmatrix} z_1 \\ z_2 \\ \cdots \\ z_m \end{pmatrix}$$

the matrix form of (17.4) turns out to be

$$\boldsymbol{A}\boldsymbol{\beta} \approx \boldsymbol{z}\,. \tag{17.6}$$

The orthogonal projection produces the solution vector

$$\bar{\boldsymbol{\beta}} = \boldsymbol{B}^{\mathrm{T}}\boldsymbol{z}\,, \qquad \bar{\beta}_k = \sum_{i=1}^{m} b_{ik}\, z_i\,; \qquad k = 1, 2, 3 \tag{17.7}$$

with

$$\boldsymbol{B} = \boldsymbol{A}\left(\boldsymbol{A}^{\mathrm{T}}\boldsymbol{A}\right)^{-1} = (b_{ik})\,; \quad i = 1, \ldots, m; \quad k = 1, 2, 3\,. \tag{17.8}$$

The components $\bar{\beta}_k$ of the solution vector $\bar{\boldsymbol{\beta}}$ issue the least squares plane (17.5).

17.3 Uncertainties of the Input Data

According to the assumptions made, the minimized sum of squared residuals

$$\bar{Q} = \left(\boldsymbol{z} - \boldsymbol{A}\bar{\boldsymbol{\beta}}\right)^{\mathrm{T}}\left(\boldsymbol{z} - \boldsymbol{A}\bar{\boldsymbol{\beta}}\right)$$

provides an estimate

$$s_z^2 = \frac{\bar{Q}}{m-3} \tag{17.9}$$

of the common theoretical variance

$$\sigma_z^2 = E\{S_z^2\}$$

of the input data. The empirical variance s_z^2 has degrees of freedom $\nu = m-3$.

The extra information (17.9) obviously stems from our requesting the data set (17.1) to establish a least squares plane. Hence, we may firstly localize the expectations

$$\mu_{z_i} = z_{0,i} + f_z\,; \quad i = 1, \ldots, m \tag{17.10}$$

by means of confidence intervals

$$z_i - t_P(m-3)\, s_z \le \mu_{z_i} \le z_i + t_P(m-3)\, s_z\,; \quad i = 1, \dots, m\,, \qquad (17.11)$$

Appendix H, and secondly provide the z_i with overall uncertainties

$$u_{z_i} = t_P(m-3)\, s_z + f_{s,z}\,; \quad i = 1, \dots, m\,. \qquad (17.12)$$

After all, we expect the intervals

$$z_i - u_{z_i} \le z_{0,i} \le z_i + u_{z_i}\,; \quad i = 1, \dots, m \qquad (17.13)$$

to localize the true values $z_{0,i}$ of the measured z_i.

17.4 Uncertainties of the Components of the Solution Vector

Random Errors

Albeit there are no repeated measurements, we may nevertheless assign an empirical variance–covariance matrix to the components of the solution vector $\bar{\boldsymbol{\beta}}$. Deploying the expectations

$$E\{\bar{\beta}_k\} = \mu_{\bar{\beta}_k} = \sum_{i=1}^{m} b_{ik}\mu_{z_i}\,; \quad k = 1, 2, 3$$

of the $\bar{\beta}_k$, as a start, we formalize the theoretical variances and the theoretical covariance

$$\sigma_{\bar{\beta}_k \bar{\beta}_k{}'} = E\left\{(\bar{\beta}_k - \mu_{\bar{\beta}_k})(\bar{\beta}_k{}' - \mu_{\bar{\beta}_k{}'})\right\}$$

$$= E\left\{\left[\sum_{i=1}^{m} b_{ik}\left(Z_i - \mu_{z_i}\right)\right]\left[\sum_{j=1}^{m} b_{jk'}\left(Z_j - \mu_{z_j}\right)\right]\right\} = \sigma_z^2 \sum_{i=1}^{m} b_{ik} b_{ik'}\,;$$

$$k, k' = 1, 2, 3$$

their empirical counterparts being

$$s_{\bar{\beta}_k \bar{\beta}_{k'}} = s_z^2 \sum_{i=1}^{m} b_{ik} b_{ik'}\,; \qquad k, k' = 1, 2, 3$$

each of degrees of freedom $\nu = m - 3$. Hence, the empirical variance–covariance matrix of the solution vector reads

$$\boldsymbol{s}_{\bar{\beta}} = \begin{pmatrix} s_{\bar{\beta}_1\bar{\beta}_1} & s_{\bar{\beta}_1\bar{\beta}_2} & s_{\bar{\beta}_1\bar{\beta}_3} \\ s_{\bar{\beta}_2\bar{\beta}_1} & s_{\bar{\beta}_2\bar{\beta}_2} & s_{\bar{\beta}_2\bar{\beta}_3} \\ s_{\bar{\beta}_3\bar{\beta}_1} & s_{\bar{\beta}_3\bar{\beta}_2} & s_{\bar{\beta}_3\bar{\beta}_3} \end{pmatrix}. \qquad (17.14)$$

Systematic Errors

The formal decomposition

$$\bar{\beta}_k = \sum_{i=1}^{m} b_{ik} \left[z_{0,i} + (z_i - \mu_{z_i}) + f_z \right]; \qquad k = 1, 2, 3$$

suggests the systematic errors

$$f_{\bar{\beta}_k} = f_z \sum_{i=1}^{m} b_{ik}; \qquad k = 1, 2, 3.$$

As

$$\sum_{i=1}^{m} b_{i1} = 1, \qquad \sum_{i=1}^{m} b_{i2} = 0, \qquad \sum_{i=1}^{m} b_{i3} = 0 \tag{17.15}$$

we have

$$\begin{aligned} f_{\bar{\beta}_1} &= f_z \\ f_{\bar{\beta}_2} &= 0 \\ f_{\bar{\beta}_3} &= 0 \end{aligned} \tag{17.16}$$

which obviously is what we expect.

Confidence Intervals and Overall Uncertainties

Given we localize the expectations

$$\begin{aligned} E\left\{\bar{\beta}_1\right\} &= \mu_{\bar{\beta}_1} = \beta_{0,1} + f_z \\ E\left\{\bar{\beta}_2\right\} &= \mu_{\bar{\beta}_2} = \beta_{0,2} \\ E\left\{\bar{\beta}_3\right\} &= \mu_{\bar{\beta}_3} = \beta_{0,3} \end{aligned} \tag{17.17}$$

of the components $\bar{\beta}_k$ by confidence intervals

$$\begin{aligned} \bar{\beta}_1 - t_P(m-3)\, s_{\bar{\beta}_1} &\le \mu_{\bar{\beta}_1} \le \bar{\beta}_1 + t_P(m-3)\, s_{\bar{\beta}_1} \\ \bar{\beta}_2 - t_P(m-3)\, s_{\bar{\beta}_2} &\le \mu_{\bar{\beta}_2} \le \bar{\beta}_2 + t_P(m-3)\, s_{\bar{\beta}_2} \\ \bar{\beta}_3 - t_P(m-3)\, s_{\bar{\beta}_3} &\le \mu_{\bar{\beta}_3} \le \bar{\beta}_3 + t_P(m-3)\, s_{\bar{\beta}_3} \end{aligned} \tag{17.18}$$

the uncertainties of the estimators $\bar{\beta}_1, \bar{\beta}_2, \bar{\beta}_3$ take the form

$$\begin{aligned} u_{\bar{\beta}_1} &= t_P(m-3)\, s_{\bar{\beta}_1} + f_{s,z} \\ u_{\bar{\beta}_2} &= t_P(m-3)\, s_{\bar{\beta}_2} \\ u_{\bar{\beta}_3} &= t_P(m-3)\, s_{\bar{\beta}_3}\,. \end{aligned} \tag{17.19}$$

Neither $u_{\bar{\beta}_2}$ nor $u_{\bar{\beta}_3}$ holds a systematic error, $u_{\bar{\beta}_1}$ however includes $f_{s,z}$ as f_z shifts the plane parallel to itself either down or up the z-axis.

Uncertainty Bowls

Inserting the

$$\begin{aligned}\bar{\beta}_k &= \sum_{i=1}^{m} b_{ik} z_i \\ &= \beta_{k,0} + \sum_{i=1}^{m} b_{ik}(z_i - \mu_{z_i}) + f_z \sum_{i=1}^{m} b_{ik}\,; \qquad k = 1,2,3\end{aligned} \tag{17.20}$$

into the least squares plane

$$z(x,y) = \bar{\beta}_1 + \bar{\beta}_2\, x + \bar{\beta}_3\, y$$

produces

$$z(x,y) = \beta_{0,1} + \beta_{0,2}\, x + \beta_{0,3}\, y + \sum_{i=1}^{m} (b_{i1} + b_{i2}x + b_{i3}y)(z_i - \mu_{z_i}) + f_z\,.$$

For any fixed pair (x,y) the expectation of $Z(x,y)$ is given by

$$E\left\{Z(x,y)\right\} = \mu_{z(x,y)} = \beta_{0,1} + \beta_{0,2}\, x + \beta_{0,3}\, y + f_z\,.$$

Thus, the theoretical variance in (x,y) is given by

$$\begin{aligned}\sigma^2_{z(x,y)} &= E\left\{\left(Z(x,y) - \mu_{z(x,y)}\right)^2\right\} \\ &= \sigma_z^2 \sum_{i=1}^{m} (b_{i1} + b_{i2}x + b_{i3}y)^2\,, \qquad \sigma_z^2 = E\left\{(Z_i - \mu_{z_i})^2\right\}.\end{aligned}$$

Substituting s_z^2 for σ_z^2 and submitting f_z to a worst-case estimation brings forth the uncertainty surfaces

$$z(x,y) \pm u_{z(x,y)}$$

$$u_{z(x,y)} = t_P(m-3)\, s_z \sqrt{\sum_{i=1}^{m} (b_{i1} + b_{i2}\, x + b_{i3}\, y)^2} + f_{s,z} \tag{17.21}$$

The surfaces, to be placed in symmetry to the least squares plane (17.5), tellingly suggest to be termed "uncertainty bowls." They span a spatial region which we expect to localize the true plane (16.1).

17.5 *EPC*-Region

The *EPC*-region localizes the triple of true values $(\beta_{0,1}, \beta_{0,2}, \beta_{0,3})$ with respect to the triple of estimators $(\bar{\beta}_1, \bar{\beta}_2, \bar{\beta}_3)$.

Confidence Ellipsoid

Inserting the expectations (17.17) into the decompositions (17.20) we find

$$\bar{\beta}_k = \mu_{\bar{\beta}_k} + \sum_{i=1}^{m} b_{ik}\,(z_i - \mu_{z_i})\,; \qquad k = 1, 2, 3\,.$$

Following Appendix C, Hotelling's ellipsoid takes the form

$$(\bar{\boldsymbol{\beta}} - \boldsymbol{\mu}_{\bar{\boldsymbol{\beta}}})^{\mathrm{T}} s_{\bar{\boldsymbol{\beta}}}^{-1} (\bar{\boldsymbol{\beta}} - \boldsymbol{\mu}_{\bar{\boldsymbol{\beta}}}) = t_P^2(3, m-3)$$

with $s_{\bar{\boldsymbol{\beta}}}$ as given in (17.14). Thus, we expect the confidence ellipse

$$(\boldsymbol{\beta} - \bar{\boldsymbol{\beta}})^{\mathrm{T}} s_{\bar{\boldsymbol{\beta}}}^{-1} (\boldsymbol{\beta} - \bar{\boldsymbol{\beta}}) = t_P^2(3, m-3) \qquad (17.22)$$

to localize the point

$$\boldsymbol{\mu}_{\bar{\boldsymbol{\beta}}} = \begin{pmatrix} \mu_{\bar{\beta}_1} \\ \mu_{\bar{\beta}_2} \\ \mu_{\bar{\beta}_3} \end{pmatrix} \qquad (17.23)$$

with probability P.

Security Polygon

The measuring conditions, implying the z-coordinates to be charged by one and the same unknown systematic error, prompt the security polygon to degenerate into an interval

$$\begin{aligned} -f_{s,z} \le f_{\bar{\beta}_1} &\le f_{s,z} \\ f_{\bar{\beta}_2} &= 0 \\ f_{\bar{\beta}_3} &= 0 \end{aligned} \qquad (17.24)$$

which, for convenience, we consider a "stick" of length $2f_{s,z}$.

Mergence of Ellipse and Interval

The merging of a confidence ellipsoid with a degenerated security polyhedron is addressed in Appendix G.

Figure 17.1 depicts the lattice shaped true plane (16.1), the least squares plane (17.5), the uncertainties (17.19) of the estimators $\bar{\beta}_1$, $\bar{\beta}_2$, $\bar{\beta}_3$, the upper and lower uncertainty bowl (17.21) and, finally, the EPC-region, meant to localize the triple of true values $\beta_{0,1}$, $\beta_{0,2}$, and $\beta_{0,3}$.

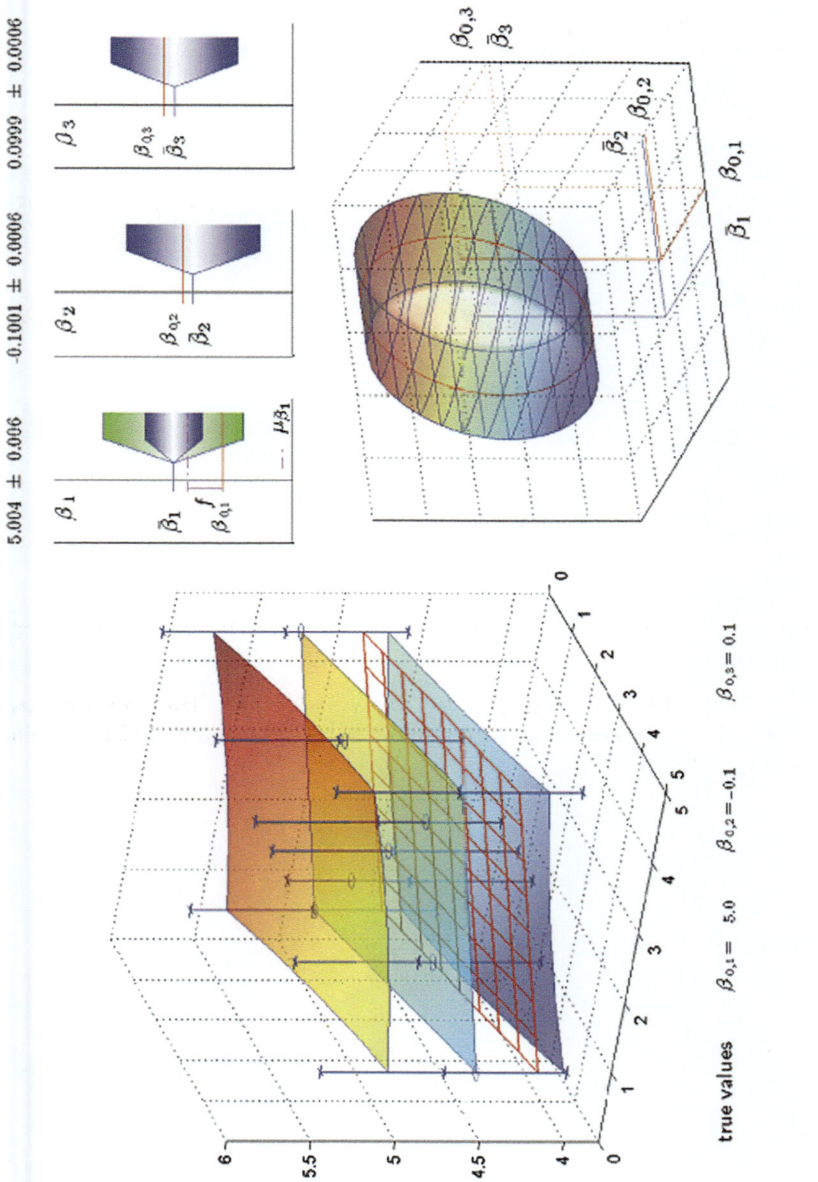

Fig. 17.1. Planes, case (i). *Left*: least squares plane with uncertainty bowls and true plane (lattice shaped). *Top right*: uncertainty intervals localizing the plane's true parameters $\beta_{0,1}$, $\beta_{0,2}$, and $\beta_{0,3}$. The $z-$s' systematic errors are assumed equal and thus do not affect $\bar{\beta}_2$ and $\bar{\beta}_3$. *Bottom right*: EP-region localizing the triple of true values $\beta_{0,1}$, $\beta_{0,2}$, and $\beta_{0,3}$

18 Planes: Case (ii)

Case (ii), Table 16.1 refers to correct x, y-coordinates and erroneous however repeatedly measured z-values. As now the empirical variances of the $z-$s are directly accessible, the scattering of the random errors may vary from z-coordinate to z-coordinate. At the same time varying unknown systematic errors are admitted. Naturally, the minimized sum of squared residuals still enters the construction of the solution vector—proves, meanwhile, no longer serviceable with respect to assessing uncertainties.

18.1 Fitting Conditions

Suppose, there are $m > 3$ data triples

$$(x_{0,1}, y_{0,1}, \bar{z}_1), \quad (x_{0,2}, y_{0,2}, \bar{z}_2), \quad \ldots \quad , (x_{0,m}, y_{0,m}, \bar{z}_m) \tag{18.1}$$

the $x_{0,i}$ and $y_{0,i}$ of which being correct and the $\bar{z}_i$ arithmetic means

$$\bar{z}_i = \frac{1}{n}\sum_{l=1}^{n} z_{il} = z_{0,i} + (\bar{z}_i - \mu_{\bar{z}_i}) + f_{\bar{z}_i}\,; \quad i = 1, \ldots, m$$

$$E\{\bar{Z}_i\} = \mu_{\bar{z}_i}\,, \quad -f_{s,\bar{z}_i} \le f_{\bar{z}_i} \le f_{s,\bar{z}_i}$$

each comprising the same number n of repeated measurements. The empirical variances

$$s_{\bar{z}_i}^2 = \frac{1}{n-1}\sum_{l=1}^{n}(z_{il} - \bar{z}_i)^2\,; \quad i = 1, \ldots, m$$

and the error limits of the unknown systematic errors issue the uncertainties $u_{\bar{z}_i}$ of the means $\bar{z}_i$; we have

$$\bar{z}_i \pm u_{\bar{z}_i}\,, \quad u_{\bar{z}_i} = \frac{t_P(n-1)}{\sqrt{n}}\, s_{\bar{z}_i} + f_{s,\bar{z}_i}\,; \quad i = 1, \ldots, m\,. \tag{18.2}$$

We expect the true values $z_{0,i}$ of the $\bar{z}_i$ to be localized by

$$\bar{z}_i - u_{\bar{z}_i} \le z_{0,i} \le \bar{z}_i + u_{\bar{z}_i}\,; \quad i = 1, \ldots, m\,.$$

18.2 Orthogonal Projection

Let us cast the inconsistent, over-determined, linear system

$$\beta_1 + \beta_2\, x_{0,i} + \beta_3\, y_{0,i} \approx \bar{z}_i\,; \qquad i = 1, \ldots, m > 3 \tag{18.3}$$

into matrix form

$$\boldsymbol{A}\boldsymbol{\beta} \approx \bar{\boldsymbol{z}}$$

where

$$\boldsymbol{A} = \begin{pmatrix} 1 & x_{0,1} & y_{0,1} \\ 1 & x_{0,2} & y_{0,2} \\ \cdots & \cdots & \cdots \\ 1 & x_{0,m} & y_{0,m} \end{pmatrix}, \qquad \boldsymbol{\beta} = \begin{pmatrix} \beta_1 \\ \beta_2 \\ \beta_3 \end{pmatrix}, \qquad \bar{\boldsymbol{z}} = \begin{pmatrix} \bar{z}_1 \\ \bar{z}_2 \\ \cdots \\ \bar{z}_m \end{pmatrix}.$$

The orthogonal projection produces

$$\bar{\boldsymbol{\beta}} = \boldsymbol{B}^{\mathrm{T}} \bar{\boldsymbol{z}}\,, \qquad \bar{\beta}_k = \sum_{i=1}^{m} b_{ik} \bar{z}_i\,; \qquad k = 1, 2, 3 \tag{18.4}$$

with

$$\boldsymbol{B} = \boldsymbol{A}\left(\boldsymbol{A}^{\mathrm{T}}\boldsymbol{A}\right)^{-1} = (b_{ik})\,; \quad i = 1, \ldots, m; \quad k = 1, 2, 3\,.$$

The components $\bar{\beta}_k$; $k = 1, 2, 3$ of the solution vector $\bar{\boldsymbol{\beta}}$ devise the least squares plane

$$z(x) = \bar{\beta}_1 + \bar{\beta}_2\, x + \bar{\beta}_3\, y\,. \tag{18.5}$$

18.3 Uncertainties of the Components of the Solution Vector

Random Errors

We set out to formally establish an ensemble of least squares planes as issued by the data $(z_{1l}, z_{2l}, \ldots, z_{ml})$, $l = 1, \ldots, n$, see Sect. 7.3. The pertaining estimators are

$$\bar{\beta}_{kl} = \sum_{i=1}^{m} b_{ik}\, z_{il}\,, \quad l = 1, \ldots, n\,. \tag{18.6}$$

Inserting the

$$\bar{z}_i = \frac{1}{n}\sum_{l=1}^{n} z_{il}\,; \quad i = 1, \ldots, m$$

into (18.4) reestablishes

$$\bar{\beta}_k = \sum_{i=1}^{m} b_{ik}\left[\frac{1}{n}\sum_{l=1}^{n} z_{il}\right] = \frac{1}{n}\sum_{l=1}^{n}\left[\sum_{i=1}^{m} b_{ik} z_{il}\right]$$

$$= \frac{1}{n}\sum_{l=1}^{n}\bar{\beta}_{kl}\,; \quad k = 1, 2, 3$$

which, as a matter of course, we could just as well have taken directly from (18.6). We now deploy the differences

$$\bar{\beta}_{kl} - \bar{\beta}_k = \sum_{i=1}^{m} b_{ik}\left(z_{il} - \bar{z}_i\right)$$

to formalize the elements

$$s_{\bar{\beta}_k\bar{\beta}_{k'}} = \frac{1}{n-1}\sum_{l=1}^{n}\left(\bar{\beta}_{kl} - \bar{\beta}_k\right)\left(\bar{\beta}_{k'l} - \bar{\beta}_{k'}\right)$$

$$= \frac{1}{n-1}\sum_{l=1}^{n}\left[\sum_{i=1}^{m} b_{ik}\left(z_{il} - \bar{z}_i\right)\right]\left[\sum_{j=1}^{m} b_{jk'}\left(z_{jl} - \bar{z}_j\right)\right]$$

$$= \sum_{i,j=1}^{m} b_{ik}\, b_{jk'}\, s_{ij}\,; \quad k, k' = 1, 2, 3$$

of the empirical variance–covariance matrix $\boldsymbol{s}_{\bar{\beta}}$ of the solution vector. Here the

$$s_{ij} = \frac{1}{n-1}\sum_{l=1}^{n}(z_{il} - \bar{z}_i)(z_{jl} - \bar{z}_j) \qquad i, j = 1, \ldots, m$$

denote the elements of the empirical variance–covariance matrix

$$\boldsymbol{s} = (s_{ij})\,; \quad i, j = 1, \ldots, m \tag{18.7}$$

of the input data each having degrees of freedom $\nu = n - 1$. After all, we may cast this into

$$\boldsymbol{s}_{\bar{\beta}} = \begin{pmatrix} s_{\bar{\beta}_1\bar{\beta}_1} & s_{\bar{\beta}_1\bar{\beta}_2} & s_{\bar{\beta}_1\bar{\beta}_3} \\ s_{\bar{\beta}_2\bar{\beta}_1} & s_{\bar{\beta}_2\bar{\beta}_2} & s_{\bar{\beta}_2\bar{\beta}_3} \\ s_{\bar{\beta}_3\bar{\beta}_1} & s_{\bar{\beta}_3\bar{\beta}_2} & s_{\bar{\beta}_3\bar{\beta}_3} \end{pmatrix} = \boldsymbol{B}^{\mathrm{T}}\boldsymbol{s}\,\boldsymbol{B}\,, \qquad s_{\bar{\beta}_k\bar{\beta}_k} \equiv s^2_{\bar{\beta}_k}\,. \tag{18.8}$$

Systematic Errors

The propagated systematic errors are to be read from

$$\bar{\beta}_k = \sum_{i=1}^{m} b_{ik}[z_{0,i} + (\bar{z}_i - \mu_{\bar{z}_i}) + f_{\bar{z}_i}]\,; \quad k = 1, 2, 3\,. \tag{18.9}$$

We obviously have

$$f_{\bar{\beta}_k} = \sum_{i=1}^{m} b_{ik} f_{\bar{z}_i}\,; \quad k = 1, 2, 3 \tag{18.10}$$

the worst-case estimations being

$$f_{s,\bar{\beta}_k} = \sum_{i=1}^{m} |\, b_{ik} \,|\, f_{s,\bar{z}_i}\,; \quad k = 1, 2, 3\,. \tag{18.11}$$

Assuming the $f_{\bar{z}_i}$ and the $f_{s,\bar{z}_i}$ to be equal

$$\begin{gathered} f_{\bar{z}_i} = f_z\,, \qquad f_{s,\bar{z}_i} = f_{s,z}\,; \qquad i = 1, \ldots m \\ -f_{s,z} \le f_z \le f_{s,z} \end{gathered} \tag{18.12}$$

(17.15) issues

$$\begin{aligned} f_{\bar{\beta}_1} &= f_z \\ f_{\bar{\beta}_2} &= 0 \\ f_{\bar{\beta}_3} &= 0\,. \end{aligned} \tag{18.13}$$

18.4 Confidence Intervals and Overall Uncertainties

We confine the expectations

$$\begin{aligned} E\{\bar{\beta}_1\} &= \mu_{\bar{\beta}_1} = \beta_{0,1} + f_{\bar{\beta}_1} \\ E\{\bar{\beta}_2\} &= \mu_{\bar{\beta}_2} = \beta_{0,2} + f_{\bar{\beta}_2} \\ E\{\bar{\beta}_3\} &= \mu_{\bar{\beta}_3} = \beta_{0,3} + f_{\bar{\beta}_3} \end{aligned} \tag{18.14}$$

to confidence intervals

$$\begin{aligned} \bar{\beta}_1 - \frac{t_P(n-1)}{\sqrt{n}}\, s_{\bar{\beta}_1} \le \mu_{\bar{\beta}_1} \le \bar{\beta}_1 + \frac{t_P(n-1)}{\sqrt{n}}\, s_{\bar{\beta}_1} \\ \bar{\beta}_2 - \frac{t_P(n-1)}{\sqrt{n}}\, s_{\bar{\beta}_2} \le \mu_{\bar{\beta}_2} \le \bar{\beta}_2 + \frac{t_P(n-1)}{\sqrt{n}}\, s_{\bar{\beta}_2} \\ \bar{\beta}_3 - \frac{t_P(n-1)}{\sqrt{n}}\, s_{\bar{\beta}_3} \le \mu_{\bar{\beta}_3} \le \bar{\beta}_3 + \frac{t_P(n-1)}{\sqrt{n}}\, s_{\bar{\beta}_3} \end{aligned}$$

of probability P. Hence, the uncertainties of the components $\bar{\beta}_1$, $\bar{\beta}_2$, $\bar{\beta}_3$ of the solution vector take the form

$$\begin{aligned} u_{\bar{\beta}_1} &= \frac{t_P(n-1)}{\sqrt{n}}\, s_{\bar{\beta}_1} + f_{s,\bar{\beta}_1} \\ u_{\bar{\beta}_2} &= \frac{t_P(n-1)}{\sqrt{n}}\, s_{\bar{\beta}_2} + f_{s,\bar{\beta}_2} \\ u_{\bar{\beta}_3} &= \frac{t_P(n-1)}{\sqrt{n}}\, s_{\bar{\beta}_3} + f_{s,\bar{\beta}_3}\,. \end{aligned} \tag{18.15}$$

In case of equal $f_{\bar{z}_i}$, this turns into

$$\begin{aligned} u_{\bar{\beta}_1} &= \frac{t_P(n-1)}{\sqrt{n}}\, s_{\bar{\beta}_1} + f_{s,z} \\ u_{\bar{\beta}_2} &= \frac{t_P(n-1)}{\sqrt{n}}\, s_{\bar{\beta}_2} \\ u_{\bar{\beta}_3} &= \frac{t_P(n-1)}{\sqrt{n}}\, s_{\bar{\beta}_3}\,. \end{aligned} \tag{18.16}$$

18.5 Uncertainty Bowls

Random Errors

Let us consider the estimators $\bar{\beta}_{kl}$ as given in (18.6) to produce an ensemble of n least squares planes

$$z_l(x,y) = \bar{\beta}_{1l} + \bar{\beta}_{2l}\, x + \bar{\beta}_{3l}\, y\,; \qquad l = 1,\dots,n\,.$$

Summing over l reestablishes

$$z(x,y) = \frac{1}{n}\sum_{l=1}^{n} z_l(x,y) = \bar{\beta}_1 + \bar{\beta}_2\, x + \bar{\beta}_3\, y\,.$$

Thus, for any fixed point (x, y), we have the differences

$$\begin{aligned} z_l(x,y) - z(x,y) &= \left(\bar{\beta}_{1l} - \bar{\beta}_1\right) + \left(\bar{\beta}_{2l} - \bar{\beta}_2\right) x + \left(\bar{\beta}_{3l} - \bar{\beta}_3\right) y \\ &= \sum_{i=1}^{m} \left(b_{i1} + b_{i2}x + b_{i3}y\right)\left(z_{il} - \bar{z}_i\right)\,; \quad l = 1,\dots,n \end{aligned}$$

which bring forth the empirical variance

$$s^2_{z(x,y)} = \frac{1}{n-1}\sum_{l=1}^{n} \left(z_l(x,y) - z(x,y)\right)^2 = \boldsymbol{b}^{\mathrm{T}} \boldsymbol{s}\, \boldsymbol{b}$$

where

$$\boldsymbol{b} = [(b_{11} + b_{12}x + b_{13}y) \qquad b_{21} + b_{22}x + b_{23}y) \\ \dots \quad (b_{m1} + b_{m2}x + b_{m3}y)]^T \tag{18.17}$$

denotes an auxiliary vector. The matrix $\boldsymbol{s}$ is given in (18.7). After all, the term

$$\frac{t_P(n-1)}{\sqrt{n}}\, s_{z(x,y)} \tag{18.18}$$

expresses the influence of random errors.

Systematic Errors

Inserting (18.9),

$$\bar{\beta}_k = \beta_{0,k} + \sum_{i=1}^{m} b_{ik}(\bar{z}_i - \mu_{\bar{z}_i}) + \sum_{i=1}^{m} b_{ik} f_{\bar{z}_i}\,; \quad k = 1, 2, 3\,,$$

into

$$z(x,y) = \bar{\beta}_1 + \bar{\beta}_2\, x + \bar{\beta}_3\, y$$

produces

$$\begin{aligned} z(x,y) \; &= \beta_{0,1} + \beta_{0,2}\, x + \beta_{0,3}\, y \\ &+ \sum_{i=1}^{m} (b_{i1} + b_{i2}x + b_{i3}y)(\bar{z}_i - \mu_{\bar{z}_i}) \\ &+ \sum_{i=1}^{m} (b_{i1} + b_{i2}x + b_{i3}y) f_{\bar{z}_i}\,. \end{aligned} \tag{18.19}$$

Thus, for any (x, y) the propagated systematic error reads

$$f_{z(x,y)} = \sum_{i=1}^{m} (b_{i1} + b_{i2}x + b_{i3}y) f_{\bar{z}_i} \tag{18.20}$$

its worst-case estimation being

$$f_{s,z(x,y)} = \sum_{i=1}^{m} \mid b_{i1} + b_{i2}x + b_{i3}y \mid f_{s,\bar{z}_i}\,. \tag{18.21}$$

Due to (17.15), for equal systematic errors this reduces to

$$f_{s,z(x,y)} = f_{s,z}\,. \tag{18.22}$$

Overall Uncertainty

Combining (18.18) and (18.21) issues the uncertainty bowls

$$z(x,y) \pm u_{z(x,y)}$$

$$u_{z(x,y)} = \frac{t_P(n-1)}{\sqrt{n}}\, s_{z(x,y)} + \sum_{i=1}^{m} \mid b_{i1} - b_{i2}x + b_{i3}y \mid f_{s,\bar{z}_i}\,. \qquad (18.23)$$

For equal systematic errors we have

$$z(x,y) \pm u_{z(x,y)}$$

$$u_{z(x,y)} = \frac{t_P(n-1)}{\sqrt{n}}\, s_{z(x,y)} + f_{s,z}\,. \qquad (18.24)$$

The uncertainty bowls being placed in symmetry to the least squares plane $\bar{z}(x,y)$ span a spatial region which we expect to localize the true plane (16.1).

18.6 *EPC*-Region

The *EPC*-region localizes the triple of true values $(\beta_{0,1}, \beta_{0,2}, \beta_{0,3})$ with respect to the triple of estimators $(\bar{\beta}_1, \bar{\beta}_2, \bar{\beta}_3)$.

Confidence Ellipsoid

Let us rewrite (18.9) as

$$\bar{\beta}_k = \mu_{\bar{\beta}_k} + \sum_{i=1}^{m} b_{ik}\left(\bar{z}_i - \mu_{\bar{z}_i}\right)\,; \quad k = 1,2,3$$

with

$$\mu_{\bar{\beta}_k} = \beta_{0,k} + \sum_{i=1}^{m} b_{ik} f_{\bar{z}_i}\,.$$

The empirical variance–covariance matrix of the solution vector $s_{\bar{\boldsymbol{\beta}}}$ as given in (18.8) provides Hotelling's ellipsoid

$$(\bar{\boldsymbol{\beta}} - \mu_{\bar{\boldsymbol{\beta}}})^{\mathrm{T}} s_{\bar{\boldsymbol{\beta}}}^{-1} (\bar{\boldsymbol{\beta}} - \mu_{\bar{\boldsymbol{\beta}}}) = \frac{t_P^2(3, n-1)}{n}\,.$$

Hence, we expect the confidence ellipsoid

$$(\boldsymbol{\beta} - \bar{\boldsymbol{\beta}})^{\mathrm{T}} s_{\bar{\boldsymbol{\beta}}}^{-1} (\boldsymbol{\beta} - \bar{\boldsymbol{\beta}}) = \frac{t_P^2(3, n-1)}{n} \qquad (18.25)$$

to localize the point

$$\boldsymbol{\mu}_{\bar{\boldsymbol{\beta}}} = \begin{pmatrix} \mu_{\bar{\beta}_1} \\ \mu_{\bar{\beta}_2} \\ \mu_{\bar{\beta}_3} \end{pmatrix} \tag{18.26}$$

with probability P.

Security Polyhedron

While the $f_{\bar{z}_1}, f_{\bar{z}_2}, \dots, f_{\bar{z}_m}$ "scan" the set of points lying within or on the faces of the m-dimensional hypercuboid

$$-f_{s,\bar{z}_i} \le f_{\bar{z}_i} \le f_{s,\bar{z}_i}\,; \quad i = 1, \dots, m$$

the propagated systematic errors

$$f_{\bar{\beta}_k} = \sum_{i=1}^{m} b_{ik} f_{\bar{z}_i}; \quad k = 1, 2, 3$$

span a polyhedron. However, assuming equal systematic errors as specified in (18.12), the polyhedron degenerates into an interval

$$\begin{aligned} -f_{s,z} \le f_{\bar{\beta}_1} &\le f_{s,z} \\ f_{\bar{\beta}_2} &= 0 \\ f_{\bar{\beta}_3} &= 0 \end{aligned} \tag{18.27}$$

which we notionally take as a "stick" of lengths $2f_{s,z}$.

Mergence of Ellipsoid Either with Interval or with Polyhedron

The merging of a confidence ellipsoid either with a non-degenerated or with a degenerated security polyhedron is addressed in Appendix G.

Figure 18.1 depicts the lattice shaped true plane (16.1), the least squares plane (18.5), the uncertainties (18.15) of the estimators $\bar{\beta}_1$, $\bar{\beta}_2$, $\bar{\beta}_3$ and, finally, the upper and lower uncertainty bowl (18.23).

Figure 18.2 refers to equal systematic errors and displays the lattice shaped true plane (16.1), the least squares plane (18.5), the uncertainties (18.16) of the estimators $\bar{\beta}_1$, $\bar{\beta}_2$, $\bar{\beta}_3$ and, finally, the upper and lower uncertainty bowl (18.24).

Figure 18.3 illustrates the composition of the spaces given by the confidence ellipsoid and the security polyhedron. While the ellipsoid remains fixed,

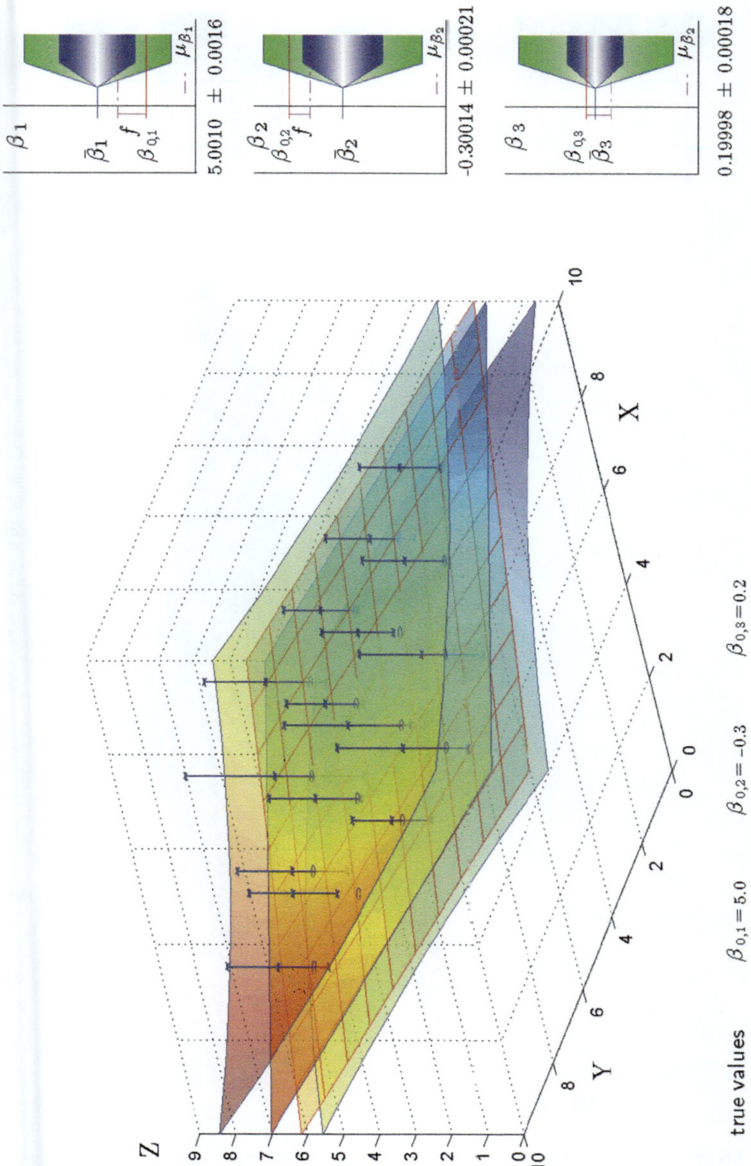

Fig. 18.1. Planes, case (ii). *Left*: least squares plane, symmetrically placed uncertainty bowls, and true plane (*lattice shaped*). *Right*: Uncertainty intervals localizing the plane's true parameters $\beta_{0,1}$, $\beta_{0,2}$, and $\beta_{0,3}$

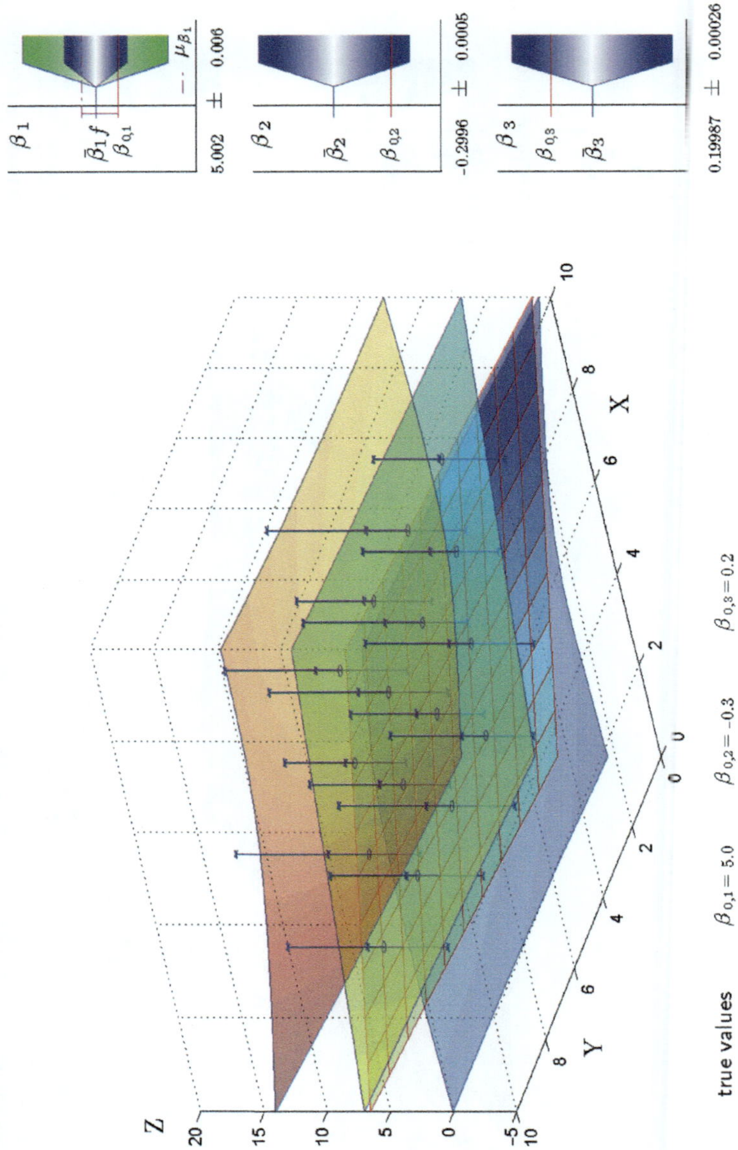

Fig. 18.2. Planes, case (ii), equal systematic errors. *Left*: least squares plane, symmetrically placed uncertainty bowls, and true plane (*lattice shaped*). *Right*: Uncertainty regions localizing the plane's true parameters $\beta_{0,1}$, $\beta_{0,2}$, and $\beta_{0,3}$. Equal systematic errors do not affect $\bar{\beta}_2$ and $\bar{\beta}_3$

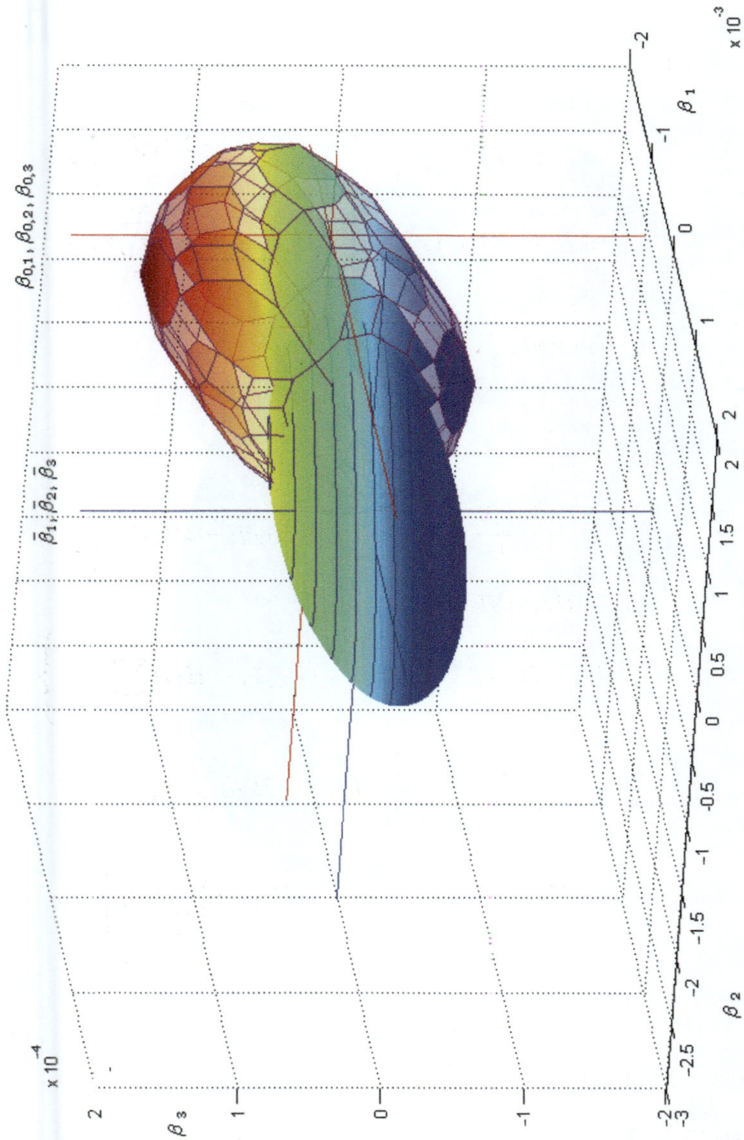

Fig. 18.3. Planes, case (ii): Composing the spaces of the confidence ellipsoid and the security polyhedron: the ellipsoid remains fixed, the center of the polyhedron moves step by step alongside the ellipsoid's skin, maintaining its spatial orientation

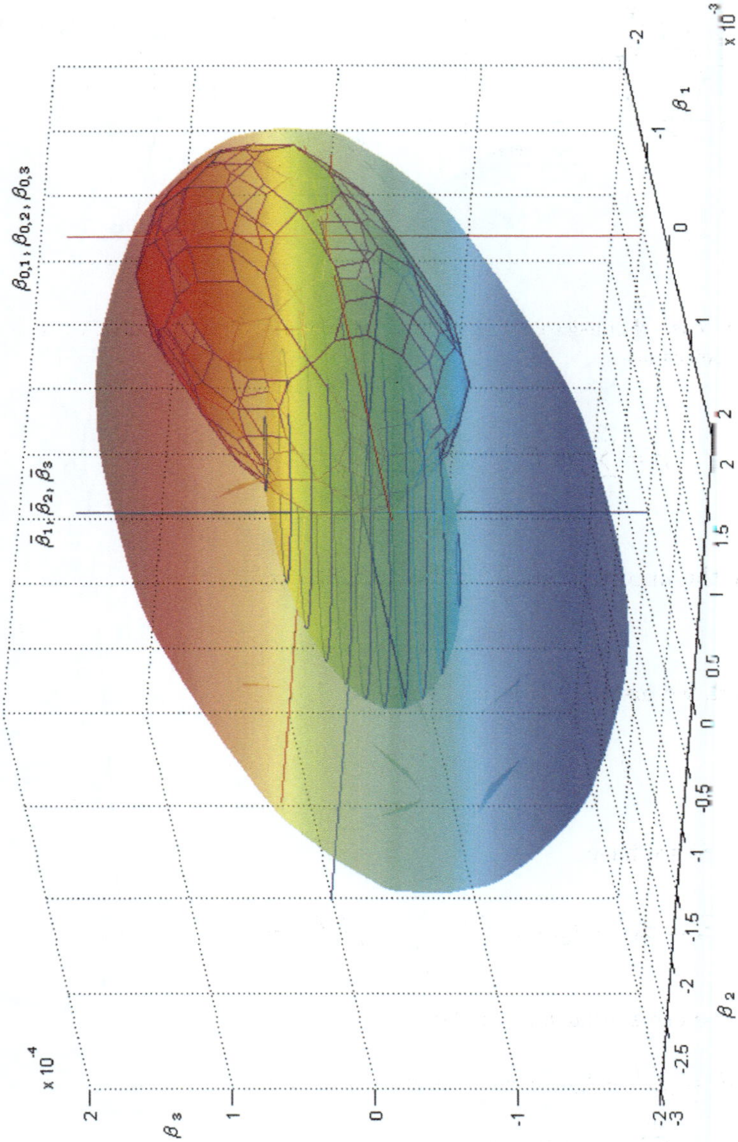

Fig. 18.4. Planes, case (ii): Uncertainty space or EPC-hull given by composing the confidence ellipsoid and the security polyhedron. The hull emerges from moving the center of the polyhedron alongside the ellipsoid's skin and is meant to localize the triple $(\beta_{0,1}, \beta_{0,2}, \beta_{0,3})$ of true values with respect to the triple of estimators $(\bar{\beta}_1, \bar{\beta}_2, \bar{\beta}_3)$

the center of the polyhedron is moved step by step along the ellipsoid's skin; while this happens, the polyhedron maintains its spatial orientation.

Figure 18.4 renders the composition of the spaces given by the confidence ellipsoid and the security polyhedron. The EPC-hull is intended to localize the triple $(\beta_{0,1}, \beta_{0,2}, \beta_{0,3})$ of true values with respect to the triple of estimators $(\bar{\beta}_1, \bar{\beta}_2, \bar{\beta}_3)$.

19 Planes: Case (iii)

Case (iii), Table 16.1 addresses measurement errors in all three coordinates. The scattering of the random errors as well as the actual values of the unknown systematic errors may vary from data triple to data triple. The now defected design matrix does not derogate the orthogonal projection, renders however the assessment of uncertainties more intricate.

19.1 Fitting Conditions

Consider $m > 3$ data triples

$$(\bar{x}_1, \bar{y}_1, \bar{z}_1), \quad (\bar{x}_2, \bar{y}_2, \bar{z}_2), \quad \cdots \quad , (\bar{x}_m, \bar{y}_m, \bar{z}_m) \tag{19.1}$$

with formal decompositions

$$\bar{x}_i = \frac{1}{n} \sum_{l=1}^{n} x_{il} = x_{0,i} + (\bar{x}_i - \mu_{\bar{x}_i}) + f_{\bar{x}_i} \, ; \quad i = 1, \ldots, m$$

$$E\{\bar{X}_i\} = \mu_{\bar{x}_i} \, , \quad -f_{s,\bar{x}_i} \le f_{\bar{x}_i} \le f_{s,\bar{x}_i}$$

$$\bar{y}_i = \frac{1}{n} \sum_{l=1}^{n} y_{il} = y_{0,i} + (\bar{y}_i - \mu_{\bar{y}_i}) + f_{\bar{y}_i} \, ; \quad i = 1, \ldots, m$$

$$E\{\bar{Y}_i\} = \mu_{\bar{y}_i} \, ; \quad -f_{s,\bar{y}_i} \le f_{\bar{y}_i} \le f_{s,\bar{y}_i}$$

and

$$\bar{z}_i = \frac{1}{n} \sum_{l=1}^{n} z_{il} = z_{0,i} + (\bar{z}_i - \mu_{\bar{z}_i}) + f_{\bar{z}_i} \, ; \quad i = 1, \ldots, m$$

$$E\{\bar{Z}_i\} = \mu_{\bar{z}_i} \, ; \quad -f_{s,\bar{z}_i} \le f_{\bar{z}_i} \le f_{s,\bar{z}_i} \, .$$

The empirical variances

$$s_{\bar{x}_i}^2 = \frac{1}{n-1}\sum_{l=1}^{n}\left(x_{il}-\bar{x}_i\right)^2$$

$$s_{\bar{y}_i}^2 = \frac{1}{n-1}\sum_{l=1}^{n}\left(y_{il}-\bar{y}_i\right)^2 \; ; \qquad i = 1,\dots,m$$

$$s_{\bar{z}_i}^2 = \frac{1}{n-1}\sum_{l=1}^{n}\left(z_{il}-\bar{z}_i\right)^2$$

and the limits $\pm f_{s,\bar{x}_i}$, $\pm f_{s,\bar{y}_i}$, $\pm f_{s,\bar{z}_i}$ of the unknown systematic errors suggest the uncertainties

$$\bar{x}_i \pm u_{\bar{x}_i}\,, \qquad u_{\bar{x}_i} = \frac{t_P(n-1)}{\sqrt{n}}\, s_{\bar{x}_i} + f_{s,\bar{x}_i}$$

$$\bar{y}_i \pm u_{\bar{y}_i}\,, \qquad u_{\bar{y}_i} = \frac{t_P(n-1)}{\sqrt{n}}\, s_{\bar{y}_i} + f_{s,\bar{y}_i}\; ; \qquad i = 1,\dots,m \tag{19.2}$$

$$\bar{z}_i \pm u_{\bar{z}_i}\,, \qquad u_{\bar{z}_i} = \frac{t_P(n-1)}{\sqrt{n}}\, s_{\bar{z}_i} + f_{s,\bar{z}_i}\,.$$

The intervals $\bar{x}_i \pm u_{\bar{x}_i}$, $\bar{y}_i \pm u_{\bar{y}_i}$, $\bar{z}_i \pm u_{\bar{z}_i}$ are meant to localize the true values $x_{0,i}$, $y_{0,i}$, $z_{0,i}$ of the input data.

19.2 Orthogonal Projection

We transfer the linear, inconsistent, and over-determined system

$$\beta_1 + \beta_2\,\bar{x}_i + \beta_3\,\bar{y}_i \approx \bar{z}_i\,; \quad i = 1,\,\dots\,, m > 3 \tag{19.3}$$

into matrix form

$$\boldsymbol{A\beta} \approx \bar{\boldsymbol{z}}$$

where

$$\boldsymbol{A} = \begin{pmatrix} 1 & \bar{x}_1 & \bar{y}_1 \\ 1 & \bar{x}_2 & \bar{y}_2 \\ \dots & \dots & \dots \\ 1 & \bar{x}_m & \bar{y}_m \end{pmatrix}, \qquad \boldsymbol{\beta} = \begin{pmatrix} \beta_1 \\ \beta_2 \\ \beta_3 \end{pmatrix}, \qquad \bar{\boldsymbol{z}} = \begin{pmatrix} \bar{z}_1 \\ \bar{z}_2 \\ \dots \\ \bar{z}_m \end{pmatrix}.$$

The orthogonal projection produces

$$\bar{\boldsymbol{\beta}} = \boldsymbol{B}^{\mathrm{T}}\bar{\boldsymbol{z}}\,, \qquad \bar{\beta}_k = \sum_{i=1}^{m} b_{ik}\bar{z}_i\,; \qquad k = 1,2,3 \tag{19.4}$$

with

$$\boldsymbol{B} = \boldsymbol{A}(\boldsymbol{A}^{\mathrm{T}}\boldsymbol{A})^{-1} = (b_{ik})\,; \qquad i = 1,\ldots,m; \quad k = 1,2,3\,.$$

The components $\bar{\beta}_k\,;\ k = 1,2,3$ of the solution vector $\bar{\boldsymbol{\beta}}$ devise the least squares plane

$$z(x) = \bar{\beta}_1 + \bar{\beta}_2\,x + \bar{\beta}_3\,y\,. \tag{19.5}$$

19.3 Series Expansion of the Solution Vector

We submit the $\bar{\beta}_k\,;\ k = 1,2,3$ to series expansions throughout a neighborhood of the point

$$(x_{0,1}\,,\ldots,x_{0,m}\,;\ y_{0,1}\,,\ldots,y_{0,m}\,;\ z_{0,1}\,,\ldots,z_{0,m})\,,$$

firstly, with respect to the n points

$$(x_{1l},x_{2l},\ldots,x_{ml}\,;\ y_{1l},y_{2l},\ldots,y_{ml}\,;\ z_{1l},z_{2l},\ldots,z_{ml})\,; \qquad l = 1,\ldots,n$$

which yields

$$\bar{\beta}_{kl}\,(x_{1l},\ldots,x_{ml}\,;\ y_{1l},\ldots,y_{ml}\,;\ z_{1l},\ldots,z_{ml})$$

$$= \bar{\beta}_k(x_{0,1},\ldots,x_{0,1}\,;\ y_{0,1},\ldots,y_{0,m}\,;\ z_{0,1},\ldots,z_{0,m})$$

$$+\sum_{i=1}^{m}\frac{\partial\bar{\beta}_k}{\partial x_{0,i}}(x_{il}-\mu_{\bar{x}_i}) + \sum_{i=1}^{m}\frac{\partial\bar{\beta}_k}{\partial y_{0,i}}(y_{il}-\mu_{\bar{y}_i}) + \sum_{i=1}^{m}\frac{\partial\bar{\beta}_k}{\partial z_{0,i}}(z_{il}-\mu_{\bar{z}_i}) + \cdots \tag{19.6}$$

$$+\sum_{i=1}^{m}\frac{\partial\bar{\beta}_k}{\partial x_{0,i}}f_{\bar{x}_i} + \sum_{i=1}^{m}\frac{\partial\bar{\beta}_k}{y_{0,i}}f_{\bar{y}_i} + \sum_{i=1}^{m}\frac{\partial\bar{\beta}_k}{\partial x_{0,i}}f_{\bar{z}_i} + \cdots$$

and, secondly, with respect to the sample-dependent means

$$(\bar{x}_1,\bar{x}_2,\ldots,\bar{x}_m\,;\ \bar{y}_1,\bar{y}_2,\ldots,\bar{y}_m\,;\ \bar{z}_1,\bar{z}_2,\ldots,\bar{z}_m)$$

which produces

$$\bar{\beta}_k(\bar{x}_1, \ldots, \bar{x}_m\,;\,\bar{y}_1, \ldots, \bar{y}_m\,;\,\bar{z}_1 \ldots, \bar{z}_m)$$

$$= \bar{\beta}_k(x_{0,1}, \ldots, x_{0,1}\,;\,y_{0,1}, \ldots, y_{0,m}\,;\,z_{0,1}, \ldots, z_{0,m})$$

$$+\sum_{i=1}^{m} \frac{\partial \bar{\beta}_k}{\partial x_{0,i}}(\bar{x}_i - \mu_{\bar{x}_i}) + \sum_{i=1}^{m} \frac{\partial \bar{\beta}_k}{\partial y_{0,i}}(\bar{y}_i - \mu_{\bar{y}_i}) + \sum_{i=1}^{m} \frac{\partial \bar{\beta}_k}{\partial z_{0,i}}(\bar{z}_i - \mu_{\bar{z}_i}) + \cdots \tag{19.7}$$

$$+\sum_{i=1}^{m} \frac{\partial \bar{\beta}_k}{\partial x_{0,i}} f_{\bar{x}_i} + \sum_{i=1}^{m} \frac{\partial \bar{\beta}_k}{y_{0,i}} f_{\bar{y}_i} + \sum_{i=1}^{m} \frac{\partial \bar{\beta}_k}{\partial x_{0,i}} f_{\bar{z}_i} + \cdots$$

where $\bar{\beta}_k(x_{0,1}, \ldots, x_{0,1}\,;\,y_{0,1}, \ldots, y_{0,m}\,;\,z_{0,1}, \ldots, z_{0,m}) \equiv \beta_{0,k}$. As there is no other choice, we approximate the derivatives in $x_{0,1}\,, \ldots, z_{0,m}$ through derivatives in $\bar{x}_1\,, \ldots, \bar{z}_m$. Furthermore, we linearize the expansions, abbreviate the partial derivatives following

$$c_{i1} = \frac{\partial \bar{\beta}_1}{\partial \bar{x}_i}\,, \quad c_{i+m,1} = \frac{\partial \bar{\beta}_1}{\partial \bar{y}_i}\,, \quad c_{i+2m,1} = \frac{\partial \bar{\beta}_1}{\partial \bar{z}_i}$$

$$c_{i2} = \frac{\partial \bar{\beta}_2}{\partial \bar{x}_i}\,, \quad c_{i+m,2} = \frac{\partial \bar{\beta}_2}{\partial \bar{y}_i}\,, \quad c_{i+2m,2} = \frac{\partial \bar{\beta}_2}{\partial \bar{z}_i}\,; \qquad i = 1, \ldots m$$

$$c_{i3} = \frac{\partial \bar{\beta}_3}{\partial \bar{x}_i}\,, \quad c_{i+m,3} = \frac{\partial \bar{\beta}_3}{\partial \bar{y}_i}\,, \quad c_{i+2m,3} = \frac{\partial \bar{\beta}_3}{\partial \bar{z}_i}$$

and transfer the coefficients c_{ik} to an auxiliary matrix

$$\boldsymbol{C}^{\mathrm{T}} = \begin{pmatrix} c_{11} & c_{21} & \cdots & c_{3m,1} \\ c_{12} & c_{22} & \cdots & c_{3m,2} \\ c_{13} & c_{23} & \cdots & c_{3m,3} \end{pmatrix}. \tag{19.8}$$

The c_{ik} are given in Appendix B. Ultimately, deploying the notations

$$\begin{aligned} v_{il} &= x_{il}\,, & v_{i+m,l} &= y_{il} & v_{i+2m,l} &= z_{il} \\ \bar{v}_i &= \bar{x}_i\,, & \bar{v}_{i+m} &= \bar{y}_i & \bar{v}_{i+2m} &= \bar{z}_i \\ \mu_i &= \mu_{\bar{x}_i}\,, & \mu_{i+m} &= \mu_{\bar{y}_i} & \mu_{i+2m} &= \mu_{\bar{z}_i} \\ f_i &= f_{\bar{x}_i}\,, & f_{i+m} &= f_{\bar{y}_i} & f_{i+2m} &= f_{\bar{z}_i} \\ f_{s,i} &= f_{s,\bar{x}_i}\,, & f_{s,i+m} &= f_{s,\bar{y}_i} & f_{s,i+2m} &= f_{s,\bar{z}_i} \end{aligned} \tag{19.9}$$

we may cast (19.6) and (19.7) into

$$\begin{aligned} \bar{\beta}_{kl} &= \beta_{0,k} + \sum_{i=1}^{3m} c_{ik}\,(v_{il} - \mu_i) + \sum_{i=1}^{3m} c_{ik} f_i \quad l = 1, \ldots, n \\ \bar{\beta}_k &= \beta_{0,k} + \sum_{i=1}^{3m} c_{ik}\,(\bar{v}_i - \mu_i) + \sum_{i=1}^{3m} c_{ik} f_i\,; \quad k = 1, 2, 3\,. \end{aligned} \tag{19.10}$$

Subtraction yields

$$\bar{\beta}_{kl} - \bar{\beta}_k = \sum_{i=1}^{3m} c_{ik}\,(v_{il} - \bar{v}_i)\;; \quad k = 1,2,3 \tag{19.11}$$

which implies

$$\bar{\beta}_k = \frac{1}{n}\sum_{l=1}^{n} \bar{\beta}_{kl}\;; \quad k = 1,2,3\,. \tag{19.12}$$

We take the propagation of random errors from (19.11) and the propagation of systematic errors from (19.10) spawning

$$f_{\bar{\beta}_k} = \sum_{i=1}^{3m} c_{ik} f_i; \quad k = 1,2,3\,. \tag{19.13}$$

19.4 Uncertainties of the Components of the Solution Vector

Random Errors

Equation (19.11) issues the elements

$$\begin{aligned} s_{\bar{\beta}_k \bar{\beta}_{k'}} &= \frac{1}{n-1}\sum_{l=1}^{n}\left[\sum_{i=1}^{3m} c_{ik}(v_{il} - \bar{v}_i)\right]\left[\sum_{j=1}^{3m} c_{jk'}(v_{jl} - \bar{v}_j)\right] \\ &= \sum_{i,j=1}^{3m} c_{ik} c_{jk'} s_{ij}\;; \quad k, k' = 1,2,3 \end{aligned}$$

of the empirical variance–covariance matrix

$$\boldsymbol{s}_{\bar{\boldsymbol{\beta}}} = \begin{pmatrix} s_{\bar{\beta}_1\bar{\beta}_1} & s_{\bar{\beta}_1\bar{\beta}_2} & s_{\bar{\beta}_1\bar{\beta}_3} \\ s_{\bar{\beta}_2\bar{\beta}_1} & s_{\bar{\beta}_2\bar{\beta}_2} & s_{\bar{\beta}_2\bar{\beta}_3} \\ s_{\bar{\beta}_3\bar{\beta}_1} & s_{\bar{\beta}_3\bar{\beta}_2} & s_{\bar{\beta}_3\bar{\beta}_3} \end{pmatrix}; \qquad s_{\bar{\beta}_k\bar{\beta}_k} \equiv s^2_{\bar{\beta}_k}$$

of the solution vector. The

$$s_{ij} = \frac{1}{n-1}\sum_{l=1}^{n}(v_{il} - \bar{v}_i)(v_{jl} - \bar{v}_j); \quad i,j = 1,\ldots,3m$$

denote the empirical variances and covariances of the input data each having degrees of freedom $\nu = n - 1$. For convenience, we let the s_{ij} establish a matrix

$$\boldsymbol{s} = (s_{ij}) \; ; \quad i, j = 1, \ldots, 3m \, . \tag{19.14}$$

Hence, $\boldsymbol{s}_{\bar{\boldsymbol{\beta}}}$ may be written as

$$\boldsymbol{s}_{\bar{\boldsymbol{\beta}}} = \boldsymbol{C}^T \boldsymbol{s} \, \boldsymbol{C} \, . \tag{19.15}$$

Systematic Errors

For equal systematic errors and equal error bounds

$$\begin{aligned}
&f_{\bar{x}_i} = f_x \, , \quad f_{s,\bar{x}_i} = f_{s,x} \, , \quad -f_{s,x} \le f_x \le f_{s,x} \\
&f_{\bar{y}_i} = f_y \, , \quad f_{s,\bar{y}_i} = f_{s,y} \, , \quad -f_{s,y} \le f_y \le f_{s,y} \; ; \quad i = 1, \ldots, m \\
&f_{\bar{z}_i} = f_z \, , \quad f_{s,\bar{z}_i} = f_{s,z} \, , \quad -f_{s,z} \le f_z \le f_{s,z}
\end{aligned} \tag{19.16}$$

(19.13) yields

$$\begin{aligned}
f_{\bar{\beta}_k} &= \sum_{i=1}^{m} c_{ik} f_{\bar{x}_i} + \sum_{i=1}^{m} c_{i+m,k} f_{\bar{y}_i} + \sum_{i=1}^{m} c_{i+2m,k} f_{\bar{z}_i} \\
&= f_x \sum_{i=1}^{m} c_{ik} + f_y \sum_{i=1}^{m} c_{i+m,k} + f_z \sum_{i=1}^{m} c_{i+2m,k} \, .
\end{aligned}$$

But as

$$\begin{aligned}
&\sum_{i=1}^{m} c_{i1} = -\bar{\beta}_2 \, , \quad \sum_{i=1}^{m} c_{i+m,1} = -\bar{\beta}_3 \, , \quad \sum_{i=1}^{m} c_{i+2m,1} = 1 \\
&\sum_{i=1}^{m} c_{i2} = 0 \, , \quad \sum_{i=1}^{m} c_{i+m,2} = 0 \, , \quad \sum_{i=1}^{m} c_{i+2m,2} = 0 \\
&\sum_{i=1}^{m} c_{i3} = 0 \, , \quad \sum_{i=1}^{m} c_{i+m,3} = 0 \, , \quad \sum_{i=1}^{m} c_{i+2m,3} = 0
\end{aligned} \tag{19.17}$$

we have

$$\begin{aligned}
f_{\bar{\beta}_1} &= -\bar{\beta}_2 f_x - \bar{\beta}_3 f_y + f_z \\
f_{\bar{\beta}_2} &= 0 \\
f_{\bar{\beta}_3} &= 0 \, .
\end{aligned} \tag{19.18}$$

Equal systematic errors only affect $\bar{\beta}_1$ and shift the plane parallel to itself either up or down the z-axis.

Confidence Intervals and Overall Uncertainties

We consider the expectations

$$E\{\bar{\beta}_1\} = \mu_{\bar{\beta}_1} = \beta_{0,1} + f_{\bar{\beta}_1}$$

$$E\{\bar{\beta}_2\} = \mu_{\bar{\beta}_2} = \beta_{0,2} + f_{\bar{\beta}_2}$$

$$E\{\bar{\beta}_3\} = \mu_{\bar{\beta}_3} = \beta_{0,3} + f_{\bar{\beta}_3}$$

to be localized by confidence intervals

$$\bar{\beta}_1 - \frac{t_P(n-1)}{\sqrt{n}}\, s_{\bar{\beta}_1} \le \mu_{\bar{\beta}_1} \le \bar{\beta}_1 + \frac{t_P(n-1)}{\sqrt{n}}\, s_{\bar{\beta}_1}$$

$$\bar{\beta}_2 - \frac{t_P(n-1)}{\sqrt{n}}\, s_{\bar{\beta}_2} \le \mu_{\bar{\beta}_2} \le \bar{\beta}_2 + \frac{t_P(n-1)}{\sqrt{n}}\, s_{\bar{\beta}_2}$$

$$\bar{\beta}_3 - \frac{t_P(n-1)}{\sqrt{n}}\, s_{\bar{\beta}_3} \le \mu_{\bar{\beta}_3} \le \bar{\beta}_3 + \frac{t_P(n-1)}{\sqrt{n}}\, s_{\bar{\beta}_3}$$

with probability P. Hence, the overall uncertainties of the components of the solution vector turn out to be

$$u_{\bar{\beta}_k} = \frac{t_P(n-1)}{\sqrt{n}}\, s_{\bar{\beta}_k} + \sum_{i=1}^{3m} |\, c_{ik} \,|\, f_{s,i}\,; \quad k = 1, 2, 3\,. \tag{19.19}$$

For equal systematic errors, this reduces to

$$\begin{aligned} u_{\bar{\beta}_1} &= \frac{t_P(n-1)}{\sqrt{n}}\, s_{\bar{\beta}_1} + |\, \bar{\beta}_2 \,|\, f_{s,x} + |\, \bar{\beta}_3 \,|\, f_{s,y} + f_{s,z} \\ u_{\bar{\beta}_2} &= \frac{t_P(n-1)}{\sqrt{n}}\, s_{\bar{\beta}_2} \\ u_{\bar{\beta}_3} &= \frac{t_P(n-1)}{\sqrt{n}}\, s_{\bar{\beta}_3}\,. \end{aligned} \tag{19.20}$$

19.5 Uncertainty Bowls

Random Errors

Let the $\bar{\beta}_k$, constitute an ensemble of n least squares planes

$$z_l(x, y) = \bar{\beta}_{1l} + \bar{\beta}_{2l}\, x + \bar{\beta}_{3l}\, y\,; \quad l = 1, \ldots, n\,. \tag{19.21}$$

Summed over l, the $z_l(x,y)$ reproduce

$$z(x,y) = \sum_{l=1}^{n} z_l(x,y)\,.$$

Thus, for any fixed point (x,y), the differences

$$\begin{aligned} z_l(x,y) - z(x,y) &= \left(\bar{\beta}_{1l} - \bar{\beta}_1\right) + \left(\bar{\beta}_{2l} - \bar{\beta}_2\right) x + \left(\bar{\beta}_{3l} - \bar{\beta}_3\right) y \\ &= \sum_{i=1}^{3m} \left(c_{i1} + c_{i2}x + c_{i3}y\right)\left(v_{il} - \bar{v}_i\right)\,; \quad l = 1, \ldots, n \end{aligned}$$

bring forth the empirical variance

$$s_{z(x,y)}^2 = \frac{1}{n-1} \sum_{l=1}^{n} \left(z_l(x,y) - z(x,y)\right)^2 = \boldsymbol{c}^{\mathrm{T}} \boldsymbol{s}\, \boldsymbol{c} \tag{19.22}$$

in which

$$\begin{aligned} \boldsymbol{c} = (c_{11} + c_{12}\,x + c_{13}\,y \qquad & c_{21} + c_{22}\,x + c_{23}\,y \\ \ldots \qquad & c_{3m,1} + c_{3m,2}\,x + c_{3m,3}\,y)^{\mathrm{T}} \end{aligned}$$

denotes an auxiliary vector.

Systematic Errors

Inserting (19.10) into (19.5) issues

$$\begin{aligned} z(x,y) &= \beta_{0,1} + \beta_{0,2}\,x + \beta_{0,3}\,y \\ &\quad + \sum_{i=1}^{3m} \left(c_{i1} + c_{i2}\,x + c_{i3}\,y\right)\left(\bar{v}_i - \mu_i\right) + \sum_{i=1}^{3m} \left(c_{i1} + c_{i2}\,x + c_{i3}\,y\right) f_i\,. \end{aligned}$$

From this we read the propagated systematic error

$$f_{z(x,y)} = \sum_{i=1}^{3m} (c_{i1} + c_{i2}\,x + c_{i3}\,y) f_i \tag{19.23}$$

its worst case estimation being

$$f_{s,z(x,y)} = \sum_{i=1}^{3m} \left|\, c_{i1} + c_{i2}\,x + c_{i3}\,y \,\right| f_{s,i}\,. \tag{19.24}$$

In case of equal systematic errors we have

$$f_{z(x,y)} = -\bar{\beta}_2 f_x - \bar{\beta}_3 f_y + f_z\,. \tag{19.25}$$

Overall Uncertainty

After all, we expect the uncertainty bowls

$$z(x,y) \pm u_{z(x,y)}$$

$$u_{z(x,y)} = \frac{t_P(n-1)}{\sqrt{n}}\sqrt{\boldsymbol{c}^{\mathrm{T}}\boldsymbol{s}\,\boldsymbol{c}} + \sum_{i=1}^{3m} \mid c_{i1} + c_{i2}x + c_{i3}y \mid f_{s,i} \tag{19.26}$$

to localize the true plane (16.1). For equal systematic errors this turns into

$$z(x,y) \pm u_{z(x,y)}$$

$$u_{z(x,y)} = \frac{t_P(n-1)}{\sqrt{n}}\sqrt{\boldsymbol{c}^{\mathrm{T}}\boldsymbol{s}\,\boldsymbol{c}} + \mid \bar{\beta}_2 \mid f_{s,x} + \mid \bar{\beta}_3 \mid f_{s,y} + f_{s,z}\,. \tag{19.27}$$

19.6 *EPC*-Region

The *EPC*-region localizes the triple of true values $(\beta_{0,1}, \beta_{0,2}, \beta_{0,3})$ with respect to the triple of estimators $(\bar{\beta}_1, \bar{\beta}_2, \bar{\beta}_3)$.

Confidence Ellipsoid

We rewrite (19.10) as

$$\bar{\beta}_k = \mu_{\bar{\beta}_k} + \sum_{i=1}^{3m} c_{ik}\left(\bar{v}_i - \mu_i\right)\,; \quad \mu_{\bar{\beta}_k} = \beta_{0,k} + \sum_{i=1}^{3m} c_{ik} f_i; \quad k = 1,2,3\,.$$

With this the variance–covariance matrix $s_{\bar{\boldsymbol{\beta}}}$ of the solution vector $\bar{\boldsymbol{\beta}}$, as given in (19.15), produces Hotelling's ellipsoid

$$(\bar{\boldsymbol{\beta}} - \boldsymbol{\mu}_{\bar{\boldsymbol{\beta}}})^{\mathrm{T}} s_{\bar{\boldsymbol{\beta}}}^{-1} (\bar{\boldsymbol{\beta}} - \boldsymbol{\mu}_{\bar{\boldsymbol{\beta}}}) = \frac{t_P^2(3, n-1)}{n}\,.$$

Hence, we expect the confidence ellipsoid

$$(\boldsymbol{\beta} - \bar{\boldsymbol{\beta}})^{\mathrm{T}} s_{\bar{\boldsymbol{\beta}}}^{-1} (\boldsymbol{\beta} - \bar{\boldsymbol{\beta}}) = \frac{t_P^2(3, n-1)}{n} \tag{19.28}$$

to localize the point

$$\boldsymbol{\mu}_{\bar{\boldsymbol{\beta}}} = \begin{pmatrix} \mu_{\bar{\beta}_1} \\ \mu_{\bar{\beta}_2} \\ \mu_{\bar{\beta}_3} \end{pmatrix} \tag{19.29}$$

with probability P.

Security Polyhedron

While the errors $f_1, f_2, \ldots, f_{3m}$ "scan" the set of points lying within or on the faces of the $3m$-dimensional hypercuboid

$$-f_{s,i} \le f_i \le f_{s,i}\,; \quad i = 1, \ldots, 3m$$

the propagated systematic errors

$$f_{\bar{\beta}_k} = \sum_{i=1}^{3m} c_{ik} f_i; \quad k = 1, 2, 3$$

span a polyhedron. In case of equal systematic errors, the polyhedron degenerates into an interval

$$\begin{aligned} -(\mid \bar{\beta}_2 \mid f_{s,x} + \mid \bar{\beta}_3 \mid f_{s,y} + f_{s,z}) \le f_{\bar{\beta}_1} &\le \mid \bar{\beta}_2 \mid f_{s,x} + \mid \bar{\beta}_3 \mid f_{s,y} + f_{s,z} \\ f_{\bar{\beta}_2} &= 0 \\ f_{\bar{\beta}_3} &= 0 \end{aligned} \qquad (19.30)$$

which we notionally consider "a stick" of lengths $2(\mid \bar{\beta}_2 \mid f_{s,x} + \mid \bar{\beta}_3 \mid f_{s,y} + f_{s,z})$.

Mergence of Ellipsoid Either with Interval or with Polyhedron

The merging of a confidence ellipsoid either with a non-degenerated or with a degenerated security polyhedron is addressed in Appendix G.

Figure 19.1 depicts the lattice shaped true plane (16.1), the least squares plane (19.5), the uncertainties (19.19) of the estimators $\bar{\beta}_1$, $\bar{\beta}_2$, $\bar{\beta}_3$ and, finally, the upper and lower uncertainty bowl (19.26).

Figure 19.2 refers to equal systematic errors and displays the lattice shaped true plane (16.1), the least squares plane (19.5), the uncertainties (19.20) of the estimators $\bar{\beta}_1$, $\bar{\beta}_2$, $\bar{\beta}_3$ and, finally, the upper and lower uncertainty bowl (19.27).

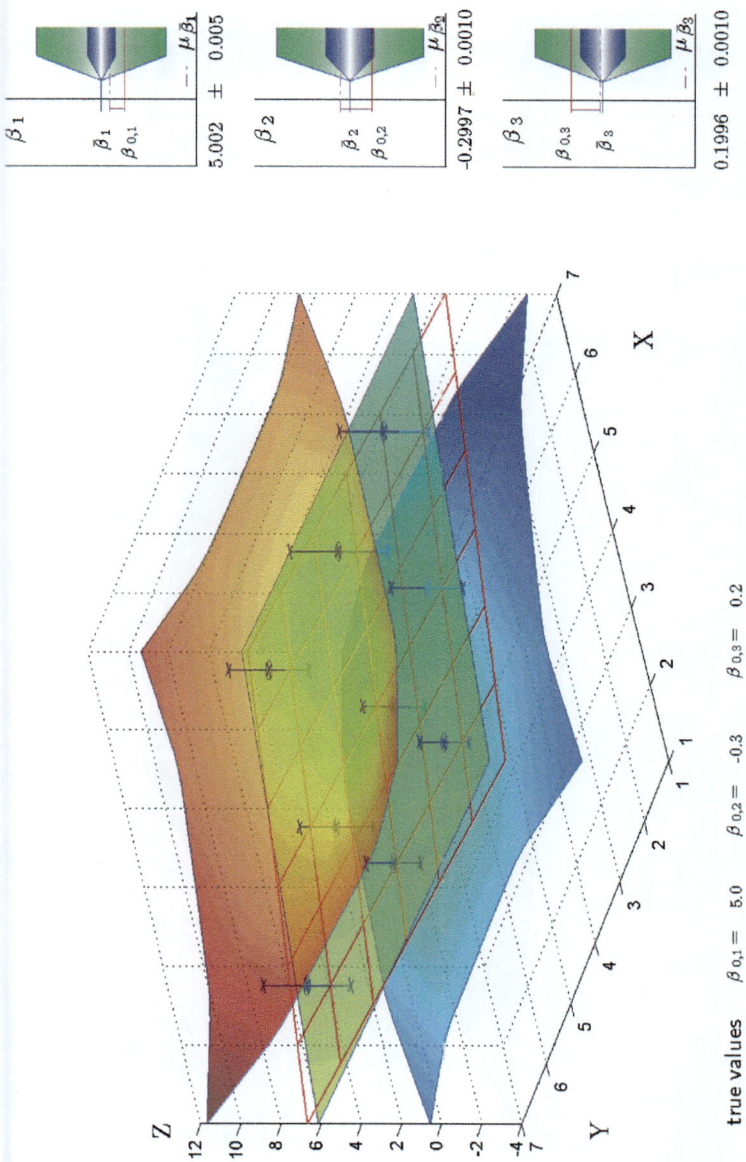

Fig. 19.1. Planes, case (iii). *Left*: least squares plane, symmetrically placed uncertainty bowls, and true plane (*lattice shaped*) . *Right*: Uncertainty intervals localizing the plane's true parameters $\beta_{0,1}$ $\beta_{0,2}$, and $\beta_{0,3}$

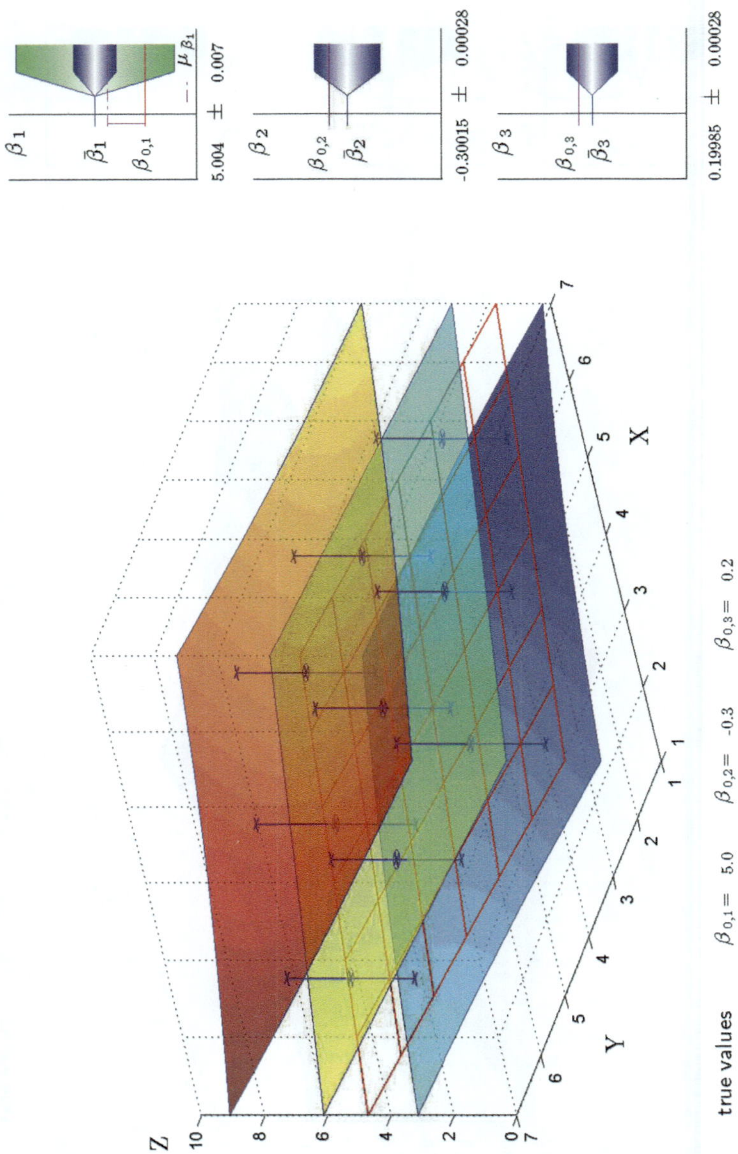

Fig. 19.2. Planes, case (iii), equal systematic errors in x, y, z. *Left*: least squares plane, symmetrically placed uncertainty bowls, and true plane (*lattice shaped*). *Right*: Uncertainty intervals localizing the true values $\beta_{0,1}$ $\beta_{0,2}$, and $\beta_{0,3}$. Equal systematic errors do not affect $\bar{\beta}_2$ and $\bar{\beta}_3$

Part VII

Fitting of Parabolas

20 Preliminaries

The fitting of parabolas relies on data pairs

$$(x_1, y_1), \quad (x_2, y_2), \quad \dots \quad , (x_m, y_m)\,; \qquad m > 3.$$

20.1 Distinction of Cases

We shall discuss three fitting situations (Table 20.1):

Table 20.1. Fitting of Parabolas, Three Cases

Case	Abscissas	Ordinates
(i)	Error-free	Individual measurements
(ii)	Error-free	Repeated measurements
(iii)	Repeated measurements	Repeated measurements

Case (i) considers correct abscissas and individually measured ordinates. Each ordinate is supposed to hold a particular random error, stemming from one and the same normal distribution, and a common unknown systematic error. Case (ii) still assumes error-free abscissas but submits the ordinates to repeated measurements. Then, the scattering of the random errors as well as the actual values of the unknown systematic errors may vary from ordinate to ordinate. Case (iii), finally, addresses erroneous abscissas and erroneous ordinates thus admitting variances and unknown systematic errors to vary from data point to data point.

20.2 True Parabola

Let the equation of a parabola

$$y(x) = \beta_{0,1} + \beta_{0,2}\, x + \beta_{0,3}\, x^2 \tag{20.1}$$

be fulfilled by the m data pairs

$$(x_{0,1}, y_{0,1}), \quad (x_{0,2}, y_{0,2}), \quad \dots \quad , (x_{0,m}, y_{0,m}) \,. \tag{20.2}$$

Written in matrices, this reads

$$\boldsymbol{A}\boldsymbol{\beta}_0 = \boldsymbol{y}_0 \tag{20.3}$$

where

$$\boldsymbol{A} = \begin{pmatrix} 1 & x_{0,1} & x_{0,1}^2 \\ 1 & x_{0,2} & x_{0,2}^2 \\ \dots & \dots & \dots \\ 1 & x_{0,m} & x_{0,m}^2 \end{pmatrix}, \quad \boldsymbol{\beta_0} = \begin{pmatrix} \beta_{0,1} \\ \beta_{0,2} \\ \beta_{0,3} \end{pmatrix}, \quad \boldsymbol{y_0} = \begin{pmatrix} y_{0,1} \\ y_{0,2} \\ \dots \\ y_{0,m} \end{pmatrix}. \tag{20.4}$$

Given the design matrix $\boldsymbol{A}$ has rank 3, the linear system (20.3) formally reproduces

$$\boldsymbol{\beta}_0 = \boldsymbol{B}^{\mathrm{T}}\boldsymbol{z_0} \,, \qquad \boldsymbol{B} = \boldsymbol{A}\left(\boldsymbol{A}^{\mathrm{T}}\boldsymbol{A}\right)^{-1} . \tag{20.5}$$

We address (20.1) as the true parabola, (20.2) as the true input data, and $\boldsymbol{\beta}_0$ as the true solution vector.

On substituting a set of measured data for the true ones, (20.3) becomes inconsistent. This observation suggests to fit a least squares parabola to the defective input and to assess the true parameters $\beta_{0,1}$, $\beta_{0,2}$, and $\beta_{0,3}$ by means of suitably shaped uncertainty intervals.

21 Parabolas: Case (i)

Case (i), Table 20.1 considers error-free abscissas and erroneous, individually measured ordinates.

21.1 Fitting Conditions

The adjustment relates to $m > 3$ data pairs

$$(x_{0,1}, y_1), \quad (x_{0,2}, y_2), \quad \ldots \quad , (x_{0,m}, y_m) \tag{21.1}$$

assuming the $x_{0,i}$ correct and the y_i erroneous

$$y_i = y_{0,i} + (y_i - \mu_{y_i}) + f_{y_i} \, ; \quad i = 1, \, \ldots \, , m \tag{21.2}$$

with expectations $\mu_{y_i} = E\{Y_i\}$. We assume the random errors $(y_i - \mu_{y_i})$ to come from the same normal density and thus to have a common theoretical variance. Further, we suppose the y_i to be burdened by the same unknown systematic error

$$\begin{aligned} f_{y_i} = f_y \, ; \qquad i = 1, \, \ldots \, , m \\ -f_{s,y} \le f_y \le f_{s,y} \, . \end{aligned} \tag{21.3}$$

For the time being, we are obviously not in a position to quote the uncertainties of the y_i.

21.2 Orthogonal Projection

We are faced with an inconsistent, over-determined, linear system

$$\beta_1 + \beta_2 \, x_{0,i} + \beta_3 \, x_{0,i}^2 \approx y_i \, ; \qquad i = 1, \, \ldots \, , m > 3 \tag{21.4}$$

and ask for a parabola

$$y(x) = \bar{\beta}_1 + \bar{\beta}_2 \, x + \bar{\beta}_3 \, x^2 \tag{21.5}$$

fitting the input data in terms of least squares. Introducing

$$\boldsymbol{A} = \begin{pmatrix} 1 & x_{0,1} & x_{0,1}^2 \\ 1 & x_{0,2} & x_{0,2}^2 \\ \dots & \dots & \dots \\ 1 & x_{0,m} & x_{0,m}^2 \end{pmatrix}, \qquad \boldsymbol{\beta} = \begin{pmatrix} \beta_1 \\ \beta_2 \\ \beta_3 \end{pmatrix}, \qquad \boldsymbol{y} = \begin{pmatrix} y_1 \\ y_2 \\ \dots \\ y_m \end{pmatrix}$$

the matrix form of (21.4) reads

$$\boldsymbol{A}\boldsymbol{\beta} \approx \boldsymbol{y} . \tag{21.6}$$

The orthogonal projection produces

$$\bar{\boldsymbol{\beta}} = \boldsymbol{B}^{\mathrm{T}}\boldsymbol{y} , \qquad \bar{\beta}_k = \sum_{i=1}^{m} b_{ik}\, y_i \, ; \qquad k = 1, 2, 3 \tag{21.7}$$

where

$$\boldsymbol{B} = \boldsymbol{A}\left(\boldsymbol{A}^{\mathrm{T}}\boldsymbol{A}\right)^{-1} = (b_{ik}) \, ; \quad i = 1, \dots, m; \quad k = 1, 2, 3 . \tag{21.8}$$

The components $\bar{\beta}_k$; $k = 1, 2, 3$ of the solution vector (21.7) establish the least squares parabola (21.5).

21.3 Uncertainties of the Input Data

According to the assumption made, the minimized sum of squared residuals

$$\bar{Q} = \left(\boldsymbol{y} - \boldsymbol{A}\bar{\boldsymbol{\beta}}\right)^{\mathrm{T}} \left(\boldsymbol{y} - \boldsymbol{A}\bar{\boldsymbol{\beta}}\right)$$

issues an estimate

$$s_y^2 = \frac{\bar{Q}}{m - 3} \tag{21.9}$$

of the common theoretical variance

$$\sigma_y^2 = E\{S_y^2\}$$

of the input data. The empirical variance s_y^2 has degrees of freedom $\nu = m - 3$. Obviously, this information is due to our requesting the input data to devise a parabola. Thus, we are in a position to localize the expectations

$$\mu_{y_i} = y_{0,i} + f_y \, ; \quad i = 1, \dots, m \tag{21.10}$$

through confidence intervals

$$y_i - t_P(m-3)\, s_y \le \mu_{y_i} \le y_i + t_P(m-3)\, s_y \, ; \quad i = 1, \dots, m , \tag{21.11}$$

Appendix H, so that the overall uncertainties of the input data take the form

$$u_{y_i} = t_P(m-3)\, s_y + f_{s,y}\,; \qquad i = 1, \ldots, m\,. \tag{21.12}$$

Hence, we consider the intervals

$$y_i - u_{y_i} \le y_{0,i} \le y_i + u_{y_i}\,; \qquad i = 1, \ldots, m$$

to localize the true values $y_{0,i}$ of the ordinates y_i ; $i = 1, \ldots, m$.

21.4 Uncertainties of the Components of the Solution Vector

Random Errors

Though we are not endued with repeated measurements, we may nevertheless assign an empirical variance–covariance matrix to the components $\bar{\beta}_k$; $k = 1, 2, 3$ of the solution vector. To begin with, we refer to the expectations

$$E\{\bar{\beta}_k\} = \mu_{\bar{\beta}_k} = \sum_{i=1}^{m} b_{ik}\mu_{y_i}\,; \quad k = 1, 2, 3\,.$$

Thus, the theoretical variances and the theoretical covariances of the components of the solution vector take the form

$$\sigma_{\bar{\beta}_k \bar{\beta}_k{}'} = E\left\{(\bar{\beta}_k - \mu_{\bar{\beta}_k})(\bar{\beta}_k{}' - \mu_{\bar{\beta}_k{}'})\right\}$$

$$= E\left\{\left[\sum_{i=1}^{m} b_{ik}\left(Y_i - \mu_{y_i}\right)\right]\left[\sum_{j=1}^{m} b_{jk'}\left(Y_j - \mu_{y_j}\right)\right]\right\} = \sigma_y^2 \sum_{i=1}^{m} b_{ik} b_{ik}{}'\,;$$

$$k = 1, 2, 3\,.$$

Substituting s_y^2 for σ_y^2 issues the empirical counterparts

$$s_{\bar{\beta}_k \bar{\beta}_k{}'} = s_y^2 \sum_{i=1}^{m} b_{ik} b_{ik}{}'\,.$$

Hence, the empirical variance–covariance matrix of the solution vector reads

$$\boldsymbol{s}_{\bar{\beta}} = \begin{pmatrix} s_{\bar{\beta}_1\bar{\beta}_1} & s_{\bar{\beta}_1\bar{\beta}_2} & s_{\bar{\beta}_1\bar{\beta}_3} \\ s_{\bar{\beta}_2\bar{\beta}_1} & s_{\bar{\beta}_2\bar{\beta}_2} & s_{\bar{\beta}_2\bar{\beta}_3} \\ s_{\bar{\beta}_3\bar{\beta}_1} & s_{\bar{\beta}_3\bar{\beta}_2} & s_{\bar{\beta}_3\bar{\beta}_3} \end{pmatrix} \tag{21.13}$$

the elements having degrees of freedom $\nu = m - 3$.

Whenever appropriate, we denote

$$s_{\bar{\beta}_1\bar{\beta}_1} = s_{\bar{\beta}_1}^2\,, \qquad s_{\bar{\beta}_2\bar{\beta}_2} = s_{\bar{\beta}_2}^2\,, \qquad s_{\bar{\beta}_3\bar{\beta}_3} = s_{\bar{\beta}_3}^2\,.$$

Systematic Errors

The formal decompositions

$$\bar{\beta}_k = \sum_{i=1}^{m} b_{ik} \left[y_{0,i} + (y_i - \mu_{y_i}) + f_y \right] ; \qquad k = 1, 2, 3 \tag{21.14}$$

suggest the propagated systematic errors

$$f_{\bar{\beta}_k} = f_y \sum_{i=1}^{m} b_{ik} ; \quad k = 1, 2, 3 .$$

As

$$\sum_{i=1}^{m} b_{i1} = 1 , \qquad \sum_{i=1}^{m} b_{i2} = 0 , \qquad \sum_{i=1}^{m} b_{i3} = 0 \tag{21.15}$$

we have

$$\begin{aligned} f_{\bar{\beta}_1} &= f_y \\ f_{\bar{\beta}_2} &= 0 \\ f_{\bar{\beta}_3} &= 0 . \end{aligned} \tag{21.16}$$

Confidence Intervals and Overall Uncertainties

We localize the expectations

$$\begin{aligned} E\{\bar{\beta}_1\} &= \mu_{\bar{\beta}_1} = \beta_{0,1} + f_y \\ E\{\bar{\beta}_2\} &= \mu_{\bar{\beta}_2} = \beta_{0,2} \\ E\{\bar{\beta}_3\} &= \mu_{\bar{\beta}_3} = \beta_{0,3} \end{aligned} \tag{21.17}$$

through confidence intervals

$$\begin{aligned} \bar{\beta}_1 - t_P(m-3)\, s_{\bar{\beta}_1} \le \mu_{\bar{\beta}_1} \le \bar{\beta}_1 + t_P(m-3)\, s_{\bar{\beta}_1} \\ \bar{\beta}_2 - t_P(m-3)\, s_{\bar{\beta}_2} \le \mu_{\bar{\beta}_2} \le \bar{\beta}_2 + t_P(m-3)\, s_{\bar{\beta}_2} \\ \bar{\beta}_3 - t_P(m-3)\, s_{\bar{\beta}_3} \le \mu_{\bar{\beta}_3} \le \bar{\beta}_3 + t_P(m-3)\, s_{\bar{\beta}_3} \end{aligned} \tag{21.18}$$

with probability P. Hence, the uncertainties of the estimators $\bar{\beta}_1, \bar{\beta}_2, \bar{\beta}_3$ are given by

$$\begin{aligned} u_{\bar{\beta}_1} &= t_P(m-3)\, s_{\bar{\beta}_1} + f_{s,y} \\ u_{\bar{\beta}_2} &= t_P(m-3)\, s_{\bar{\beta}_2} \\ u_{\bar{\beta}_3} &= t_P(m-3)\, s_{\bar{\beta}_3} . \end{aligned} \tag{21.19}$$

While $u_{\bar{\beta}_2}$ and $u_{\bar{\beta}_3}$ do not exhibit systematic uncertainty components, $u_{\bar{\beta}_1}$ is burdened by $f_{s,y}$. This is obvious, as f_y shifts the parabola parallel to itself either down or up the y-axis.

21.5 Uncertainty Band

Inserting the formal decompositions

$$\begin{aligned}\bar{\beta}_k &= \sum_{i=1}^{m} b_{ik} y_i \\ &= \beta_{k,0} + \sum_{i=1}^{m} b_{ik}(y_i - \mu_{y_i}) + f_y \sum_{i=1}^{m} b_{ik}\,; \quad k = 1,2,3\end{aligned} \tag{21.20}$$

into the least squares parabola

$$y(x) = \bar{\beta}_1 + \bar{\beta}_2\, x + \bar{\beta}_3\, x^2$$

yields

$$y(x) = \beta_{0,1} + \beta_{0,2}\, x + \beta_{0,3}\, x^2 + \sum_{i=1}^{m} (b_{i1} + b_{i2}\, x + b_{i3}\, x^2)(y_i - \mu_{y_i}) + f_y\,.$$

For any fixed x, the expectation of $Y(x)$ brings forth

$$E\{Y(x)\} = \mu_{y(x)} = \beta_{0,1} + \beta_{0,2}\, x + \beta_{0,3}\, x^2 + f_y\,.$$

Thus, there is an x-dependent theoretical variance

$$\begin{aligned}\sigma^2_{y(x)} &= E\{(Y(x) - \mu_{y(x)})^2\} \\ &= \sigma_y^2 \sum_{i=1}^{m} (b_{i1} + b_{i2}x + b_{i3}x^2)^2\,, \quad \sigma_y^2 = E\left\{(Y_i - \mu_{y_i})^2\right\}\,.\end{aligned}$$

Substituting s_y^2 for σ_y^2 and submitting f_y to a worst-case estimation produces the uncertainty band

$$\begin{aligned} & y(x) \pm u_{y(x)} \\ & u_{z(x)} = t_P(m-3)\, s_y \sqrt{\sum_{i=1}^{m} (b_{i1} + b_{i2}\, x + b_{i3}\, x^2)^2} + f_{s,y}\,. \end{aligned} \tag{21.21}$$

The boundary lines, running in symmetry to $\bar{y}(x)$, are expected to localize the true parabola (20.1).

21.6 *EPC*-Region

The *EPC*-region confines the triple of true values $(\beta_{0,1}, \beta_{0,2}, \beta_{0,3})$ with respect to the triple of estimators $(\bar{\beta}_1, \bar{\beta}_2, \bar{\beta}_3)$.

Confidence Ellipsoid

Let us consider the formal decompositions

$$\bar{\beta}_k = \mu_{\bar{\beta}_k} + \sum_{i=1}^{m} b_{ik}\,(y_i - \mu_{y_i})\,; \qquad k = 1, 2, 3$$

with $\mu_{\bar{\beta}_k}$ as given in (21.17). As $s_{\bar{\boldsymbol{\beta}}}$ is known from (21.13), Hotelling's ellipsoid reads

$$(\bar{\boldsymbol{\beta}} - \mu_{\bar{\boldsymbol{\beta}}})^{\mathrm{T}} s_{\bar{\boldsymbol{\beta}}}^{-1} (\bar{\boldsymbol{\beta}} - \mu_{\bar{\boldsymbol{\beta}}}) = t_P^2(3, m-3)\,,$$

Appendix C. The pertaining confidence ellipsoid

$$(\boldsymbol{\beta} - \bar{\boldsymbol{\beta}})^{\mathrm{T}} s_{\bar{\boldsymbol{\beta}}}^{-1} (\boldsymbol{\beta} - \bar{\boldsymbol{\beta}}) = t_P^2(3, m-3) \qquad (21.22)$$

is expected to localize the point

$$\boldsymbol{\mu}_{\bar{\boldsymbol{\beta}}} = \begin{pmatrix} \mu_{\bar{\beta}_1} \\ \mu_{\bar{\beta}_2} \\ \mu_{\bar{\beta}_3} \end{pmatrix} \qquad (21.23)$$

with probability P.

Security Polygon

The measuring conditions referred to entail the security polygon to degenerate into an interval

$$\begin{aligned} -f_{s,y} \le f_{\bar{\beta}_1} &\le f_{s,y} \\ f_{\bar{\beta}_2} &= 0 \\ f_{\bar{\beta}_3} &= 0\,. \end{aligned} \qquad (21.24)$$

For convenience, we consider the interval a "stick" of length $2f_{s,y}$

Mergence of Ellipsoid and Interval

The merging of a confidence ellipsoid with a degenerated security polyhedron is addressed in Appendix G.

Figure 21.1 displays the true parabola (20.1), the fitted parabola (21.5), the uncertainties (21.19) of the estimators $\bar{\beta}_1, \bar{\beta}_2$, and $\bar{\beta}_3$ and, finally, the uncertainty band (21.21).

The illustrations are based on simulated data implying known true values and, in particular, extensive graphical scale transformations, Appendix A.

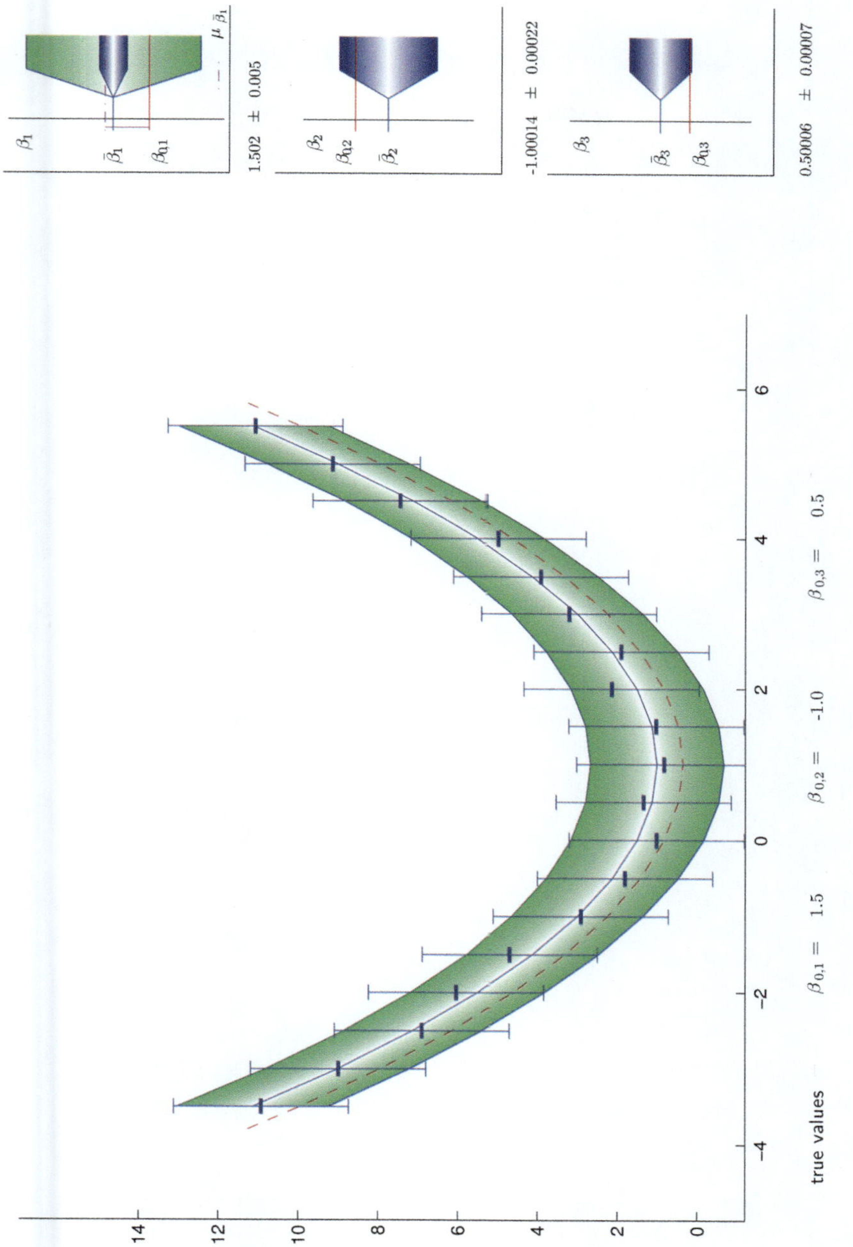

Fig. 21.1. Parabolas, case (i). *Left*: adjusted parabola, uncertainty band, and true parabola (*dashed line*). *Right*: Uncertainty intervals localizing the parabola's true parameters $\beta_{0,1}$, $\beta_{0,2}$, and $\beta_{0,3}$. Equal systematic errors do not affect $\bar{\beta}_2$ and $\bar{\beta}_3$

22 Parabolas: Case (ii)

Case (ii), Table 20.1 still invokes correct abscissas, introduces, meanwhile, repeated measurements of the ordinates. As the empirical variances of the ordinates are directly accessible, the scattering of the random errors may vary from ordinate to ordinate; at the same time varying unknown systematic errors are admitted. It goes without saying that the minimized sum of squared residuals still enters the construction of the solution vector proves, however, no longer serviceable with respect to assessing uncertainties.

22.1 Fitting Conditions

Given $m > 3$ data pairs

$$(x_{0,1}, \bar{y}_1), \quad (x_{0,2}, \bar{y}_2), \quad \ldots \quad , (x_{0,m}, \bar{y}_m) \tag{22.1}$$

the x-coordinates being error-free and the y-coordinates measured quantities

$$\bar{y}_i = \frac{1}{n}\sum_{l=1}^{n} y_{il} = y_{0,i} + (\bar{y}_i - \mu_{\bar{y}_i}) + f_{\bar{y}_i}\,; \quad i = 1, \ldots, m$$

$$E\{\bar{Y}_i\} = \mu_{\bar{y}_i}\,, \quad -f_{s,\bar{y}_i} \le f_{\bar{y}_i} \le f_{s,\bar{y}_i}$$

each comprising the same number n of repeated measurements. Combining the empirical variances of the ordinates

$$s_{\bar{y}_i}^2 = \frac{1}{n-1}\sum_{l=1}^{n}(y_{il} - \bar{y}_i)^2\,; \quad i = 1, \ldots, m$$

with the boundaries of the unknown systematic errors the uncertainties of the input data are given by

$$\bar{y}_i \pm u_{\bar{y}_i}\,, \quad u_{\bar{y}_i} = \frac{t_P(n-1)}{\sqrt{n}}\, s_{\bar{y}_i} + f_{\bar{y}_i}\,; \quad i = 1, \ldots, m\,. \tag{22.2}$$

We expect the intervals

$$\bar{y}_i - u_{\bar{y}_i} \le y_{0,i} \le \bar{y}_i + u_{\bar{y}_i}\,; \quad i = 1, \ldots, m$$

to localize the true ordinates $y_{0,i}$ of the measured data $\bar{y}_i$.

22.2 Orthogonal Projection

Let us rewrite the linear, inconsistent, over-determined system

$$\beta_1 + \beta_2 \, x_{0,i} + \beta_3 \, x_{0,i}^2 \approx \bar{y}_i \, ; \qquad i = 1, \ldots, m > 3 \tag{22.3}$$

in matrix form. Defining

$$\boldsymbol{A} = \begin{pmatrix} 1 & x_{0,1} & x_{0,1}^2 \\ 1 & x_{0,2} & x_{0,2}^2 \\ \cdots & \cdots & \cdots \\ 1 & x_{0,m} & x_{0,m}^2 \end{pmatrix}, \qquad \boldsymbol{\beta} = \begin{pmatrix} \beta_1 \\ \beta_2 \\ \beta_3 \end{pmatrix}, \qquad \bar{\boldsymbol{y}} = \begin{pmatrix} \bar{y}_1 \\ \bar{y}_2 \\ \cdots \\ \bar{y}_m \end{pmatrix}$$

we have

$$\boldsymbol{A}\boldsymbol{\beta} \approx \bar{\boldsymbol{y}} \, .$$

The orthogonal projection yields

$$\bar{\boldsymbol{\beta}} = \boldsymbol{B}^{\mathrm{T}} \bar{\boldsymbol{y}} \, , \qquad \bar{\beta}_k = \sum_{i=1}^{m} b_{ik} \bar{y}_i \, ; \qquad k = 1, 2, 3 \tag{22.4}$$

where

$$\boldsymbol{B} = \boldsymbol{A} \left(\boldsymbol{A}^{\mathrm{T}} \boldsymbol{A} \right)^{-1} = (b_{ik}) \, ; \quad i = 1, \ldots, m; \quad k = 1, 2, 3 \, .$$

The components $\bar{\beta}_k$ of the solution vector $\bar{\boldsymbol{\beta}}$ establish the fitted parabola

$$y(x) = \bar{\beta}_1 + \bar{\beta}_2 \, x + \bar{\beta}_3 \, x^2 \, . \tag{22.5}$$

22.3 Uncertainties of the Components of the Solution Vector

Random Errors

We notionally consider an ensemble of least squares parabolas taken from the data sets $(y_{1l}, y_{2l}, \ldots, y_{ml})$; $l = 1, \ldots, n$, Sect. 7.3. Obviously, the belonging estimators are

$$\bar{\beta}_{kl} = \sum_{i=1}^{m} b_{ik} \, y_{il} \, ; \quad l = 1, \ldots, n \, . \tag{22.6}$$

Inserting the means

$$\bar{y}_i = \frac{1}{n} \sum_{l=1}^{n} y_{il} \, ; \quad i = 1, \ldots, m$$

into (22.4) provides the same result, namely

$$\bar{\beta}_k = \sum_{i=1}^{m} b_{ik}\left[\frac{1}{n}\sum_{l=1}^{n} y_{il}\right] = \frac{1}{n}\sum_{l=1}^{n}\left[\sum_{i=1}^{m} b_{ik} y_{il}\right]$$

$$= \frac{1}{n}\sum_{l=1}^{n} \bar{\beta}_{kl}\,; \qquad k = 1, 2, 3\,.$$

The differences

$$\bar{\beta}_{kl} - \bar{\beta}_k = \sum_{i=1}^{m} b_{ik}\left(y_{il} - \bar{y}_i\right)$$

bring forth the elements

$$s_{\bar{\beta}_k \bar{\beta}_{k'}} = \frac{1}{n-1}\sum_{l=1}^{n}\left(\bar{\beta}_{kl} - \bar{\beta}_k\right)\left(\bar{\beta}_{k'l} - \bar{\beta}_{k'}\right)$$

$$= \frac{1}{n-1}\sum_{l=1}^{n}\left[\sum_{i=1}^{m} b_{ik}\left(y_{il} - \bar{y}_i\right)\right]\left[\sum_{j=1}^{m} b_{jk'}\left(y_{jl} - \bar{y}_j\right)\right]$$

$$= \sum_{i,j=1}^{m} b_{ik}\, b_{jk'}\, s_{ij}\,; \qquad k, k' = 1, 2, 3$$

of the empirical variance–covariance matrix of the solution vector. Here, the

$$s_{ij} = \frac{1}{n-1}\sum_{l=1}^{n}(y_{il} - \bar{y}_i)(y_{jl} - \bar{y}_j)\,; \quad i, j = 1, \ldots, m$$

denote the elements of the empirical variance–covariance matrix

$$\boldsymbol{s} = (s_{ij})\,; \quad i, j = 1, \ldots, m \tag{22.7}$$

of the input data, each having degrees of freedom $\nu = n - 1$. Thus, the empirical variance–covariance matrix of the solution vector $\bar{\boldsymbol{\beta}}$ reads

$$\boldsymbol{s}_{\bar{\beta}} = \begin{pmatrix} s_{\bar{\beta}_1\bar{\beta}_1} & s_{\bar{\beta}_1\bar{\beta}_2} & s_{\bar{\beta}_1\bar{\beta}_3} \\ s_{\bar{\beta}_2\bar{\beta}_1} & s_{\bar{\beta}_2\bar{\beta}_2} & s_{\bar{\beta}_2\bar{\beta}_3} \\ s_{\bar{\beta}_3\bar{\beta}_1} & s_{\bar{\beta}_3\bar{\beta}_2} & s_{\bar{\beta}_3\bar{\beta}_3} \end{pmatrix} = \boldsymbol{B}^{\mathrm{T}}\boldsymbol{s}\,\boldsymbol{B}; \qquad s_{\bar{\beta}_k\bar{\beta}_k} \equiv s^2_{\bar{\beta}_k}\,. \tag{22.8}$$

Systematic Errors

The propagated systematic errors as issued by the decompositions

$$\bar{\beta}_k = \sum_{i=1}^{m} b_{ik}[\, y_{0,i} + (\bar{y}_i - \mu_{\bar{y}_i}) + f_{\bar{y}_i}\,]\,; \qquad k = 1,2,3 \tag{22.9}$$

are

$$f_{\bar{\beta}_k} = \sum_{i=1}^{m} b_{ik} f_{\bar{y}_i}\,; \qquad k = 1,2,3\,. \tag{22.10}$$

This suggests the worst-case estimations

$$f_{s,\bar{\beta}_k} = \sum_{i=1}^{m} |\, b_{ik} \,|\, f_{s,\bar{y}_i}\,; \qquad k = 1,2,3\,. \tag{22.11}$$

Should the $f_{\bar{y}_i}$ and $f_{s,\bar{y}_i}$ be equal

$$\begin{gathered} f_{\bar{y}_i} = f_y\,, \qquad f_{s,\bar{y}_i} = f_{s,y}\,; \qquad i = 1,\dots m \\ -f_{s,y} \le f_y \le f_{s,y} \end{gathered} \tag{22.12}$$

(21.15) produces

$$\begin{aligned} f_{\bar{\beta}_1} &= f_y \\ f_{\bar{\beta}_2} &= 0 \\ f_{\bar{\beta}_3} &= 0\,. \end{aligned} \tag{22.13}$$

Confidence Intervals and Overall Uncertainties

Confining the expectations

$$\begin{aligned} E\{\bar{\beta}_1\} &= \mu_{\bar{\beta}_1} = \beta_{0,1} + f_{\bar{\beta}_1} \\ E\{\bar{\beta}_2\} &= \mu_{\bar{\beta}_2} = \beta_{0,2} + f_{\bar{\beta}_2} \\ E\{\bar{\beta}_3\} &= \mu_{\bar{\beta}_3} = \beta_{0,3} + f_{\bar{\beta}_3} \end{aligned} \tag{22.14}$$

with probability P to confidence intervals

$$\begin{aligned} \bar{\beta}_1 - \frac{t_P(n-1)}{\sqrt{n}}\, s_{\bar{\beta}_1} &\le \mu_{\bar{\beta}_1} \le \bar{\beta}_1 + \frac{t_P(n-1)}{\sqrt{n}}\, s_{\bar{\beta}_1} \\ \bar{\beta}_2 - \frac{t_P(n-1)}{\sqrt{n}}\, s_{\bar{\beta}_2} &\le \mu_{\bar{\beta}_2} \le \bar{\beta}_2 + \frac{t_P(n-1)}{\sqrt{n}}\, s_{\bar{\beta}_2} \\ \bar{\beta}_3 - \frac{t_P(n-1)}{\sqrt{n}}\, s_{\bar{\beta}_3} &\le \mu_{\bar{\beta}_3} \le \bar{\beta}_3 + \frac{t_P(n-1)}{\sqrt{n}}\, s_{\bar{\beta}_3} \end{aligned}$$

suggests to assess the uncertainties of the components of the solution vector as

$$\begin{aligned} u_{\bar{\beta}_1} &= \frac{t_P(n-1)}{\sqrt{n}}\, s_{\bar{\beta}_1} + f_{s,\bar{\beta}_1} \\ u_{\bar{\beta}_2} &= \frac{t_P(n-1)}{\sqrt{n}}\, s_{\bar{\beta}_2} + f_{s,\bar{\beta}_2} \\ u_{\bar{\beta}_3} &= \frac{t_P(n-1)}{\sqrt{n}}\, s_{\bar{\beta}_3} + f_{s,\bar{\beta}_3}\,. \end{aligned} \tag{22.15}$$

Given the $f_{\bar{y}_i}$ are equal, this gives way to

$$\begin{aligned} u_{\bar{\beta}_1} &= \frac{t_P(n-1)}{\sqrt{n}}\, s_{\bar{\beta}_1} + f_{s,z} \\ u_{\bar{\beta}_2} &= \frac{t_P(n-1)}{\sqrt{n}}\, s_{\bar{\beta}_2} \\ u_{\bar{\beta}_3} &= \frac{t_P(n-1)}{\sqrt{n}}\, s_{\bar{\beta}_3}\,. \end{aligned} \tag{22.16}$$

22.4 Uncertainty Band

Random Errors

By means of the estimators $\bar{\beta}_{kl}$ as given in (22.6) we may visualize an ensemble of n least squares parabolas

$$y_l(x) = \bar{\beta}_{1l} + \bar{\beta}_{2l}\, x + \bar{\beta}_{3l}\, x^2\,; \qquad l = 1, \ldots, n\,.$$

On the average, the members of the ensemble reissue the least squares parabola

$$\begin{aligned} y(x) &= \frac{1}{n} \sum_{l=1}^{n} y_l(x) \\ &= \bar{\beta}_1 + \bar{\beta}_2\, x + \bar{\beta}_3\, x^2\,. \end{aligned}$$

Hence, for any fixed x, the differences

$$\begin{aligned} y_l(x) - y(x) &= \left(\bar{\beta}_{1l} - \bar{\beta}_1\right) + \left(\bar{\beta}_{2l} - \bar{\beta}_2\right) x + \left(\bar{\beta}_{3l} - \bar{\beta}_3\right) x^2 \\ &= \sum_{i=1}^{m} \left(b_{i1} + b_{i2}x + b_{i3}x^2\right) \left(y_{il} - \bar{y}_i\right)\,; \quad l = 1, \ldots, n \end{aligned}$$

bring forth the empirical variance

$$s_{y(x)}^2 = \frac{1}{n-1}\sum_{l=1}^{n}(y_l(x) - y(x))^2 = \boldsymbol{b}^T\boldsymbol{s}\,\boldsymbol{b} \tag{22.17}$$

where

$$\begin{aligned}\boldsymbol{b} = [(b_{11} + b_{12}x + b_{13}x^2) \qquad (b_{21} + b_{22}x + b_{23}x^2) \\ \ldots \quad (b_{m1} + b_{m3}x + b_{m3}x^2)]^T\end{aligned} \tag{22.18}$$

designates an auxiliary vector and $\boldsymbol{s}$ the empirical variance–covariance matrix of the input data as quoted in (22.7). Hence, the term

$$\frac{t_P(n-1)}{\sqrt{n}}\,s_{y(x)} \tag{22.19}$$

assesses the influence of random errors.

Systematic Errors

Inserting (22.9),

$$\bar{\beta}_k = \beta_{0,k} + \sum_{i=1}^{m} b_{ik}[(\bar{y}_i - \mu_{\bar{y}_i}) + f_{\bar{y}_i}]\,; \quad k = 1,2,3$$

into

$$y(x) = \bar{\beta}_1 + \bar{\beta}_2\,x + \bar{\beta}_3\,x^2$$

produces

$$\begin{aligned} y(x) &= \beta_{0,1} + \beta_{0,2}\,x + \beta_{0,3}\,x^2 \\ &+ \sum_{i=1}^{m}(b_{i1} + b_{i2}\,x + b_{i3}\,x^2)(\bar{y}_i - \mu_{\bar{y}_i}) \\ &+ \sum_{i=1}^{m}(b_{i1} + b_{i2}\,x + b_{i3}\,x^2)f_{\bar{y}_i}\,. \end{aligned} \tag{22.20}$$

Hence, we consider

$$f_{y(x)} = \sum_{i=1}^{m}(b_{i1} + b_{i2}\,x + b_{i3}\,x^2)f_{\bar{y}_i} \tag{22.21}$$

the propagated systematic error, its worst-case estimation being

$$f_{s,z(x,y)} = \sum_{i=1}^{m} \mid b_{i1} + b_{i2}\,x + b_{i3}\,x^2 \mid f_{s,\bar{y}_i}\,. \tag{22.22}$$

Assuming equal systematic errors, (21.15) causes

$$f_{s,y(x)} = f_{s,y}\,. \tag{22.23}$$

Overall Uncertainty

Combining (22.19) with (22.22) yields

$$y(x) \pm u_{y(x)}$$

$$u_{y(x)} = \frac{t_P(n-1)}{\sqrt{n}}\, s_{y(x)} + \sum_{i=1}^{m} \mid b_{i1} + b_{i2}x + b_{i3}x^2 \mid f_{s,\bar{y}_i} \tag{22.24}$$

which for equal systematic errors turns into

$$y(x) \pm u_{y(x)}$$

$$u_{y(x)} = \frac{t_P(n-1)}{\sqrt{n}}\, s_{y(x)} + f_{s,y}\,. \tag{22.25}$$

At any rate, we expect the uncertainty band to localize the true parabola (20.1).

22.5 *EPC*-Region

The *EPC*-region localizes the triple of true values $(\beta_{0,1}, \beta_{0,2}, \beta_{0,3})$ with respect to the triple of estimators $(\bar{\beta}_1, \bar{\beta}_2, \bar{\beta}_3)$.

Confidence Ellipsoid

Recasting (22.9), we find

$$\bar{\beta}_k = \mu_{\bar{\beta}_k} + \sum_{i=1}^{m} b_{ik}\left(\bar{y}_i - \mu_{\bar{y}_i}\right)$$

with

$$\mu_{\bar{\beta}_k} = \beta_{0,k} + \sum_{i=1}^{m} b_{ik} f_{\bar{y}_i}\,.$$

The variance–covariance matrix of the solution vector $s_{\bar{\boldsymbol{\beta}}}$ as given in (22.8) produces Hotelling's ellipsoid

$$(\bar{\boldsymbol{\beta}} - \mu_{\bar{\boldsymbol{\beta}}})^{\mathrm{T}} s_{\bar{\boldsymbol{\beta}}}^{-1} (\bar{\boldsymbol{\beta}} - \mu_{\bar{\boldsymbol{\beta}}}) = \frac{t_P^2(3, n-1)}{n}\,.$$

The pertaining confidence ellipsoid

$$(\boldsymbol{\beta} - \bar{\boldsymbol{\beta}})^{\mathrm{T}} s_{\bar{\boldsymbol{\beta}}}^{-1} (\boldsymbol{\beta} - \bar{\boldsymbol{\beta}}) = \frac{t_P^2(3, n-1)}{n} \tag{22.26}$$

is expected to localize the point

$$\boldsymbol{\mu}_{\bar{\boldsymbol{\beta}}} = \begin{pmatrix} \mu_{\bar{\beta}_1} \\ \mu_{\bar{\beta}_2} \\ \mu_{\bar{\beta}_3} \end{pmatrix} \tag{22.27}$$

with probability P.

Security Polyhedron

While the $f_{\bar{y}_1}, f_{\bar{y}_2}, \ldots, f_{\bar{y}_m}$ "scan" the set of points lying within or on the faces of the m-dimensional hypercuboid

$$-f_{s,\bar{y}_i} \le f_{\bar{y}_i} \le f_{s,\bar{y}_i}\,; \quad i = 1, \ldots, m$$

the components of the propagated systematic error

$$f_{\bar{\beta}_k} = \sum_{i=1}^{m} b_{ik} f_{\bar{z}_i}; \quad k = 1, 2, 3$$

span a polyhedron. However, assuming equal systematic errors, the polyhedron degenerates into an interval

$$\begin{aligned} -f_{s,y} \le f_{\bar{\beta}_1} &\le f_{s,y} \\ f_{\bar{\beta}_2} &= 0 \\ f_{\bar{\beta}_3} &= 0 \end{aligned} \tag{22.28}$$

which we notionally consider a "stick" of lengths $2f_{s,y}$.

Mergence of Ellipsoid Either with Interval or with Polyhedron

The merging of a confidence ellipsoid either with a non-degenerated or with a degenerated security polyhedron is addressed in Appendix G.

Figure 22.1 displays the true parabola (20.1), the fitted parabola (22.5), the uncertainties (22.15) of the estimators $\bar{\beta}_1, \bar{\beta}_2, \bar{\beta}_3$ and, finally, the uncertainty band (22.24).

Figure 22.2 considers equal systematic errors and renders the true parabola (20.1), the fitted parabola (22.5), the uncertainties (22.16) of the estimators $\bar{\beta}_1, \bar{\beta}_2, \bar{\beta}_3$ and, ultimately, the uncertainty band (22.25).

The illustrations are based on simulated data implying known true values and, in particular, extensive graphical scale transformations, Appendix A.

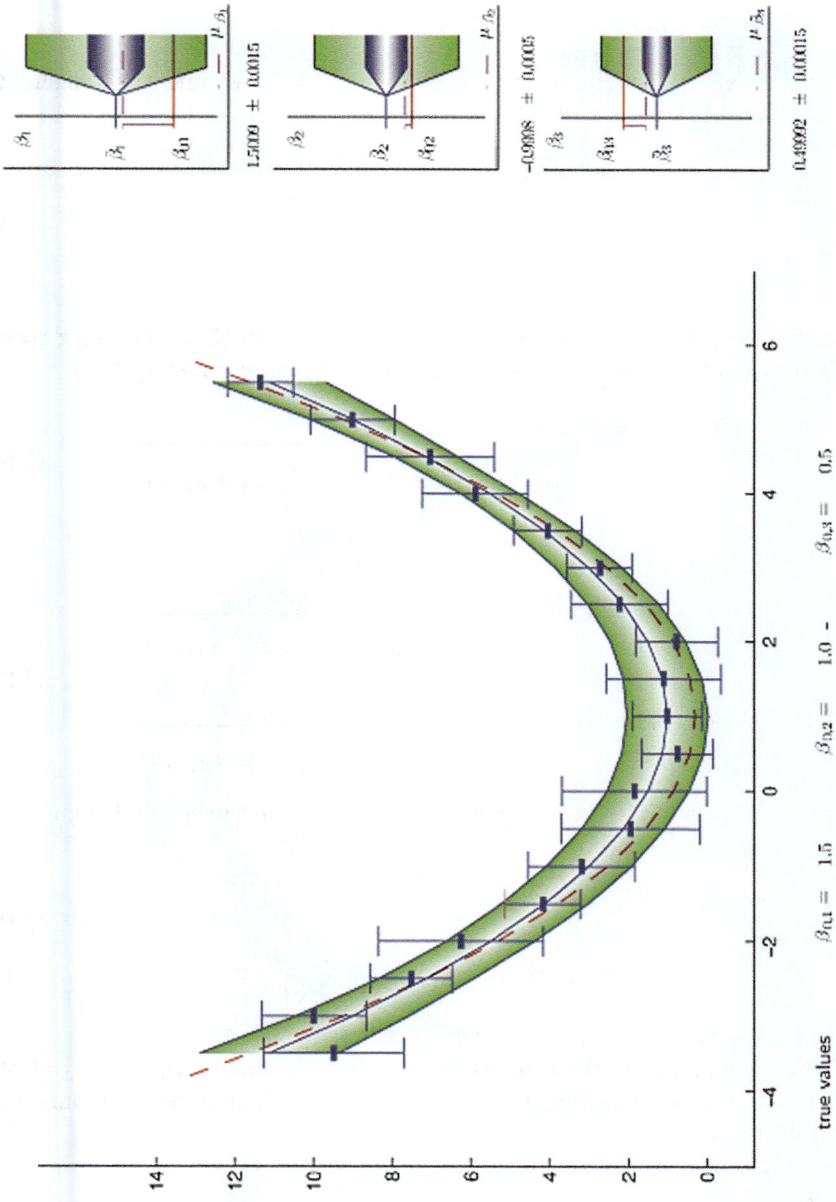

Fig. 22.1. Parabolas, case (ii). *Left*: adjusted parabola, uncertainty band, and true parabola (*dcshed line*). *Right*: Uncertainty intervals localizing the parabola's true parameters $\beta_{0,1}$ $\beta_{0,2}$, and $\beta_{0,3}$

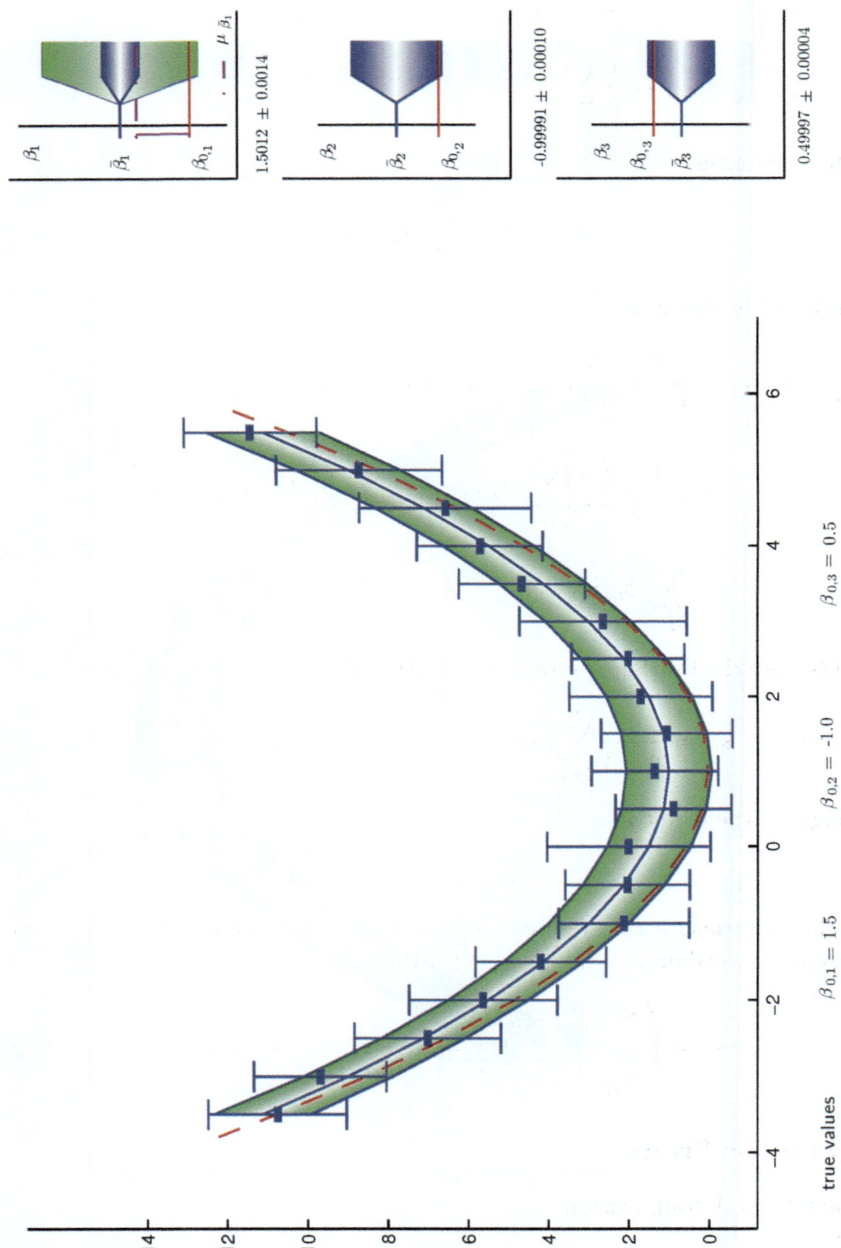

Fig. 22.2. Parabolas, case (ii), equal systematic errors. *Left*: adjusted parabola, uncertainty band, and true parabola (*dashed line*). *Right*: Uncertainty intervals localizing the parabola's true parameters $\beta_{0,1}$, $\beta_{0,2}$, and $\beta_{0,3}$. Equal systematic errors do not affect $\bar{\beta}_2$ and $\bar{\beta}_3$

23 Parabolas: Case (iii)

Case (iii), Table 20.1 suggests erroneous abscissas and erroneous ordinates blurred by varying theoretical variances and varying unknown systematic errors. The errors of the design matrix do not affect the orthogonal projection, they do, however, render the assessment of uncertainties more intricate.

23.1 Fitting Conditions

Consider $m > 3$ data pairs

$$(\bar{x}_1, \bar{y}_1), \quad (\bar{x}_2, \bar{y}_2), \quad \cdots \quad , (\bar{x}_m, \bar{y}_m) \tag{23.1}$$

with formal decompositions

$$\bar{x}_i = \frac{1}{n}\sum_{l=1}^{n} x_{il} = x_{0,i} + (\bar{x}_i - \mu_{\bar{x}_i}) + f_{\bar{x}_i}\,; \quad i = 1\,, \ldots\,, m$$

$$E\{\bar{X}_i\} = \mu_{\bar{x}_i}\,, \quad -f_{s,\bar{x}_i} \leq f_{\bar{x}_i} \leq f_{s,\bar{x}_i}$$

and

$$\bar{y}_i = \frac{1}{n}\sum_{l=1}^{n} y_{il} = y_{0,i} + (\bar{y}_i - \mu_{\bar{y}_i}) + f_{\bar{y}_i}\,; \quad i = 1\,, \ldots\,, m$$

$$E\{\bar{Y}_i\} = \mu_{\bar{y}_i}\,; \quad -f_{s,\bar{y}_i} \leq f_{\bar{y}_i} \leq f_{s,\bar{y}_i}\,.$$

The empirical variances

$$s_{\bar{x}_i}^2 = \frac{1}{n-1}\sum_{l=1}^{n}(x_{il} - \bar{x}_i)^2\,, \quad s_{\bar{y}_i}^2 = \frac{1}{n-1}\sum_{l=1}^{n}(y_{il} - \bar{y}_i)^2\,; \quad i = 1, \ldots, m$$

and the boundaries of systematic errors issue the uncertainties of the input data

$$\bar{x}_i \pm u_{\bar{x}_i}\,, \qquad u_{\bar{x}_i} = \frac{t_P(n-1)}{\sqrt{n}}\, s_{\bar{x}_i} + f_{s,\bar{x}_i}$$
$$i = 1, \ldots, m \qquad (23.2)$$
$$\bar{y}_i \pm u_{\bar{y}_i}\,, \qquad u_{\bar{y}_i} = \frac{t_P(n-1)}{\sqrt{n}}\, s_{\bar{y}_i} + f_{s,\bar{y}_i}\,.$$

We expect the intervals $\bar{x}_i \pm u_{\bar{x}_i}$ and $\bar{y}_i \pm u_{\bar{y}_i}$ to localize the true values $x_{0,i}$ and $y_{0,i}$, respectively.

23.2 Orthogonal Projection

Putting

$$\boldsymbol{A} = \begin{pmatrix} 1 & \bar{x}_1 & \bar{x}_1^2 \\ 1 & \bar{x}_2 & \bar{x}_2^2 \\ \cdots & \cdots & \cdots \\ 1 & \bar{x}_m & \bar{x}_m^2 \end{pmatrix}, \qquad \boldsymbol{\beta} = \begin{pmatrix} \beta_1 \\ \beta_2 \\ \beta_3 \end{pmatrix}, \qquad \bar{\boldsymbol{y}} = \begin{pmatrix} \bar{y}_1 \\ \bar{y}_2 \\ \cdots \\ \bar{y}_m \end{pmatrix}$$

we may transfer the linear, inconsistent, and over-determined system

$$\beta_1 + \beta_2\,\bar{x}_i + \beta_3\,\bar{x}_i^2 \approx \bar{y}_i\,; \quad i = 1, \ldots, m > 3 \qquad (23.3)$$

into

$$\boldsymbol{A}\boldsymbol{\beta} \approx \bar{\boldsymbol{y}}\,.$$

The orthogonal projection produces the solution vector

$$\bar{\boldsymbol{\beta}} = \boldsymbol{B}^{\mathrm{T}}\bar{\boldsymbol{y}}\,, \qquad \bar{\beta}_k = \sum_{i=1}^{m} b_{ik}\bar{y}_i\,; \qquad k = 1, 2, 3 \qquad (23.4)$$

with

$$\boldsymbol{B} = \boldsymbol{A}(\boldsymbol{A}^{\mathrm{T}}\boldsymbol{A})^{-1} = (b_{ik})\,; \qquad i = 1, \ldots, m; \quad k = 1, 2, 3$$

Hence, the components $\bar{\beta}_k\,;\ k = 1, 2, 3$ of the solution vector $\bar{\boldsymbol{\beta}}$ establish the least squares parabola

$$y(x) = \bar{\beta}_1 + \bar{\beta}_2\,x + \bar{\beta}_3\,x^2\,. \qquad (23.5)$$

As the elements b_{ik} are faulty, in order to assess measurement uncertainties, we are asked to carry out series expansions.

23.3 Series Expansion of the Solution Vector

We submit the $\bar{\beta}_k$; $k = 1, 2, 3$ to series expansions throughout a neighborhood of the point

$$(x_{0,1}\,, \dots , x_{0,m}\,; \quad y_{0,1}\,, \dots , y_{0,m}),$$

primarily, with respect to the n points

$$(x_{1l}, x_{2l}, \dots , x_{ml}\,;\; y_{1l}, y_{2l}, \dots , y_{nl}); \qquad l = 1, \dots , n$$

producing

$$\begin{aligned}\bar{\beta}_{kl}(x_{1l}, \dots , x_{ml}\,;\; y_{1l}, \dots , y_{ml}) &= \bar{\beta}_k(x_{0,1}, \dots , x_{0,m}\,;\; y_{0,1}, \dots , y_{0,m})\\ &+ \sum_{i=1}^{m} \frac{\partial \bar{\beta}_k}{\partial x_{0,i}}(x_{il} - \mu_{\bar{x}_i}) + \sum_{i=1}^{m} \frac{\partial \bar{\beta}_k}{\partial y_{0,i}}(y_{il} - \mu_{\bar{y}_i}) + \cdots \\ &+ \sum_{i=1}^{m} \frac{\partial \bar{\beta}_k}{\partial x_{0,i}} f_{\bar{x}_i} + \sum_{i=1}^{m} \frac{\partial \bar{\beta}_k}{\partial y_{0,i}} f_{\bar{y}_i} + \cdots\end{aligned} \tag{23.6}$$

whereat $\bar{\beta}_k(x_{0,1}, \dots , x_{0,1}\,;\; y_{0,1}, \dots , y_{0,m}) \equiv \beta_{0,k}$ and, secondly, with respect to the sample-dependent means

$$(\bar{x}_1, \bar{x}_2, \dots , \bar{x}_m\,;\; \bar{y}_1, \bar{y}_2, \dots , \bar{y}_m)$$

issuing

$$\begin{aligned}\bar{\beta}_k\,(\bar{x}_1, \dots , \bar{x}_m;\,;\; \bar{y}_1, \dots , \bar{y}_m) &= \bar{\beta}_k(x_{0,1}, \dots , x_{0,m}\,;\; y_{0,1}, \dots , y_{0,m})\\ &+ \sum_{i=1}^{m} \frac{\partial \bar{\beta}_k}{\partial x_{0,i}}(\bar{x}_i - \mu_{\bar{x}_i}) + \sum_{i=1}^{m} \frac{\partial \bar{\beta}_k}{\partial y_{0,i}}(\bar{y}_i - \mu_{\bar{y}_i}) + \cdots \\ &+ \sum_{i=1}^{m} \frac{\partial \bar{\beta}_k}{\partial x_{0,i}} f_{\bar{x}_i} + \sum_{i=1}^{m} \frac{\partial \bar{\beta}_k}{\partial y_{0,i}} f_{\bar{y}_i} + \cdots .\end{aligned} \tag{23.7}$$

As there is no other choice, we approximate the derivatives in $x_{0,1}\,, \dots , y_{0,m}$ through derivatives in $\bar{x}_1\,, \dots , \bar{y}_m$. For convenience, we abbreviate the partial derivatives following

$$\begin{aligned} c_{i1} &= \frac{\partial \bar{\beta}_1}{\partial \bar{x}_i}\,, \quad & c_{i+m,1} &= \frac{\partial \bar{\beta}_1}{\partial \bar{y}_i}\,, \\ c_{i2} &= \frac{\partial \bar{\beta}_2}{\partial \bar{x}_i}\,, \quad & c_{i+m,2} &= \frac{\partial \bar{\beta}_2}{\partial \bar{y}_i}\,, \qquad i = 1, \dots , m \\ c_{i3} &= \frac{\partial \bar{\beta}_3}{\partial \bar{x}_i}\,, \quad & c_{i+m,3} &= \frac{\partial \bar{\beta}_3}{\partial \bar{y}_i} \end{aligned} \tag{23.8}$$

and assign the coefficients c_{ik} to an auxiliary matrix

$$\boldsymbol{C}^{\mathrm{T}} = \begin{pmatrix} c_{11} & c_{21} & \cdots & c_{2m,1} \\ c_{12} & c_{22} & \cdots & c_{2m,2} \\ c_{13} & c_{23} & \cdots & c_{2m,3} \end{pmatrix} .$$

Furthermore, we put

$$\begin{aligned} v_{il} &= x_{il}\,, & v_{i+m,l} &= y_{il} \\ \bar{v}_i &= \bar{x}_i\,, & \bar{v}_{i+m} &= \bar{y}_i \\ \mu_i &= \mu_{\bar{x}_i}\,, & \mu_{i+m} &= \mu_{\bar{y}_i} \\ f_i &= f_{\bar{x}_i}\,, & f_{i+m} &= f_{\bar{y}_i} \\ f_{s,i} &= f_{s,\bar{x}_i}\,, & f_{s,i+m} &= f_{s,\bar{y}_i}\,. \end{aligned} \tag{23.9}$$

Linearizing the expansions, we find after all

$$\begin{aligned} \bar{\beta}_{kl} &= \beta_{0,k} + \sum_{i=1}^{2m} c_{ik}\left(v_{il} - \mu_i\right) + \sum_{i=1}^{2m} c_{ik} f_i\,; \quad l = 1,\,\ldots,\,n \\ \bar{\beta}_k &= \beta_{0,k} + \sum_{i=1}^{2m} c_{ik}\left(\bar{v}_i - \mu_i\right) + \sum_{i=1}^{2m} c_{ik} f_i\,; \quad k = 1,2,3\,. \end{aligned} \tag{23.10}$$

Subtraction produces

$$\bar{\beta}_{kl} - \bar{\beta}_k = \sum_{i=1}^{2m} c_{ik}(v_{il} - \bar{v}_i)\,; \quad k = 1,2,3\,; \quad l = 1,\,\ldots,\,n\,. \tag{23.11}$$

Considering an ensemble of n least squares parabolas based on the respective l-th repeated measurements of the m measuring points $i = 1, \ldots, m$

$$(x_{1l}, \ldots, x_{ml}\,; \quad y_{1l}, \ldots, y_{ml})\,; \qquad l = 1,\,\ldots,\,n$$

we have

$$\bar{\beta}_k = \frac{1}{n}\sum_{l=1}^{n} \bar{\beta}_{kl}\,; \qquad k = 1,2,3\,. \tag{23.12}$$

While (23.11) provides a means to handle the propagation of random errors the terms

$$f_{\bar{\beta}_k} = \sum_{i=1}^{2m} c_{ik} f_i\,; \qquad k = 1,2,3 \tag{23.13}$$

devise the propagated systematic errors.

23.4 Uncertainties of the Components of the Solution Vector

Random Errors

Deploying (23.11), we set up the elements

$$\begin{aligned} s_{\bar{\beta}_k \bar{\beta}_{k'}} &= \frac{1}{n-1} \sum_{l=1}^{n} \left(\bar{\beta}_{kl} - \bar{\beta}_k\right)\left(\bar{\beta}_{k'l} - \bar{\beta}_{k'}\right) \\ &= \frac{1}{n-1} \sum_{l=1}^{n} \left[\sum_{i=1}^{2m} c_{ik}\left(v_{il} - \bar{v}_i\right)\right]\left[\sum_{j=1}^{2m} c_{jk'}\left(v_{jl} - \bar{v}_j\right)\right] \\ &= \sum_{i,j=1}^{2m} c_{ik}\, c_{jk'}\, s_{ij}\,; \qquad k, k' = 1, 2, 3 \end{aligned}$$

of the empirical variance–covariance matrix

$$\boldsymbol{s}_{\bar{\boldsymbol{\beta}}} = \begin{pmatrix} s_{\bar{\beta}_1\bar{\beta}_1} & s_{\bar{\beta}_1\bar{\beta}_2} & s_{\bar{\beta}_1\bar{\beta}_3} \\ s_{\bar{\beta}_2\bar{\beta}_1} & s_{\bar{\beta}_2\bar{\beta}_2} & s_{\bar{\beta}_2\bar{\beta}_3} \\ s_{\bar{\beta}_3\bar{\beta}_1} & s_{\bar{\beta}_3\bar{\beta}_2} & s_{\bar{\beta}_3\bar{\beta}_3} \end{pmatrix} \qquad s_{\bar{\beta}_k\bar{\beta}_k} \equiv s^2_{\bar{\beta}_k}$$

of the solution vector. Again, the

$$s_{ij} = \frac{1}{n-1} \sum_{l=1}^{n} (v_{il} - \bar{v}_i)(v_{jl} - \bar{v}_j)\,; \quad i, j = 1, \ldots, 2m$$

denote the elements of the empirical variance–covariance matrix

$$\boldsymbol{s} = (s_{ij})\,; \quad i, j = 1, \ldots, 2m \tag{23.14}$$

of the input data, each s_{ij} having degrees of freedom $\nu = n - 1$. As done previously, we condense the matrix $\boldsymbol{s}_{\bar{\boldsymbol{\beta}}}$ to

$$\boldsymbol{s}_{\bar{\boldsymbol{\beta}}} = \boldsymbol{C}^{\mathrm{T}} \boldsymbol{s}\, \boldsymbol{C}\,. \tag{23.15}$$

Systematic Errors

Assuming equal systematic errors and error bounds

$$\begin{aligned} f_{\bar{x}_i} &= f_x\,, \quad f_{s,\bar{x}_i} = f_{s,x}\,, \quad -f_{s,x} \le f_x \le f_{s,x} \\ & \qquad\qquad\qquad\qquad\qquad\qquad i = 1, \ldots, m \\ f_{\bar{y}_i} &= f_y\,, \quad f_{s,\bar{y}_i} = f_{s,y}\,, \quad -f_{s,y} \le f_y \le f_{s,y} \end{aligned} \tag{23.16}$$

(23.13) yields

$$f_{\bar{\beta}_k} = \sum_{i=1}^{m} c_{ik} f_{\bar{x}_i} + \sum_{i=1}^{m} c_{i+m,k} f_{\bar{y}_i}$$

$$= f_x \sum_{i=1}^{m} c_{ik} + f_y \sum_{i=1}^{m} c_{i+m,k} \,.$$

But as

$$\begin{aligned} \sum_{i=1}^{m} c_{i1} &= -\bar{\beta}_2 \,, & \sum_{i=1}^{m} c_{i+m,1} &= 1 \\ \sum_{i=1}^{m} c_{i2} &= -2\bar{\beta}_3 \,, & \sum_{i=1}^{m} c_{i+m,2} &= 0 \\ \sum_{i=1}^{m} c_{i3} &= 0 \,, & \sum_{i=1}^{m} c_{i+m,3} &= 0 \end{aligned} \qquad (23.17)$$

we have

$$\begin{aligned} f_{\bar{\beta}_1} &= -\bar{\beta}_2 f_x + f_y \\ f_{\bar{\beta}_2} &= -2\,\bar{\beta}_3 f_x \\ f_{\bar{\beta}_3} &= 0 \end{aligned} \qquad (23.18)$$

from which we take that equal systematic errors shift the parabola parallel to itself.

Confidence Intervals and Overall Uncertainties

Localizing the expectations

$$E\{\bar{\beta}_1\} = \mu_{\bar{\beta}_1} = \beta_{0,1} + f_{\bar{\beta}_1}$$

$$E\{\bar{\beta}_2\} = \mu_{\bar{\beta}_2} = \beta_{0,2} + f_{\bar{\beta}_2}$$

$$E\{\bar{\beta}_2\} = \mu_{\bar{\beta}_3} = \beta_{0,3} + f_{\bar{\beta}_3}$$

with probability P by confidence intervals

$$\bar{\beta}_1 - \frac{t_P(n-1)}{\sqrt{n}} s_{\bar{\beta}_1} \le \mu_{\bar{\beta}_1} \le \bar{\beta}_1 + \frac{t_P(n-1)}{\sqrt{n}} s_{\bar{\beta}_1}$$

$$\bar{\beta}_2 - \frac{t_P(n-1)}{\sqrt{n}} s_{\bar{\beta}_2} \le \mu_{\bar{\beta}_2} \le \bar{\beta}_2 + \frac{t_P(n-1)}{\sqrt{n}} s_{\bar{\beta}_2}$$

$$\bar{\beta}_3 - \frac{t_P(n-1)}{\sqrt{n}} s_{\bar{\beta}_3} \le \mu_{\bar{\beta}_3} \le \bar{\beta}_3 + \frac{t_P(n-1)}{\sqrt{n}} s_{\bar{\beta}_3}$$

the overall uncertainties of the components of the solution vector are seen to be

$$u_{\bar{\beta}_k} = \frac{t_P(n-1)}{\sqrt{n}}\, s_{\bar{\beta}_k} + \sum_{i=1}^{2m} \mid c_{ik} \mid f_{s,i}\,; \quad k = 1, 2, 3\,. \qquad (23.19)$$

In case of equal systematic errors are equal, this turns into

$$\begin{aligned} u_{\bar{\beta}_1} &= \frac{t_P(n-1)}{\sqrt{n}}\, s_{\bar{\beta}_1} + \mid \bar{\beta}_2 \mid f_{s,x} + f_{s,y} \\ u_{\bar{\beta}_2} &= \frac{t_P(n-1)}{\sqrt{n}}\, s_{\bar{\beta}_2} + 2 \mid \bar{\beta}_3 \mid f_{s,x} \\ u_{\bar{\beta}_3} &= \frac{t_P(n-1)}{\sqrt{n}}\, s_{\bar{\beta}_3}\,. \end{aligned} \qquad (23.20)$$

23.5 Uncertainty Band

Random Errors

Let us visualize an ensemble of n least squares parabolas

$$y_l(x) = \bar{\beta}_{1l} + \bar{\beta}_{2l}\, x + \bar{\beta}_{3l}\, x^2\,; \quad l = 1, \ldots, n\,.$$

as addressed in (23.12). Obviously, the ensemble reestablishes

$$y(x) = \sum_{l=1}^{n} y_l(x)\,.$$

For any fixed x, the differences

$$\begin{aligned} y_l(x) - y(x) &= \left(\bar{\beta}_{1l} - \bar{\beta}_1\right) + \left(\bar{\beta}_{2l} - \bar{\beta}_2\right) x + \left(\bar{\beta}_{3l} - \bar{\beta}_3\right) x^2 \\ &= \sum_{i=1}^{2m} \left(c_{i1} + c_{i2}x + c_{i3}x^2\right) \left(v_{il} - \bar{v}_i\right)\,; \quad l = 1, \ldots, n \end{aligned}$$

suggest an empirical variance

$$s^2_{y(x)} = \frac{1}{n-1} \sum_{l=1}^{n} \left(y_l(x) - y(x)\right)^2 = \boldsymbol{c}^{\mathrm{T}} \boldsymbol{s}\, \boldsymbol{c} \qquad (23.21)$$

letting

$$\boldsymbol{c} = \left(c_{11} + c_{12}x + c_{13}x^2 \quad c_{21} + c_{22}x + c_{23}x^2 \quad \ldots \quad c_{2m,1} + c_{2m,2}x + c_{2m,3}x^2\right)^{\mathrm{T}}$$

denote an auxiliary vector.

Systematic Errors

Combining (23.10) with (23.5) produces

$$y(x) = \beta_{0,1} + \beta_{0,2}\, x + \beta_{0,3}\, x^2$$
$$+ \sum_{i=1}^{2m} (c_{i1} + c_{i2}\, x + c_{i3}\, x^2)(\bar{v}_i - \mu_i) + \sum_{i=1}^{2m} (c_{i1} + c_{i2}\, x + c_{i3}\, x^2) f_i \,.$$

From this we read the propagated systematic error

$$f_{y(x)} = \sum_{i=1}^{2m} (c_{i1} + c_{i2}\, x + c_{i3}\, x^2) f_i \tag{23.22}$$

its worst-case estimation being

$$f_{s,y(x)} = \sum_{i=1}^{2m} \mid c_{i1} + c_{i2} x + c_{i3} x^2 \mid f_{s,i} \,. \tag{23.23}$$

For equal systematic errors, (23.17) and (23.18) issue

$$f_{y(x)} = -\bar{\beta}_2\, f_x - 2\, \bar{\beta}_3 f_x\, x + f_y \,. \tag{23.24}$$

Overall Uncertainty

After all, the uncertainty band is given by

$$y(x) \pm u_{y(x)}$$
$$u_{y(x)} = \frac{t_P(n-1)}{\sqrt{n}} \sqrt{\boldsymbol{c}^{\mathrm{T}} \boldsymbol{s}\, \boldsymbol{c}} + \sum_{i=1}^{2m} \mid c_{i1} + c_{i2} x + c_{i3} x^2 \mid f_{s,i} \tag{23.25}$$

In case of equal systematic errors we have

$$y(x) \pm u_{y(x)}$$
$$u_{y(x)} = \frac{t_P(n-1)}{\sqrt{n}} \sqrt{\boldsymbol{c}^{\mathrm{T}} \boldsymbol{s}\, \boldsymbol{c}} + \mid \bar{\beta}_2 + 2\bar{\beta}_3\, x \mid f_{s,x} + f_{s,y} \,. \tag{23.26}$$

At any rate, we expect the uncertainty band to hold the true parabola (20.1).

23.6 *EPC*-Region

The *EPC*-region localizes the triple of true values $(\beta_{0,1}, \beta_{0,2}, \beta_{0\,3})$ with respect to the triple of estimators $(\bar{\beta}_1, \bar{\beta}_2, \bar{\beta}_3)$.

Confidence Ellipsoid

From (23.10) we take

$$\bar{\beta}_k = \mu_{\bar{\beta}_k} + \sum_{i=1}^{2m} c_{ik}\,(\bar{v}_i - \mu_i)\,; \quad \mu_{\bar{\beta}_k} = \beta_{0,k} + \sum_{i=1}^{2m} c_{ik} f_i; \quad k = 1, 2, 3\,.$$

The variance–covariance matrix of the solution vector $s_{\bar{\boldsymbol{\beta}}}$ as given in (23.15) produces Hotelling's ellipsoid

$$(\bar{\boldsymbol{\beta}} - \boldsymbol{\mu}_{\bar{\boldsymbol{\beta}}})^{\mathrm{T}} s_{\bar{\boldsymbol{\beta}}}^{-1} (\bar{\boldsymbol{\beta}} - \boldsymbol{\mu}_{\bar{\boldsymbol{\beta}}}) = \frac{t_P^2(3, n-1)}{n}$$

from which we draw the confidence ellipsoid

$$(\boldsymbol{\beta} - \bar{\boldsymbol{\beta}})^{\mathrm{T}} s_{\bar{\boldsymbol{\beta}}}^{-1} (\boldsymbol{\beta} - \bar{\boldsymbol{\beta}}) = \frac{t_P^2(3, n-1)}{n}\,. \tag{23.27}$$

We expected the latter to localize the point

$$\boldsymbol{\mu}_{\bar{\boldsymbol{\beta}}} = \begin{pmatrix} \mu_{\bar{\beta}_1} \\ \mu_{\bar{\beta}_2} \\ \mu_{\bar{\beta}_3} \end{pmatrix} \tag{23.28}$$

with probability P.

Security Polyhedron

While the errors $f_1, f_2, \ldots, f_{2m}$ "scan" the set of points lying within or on the faces of the $2m$-dimensional hypercuboid

$$-f_{s,i} \le f_i \le f_{s,i}\,; \quad i = 1, \ldots, 2m$$

the components of the propagated systematic error

$$f_{\bar{\beta}_k} = \sum_{i=1}^{2m} c_{ik} f_i; \quad k = 1, 2, 3$$

span a polyhedron. For equal systematic errors the polyhedron degenerates into a polygon

$$\begin{aligned} f_{\bar{\beta}_1} &= -\bar{\beta}_2 f_x + f_y \\ f_{\bar{\beta}_2} &= -2\bar{\beta}_3 f_x \\ f_{\bar{\beta}_3} &= 0\,. \end{aligned} \tag{23.29}$$

Mergence of Ellipsoid Either with Interval or with Polyhedron

The merging of a confidence ellipsoid either with a non-degenerated or with a degenerated security polyhedron is addressed in Appendix G.

Figure 23.1 displays the true parabola (20.1), the fitted parabola (23.5), the uncertainties (23.19) of the estimators $\bar{\beta}_1, \bar{\beta}_2, \bar{\beta}_3$ and, finally, the uncertainty band (23.25).

Figure 23.2 considers equal systematic errors and renders the true parabola (20.1), the fitted parabola (23.5), the uncertainties (23.20) of the estimators $\bar{\beta}_1, \bar{\beta}_2, \bar{\beta}_3$ and, ultimately, the uncertainty band (23.26).

The illustrations are based on simulated data implying known true values and, in particular, extensive graphical scale transformations, Appendix A.

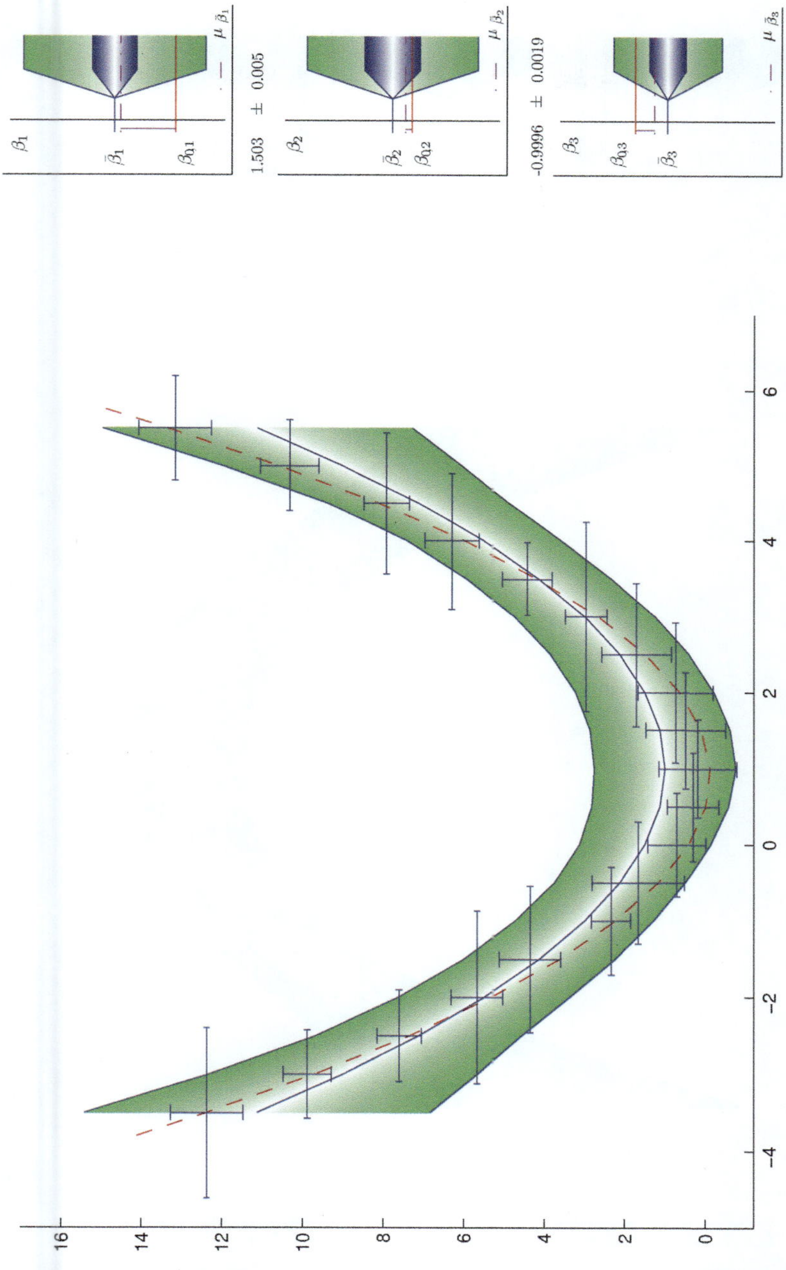

Fig. 23.1. Parabolas, case (iii). *Left*: least squares parabola, uncertainty band, and true parabola (*dashed line*). *Right*: Uncertainty intervals localizing the parabola's true parameters $\beta_{0,1}$, $\beta_{0,2}$, and $\beta_{0,3}$

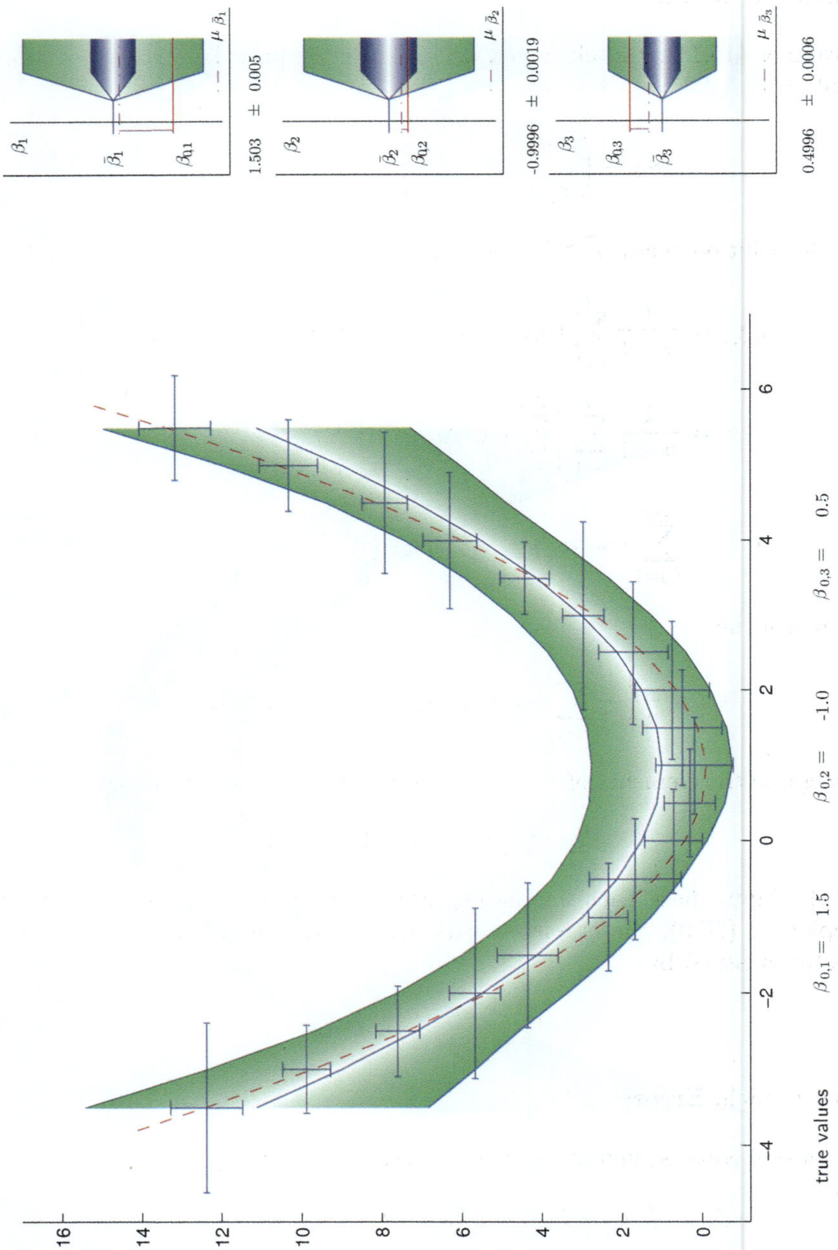

Fig. 23.2. Parabolas, case (iii), equal systematic errors in x and y. *Left*: least squares parabola, uncertainty band, and true parabola (*dashed line*). *Right*: Uncertainty intervals localizing the parabola's true parameters $\beta_{0,1}$, $\beta_{0,2}$, and $\beta_{0,3}$. Equal systematic errors do not affect $\bar{\beta}_3$

Part VIII

Non-linear Fitting

24 Series Truncation

Considering functional relationships being non-linear in the unknown parameters asks for a deviant strategy [7]. To proceed via series truncation, we firstly suppose the relationships to behave adequately smooth with respect to the unknown parameters and secondly the magnitudes of the measuring errors to remain sufficiently small. It is, however, to be observed that the truncation errors and the measuring errors jointly enter the least squares estimators and the respective uncertainties.

24.1 Homologous True Function

Let some function $\phi(x, y; a, b) = 0$ be non-linear with respect to the parameters a and b. Given a, b, the relationship involves a well-defined set of data pairs x, y. Vice versa, given the outcomes x, y of some physical experiment are known to follow $\phi(x, y; a, b) = 0$, the measured data, though erroneous, may be deployed to assess the unknown parameters a and b.

To linearize $\phi(x, y; a, b) = 0$, we rely on guessed starting values, say, a^* and b^*

$$\phi(x, y;\, a, b) = \phi(x, y;\, a^*, b^*)$$
$$+\frac{\partial \phi(x, y;\, a^*, b^*)}{\partial a^*}(a - a^*) + \frac{\partial \phi(x, y;\, a^*, b^*)}{\partial b^*}(b - b^*) + \cdots = 0 \tag{24.1}$$

and consider the differences

$$\beta_1 = a - a^* \quad \text{and} \quad \beta_2 = b - b^* \tag{24.2}$$

as unknowns. Putting

$$\xi = \frac{\dfrac{\partial \phi(x, y;\, a^*, b^*)}{\partial b^*}}{\dfrac{\partial \phi(x, y;\, a^*, b^*)}{\partial a^*}}\,, \qquad \eta = -\frac{\phi(x, y;\, a^*, b^*)}{\dfrac{\partial \phi(x, y;\, a^*, b^*)}{\partial a^*}} \tag{24.3}$$

(24.1) turns into an apparently linear model

$$\beta_1 + \beta_2 \xi \approx \eta\,; \quad i = 1, \ldots, m\,. \tag{24.4}$$

The linearization, being the result of a series truncation, is in no way connected to measurement errors. The closer the starting values a^*, b^* to the true parameters a, b, the more (24.4) resembles an identity in zero as induced via

$$\beta_1 \to 0, \qquad \beta_2 \to 0, \qquad \phi(x, y;\, a^*, b^*) \to 0\,. \tag{24.5}$$

Obviously, the identity in zero substitutes the true function $\phi(x, y; a, b) = 0$. For convenience, let us call this identity the *homologous true function.*

Should the starting values a^*, b^* be not close enough to the true values a, b, the proceeding might break down.

24.2 Fitting Conditions

Let there be $m > 2$ data pairs

$$(x_{0,1}, \bar{y}_1), \quad (x_{0,2}, \bar{y}_2), \quad \ldots \quad (x_{0,m}, \bar{y}_m) \tag{24.6}$$

the abscissas of which being error-free and ordinates arithmetic means each covering n repeated measurements

$$\bar{y}_i = \frac{1}{n}\sum_{l=1}^{n} y_{il} = y_{0,i} + (\bar{y}_i - \mu_{\bar{y}_i}) + f_{\bar{y}_i}\,; \quad i = 1, \ldots, m$$

$$E\{\bar{Y}_i\} = \mu_{\bar{y}_i}\,; \qquad -f_{s,\bar{y}_i} \le f_{\bar{y}_i} \le f_{s,\bar{y}_i}\,.$$

The empirical variances

$$s_{\bar{y}_i}^2 = \frac{1}{n-1}\sum_{l=1}^{n}(y_{il} - \bar{y}_i)^2\,, \quad i = 1, \ldots, m$$

and the error limits $\pm f_{s,\bar{y}_1}$ issue the uncertainties of the input data

$$\bar{y}_i \pm u_{\bar{y}_i}\,, \quad u_{\bar{x}_i} = \frac{t_P(n-1)}{\sqrt{n}}\, s_{\bar{y}_i} + f_{s,\bar{y}_i}\,; \quad i = 1, \ldots, m\,. \tag{24.7}$$

The intervals $\bar{y}_i \pm u_{\bar{y}_i}$ are seen to localize the true values $y_{0,i}$ of the input data.

24.3 Orthogonal Projection

The data (24.6) produce coefficients

$$\bar{\xi}_i = \frac{\dfrac{\partial\phi(x_{0,i}, \bar{y}_i\,; a^*, b^*)}{\partial b^*}}{\dfrac{\partial\phi(x_{0,i}, \bar{y}_i;\, a^*, b^*)}{\partial a^*}}\,, \qquad \bar{\eta}_i = -\frac{\phi(x_{0,i}, \bar{y}_i;\, a^*, b^*)}{\dfrac{\partial\phi(x_{0,i}, \bar{y}_i;\, a^*, b^*)}{\partial a^*}} \tag{24.8}$$

so that

$$\beta_1 + \beta_2\,\bar{\xi}_i \approx \bar{\eta}_i\,; \quad i = 1, \ldots, m\,. \tag{24.9}$$

Cast in matrices this reads

$$\boldsymbol{A}\boldsymbol{\beta} \approx \bar{\boldsymbol{\eta}}$$

given

$$\boldsymbol{A} = \begin{pmatrix} 1 & \bar{\xi}_1 \\ 1 & \bar{\xi}_2 \\ \cdots & \cdots \\ 1 & \bar{\xi}_m \end{pmatrix}, \qquad \boldsymbol{\beta} = \begin{pmatrix} \beta_1 \\ \beta_2 \end{pmatrix}, \qquad \bar{\boldsymbol{\eta}} = \begin{pmatrix} \bar{\eta}_1 \\ \bar{\eta}_2 \\ \cdots \\ \bar{\eta}_m \end{pmatrix}.$$

The orthogonal projection yields

$$\bar{\boldsymbol{\beta}} = \boldsymbol{B}^{\mathrm{T}}\bar{\boldsymbol{\eta}}\,, \qquad \bar{\beta}_k = \sum_{i=1}^{m} b_{ik}\bar{\eta}_i\,; \qquad k = 1, 2 \tag{24.10}$$

where

$$\boldsymbol{B} = \boldsymbol{A}\left(\boldsymbol{A}^{\mathrm{T}}\boldsymbol{A}\right)^{-1} = (b_{ik})\,; \quad i = 1, \ldots, m; \quad k = 1, 2\,.$$

As the coefficients b_{ik} of the matrix $\boldsymbol{B}$ are charged by measuring errors, in order to assign measurement uncertainties, the components $\bar{\beta}_1$ and $\bar{\beta}_2$ of the solution vector have to be submitted to series expansions.

Submitting (24.9) to least squares has brought forth the straight line

$$\eta(\xi) = \bar{\beta}_1 + \bar{\beta}_2\,\xi\,. \tag{24.11}$$

Hence, there are estimators

$$\bar{a} = a^* + \bar{\beta}_1\,, \qquad \bar{b} = b^* + \bar{\beta}_2 \tag{24.12}$$

of the unknown parameters a and b.

To disburden the discussion, we confine ourselves to a simplifying

Example

Given a function

$$\phi(x, y; a, b) = y - a\,\exp\left(-b/x\right) = 0\,; \quad a, b > 0\,, \quad x > 0 \tag{24.13}$$

with arbitrarily chosen parameters $a = 3e$ and $b = 1$ and $m > 2$ data pairs (24.6) holding error-free abscissas and ordinates being arithmetic means. As

$$\xi(x, a^*) = -\frac{a^*}{x}, \qquad \eta = y \exp(b^*/x) - a^* \tag{24.14}$$

the least squares proceeding (24.9) brings forth

$$\boldsymbol{A} = \begin{pmatrix} 1 & -a^*/x_{0,1} \\ 1 & -a^*/x_{0,2} \\ \dots & \dots \\ 1 & -a^*/x_{0,m} \end{pmatrix}, \quad \boldsymbol{\beta} = \begin{pmatrix} \beta_1 \\ \beta_2 \end{pmatrix}, \quad \bar{\boldsymbol{\eta}} = \begin{pmatrix} \bar{y}_1 \exp(b^*/x_{0,1}) - a^* \\ \bar{y}_2 \exp(b^*/x_{0,2}) - a^* \\ \dots \\ \bar{y}_m \exp(b^*/x_{0,m}) - a^* \end{pmatrix}$$

where rank $(\boldsymbol{A}) = 2$ is assumed. Again, the components of the solution vector are given by

$$\bar{\beta}_k = \sum_{i=1}^{m} b_{ik}\bar{\eta}_i \, ; \qquad k = 1, 2 \, . \tag{24.15}$$

Let us now consider the homologous true system

$$\boldsymbol{A}\boldsymbol{\beta}_0 = \boldsymbol{\eta}_0 \, ; \qquad \boldsymbol{\beta}_0 = \boldsymbol{B}^{\mathrm{T}}\boldsymbol{\eta}_0 \tag{24.16}$$

with

$$\boldsymbol{A} = \begin{pmatrix} 1 & -a/x_{0,1} \\ 1 & -a/x_{0,2} \\ \dots & \dots \\ 1 & -a/x_{0,m} \end{pmatrix}, \quad \boldsymbol{\beta}_0 = \begin{pmatrix} \beta_{0,1} \\ \beta_{0,2} \end{pmatrix}, \quad \boldsymbol{\eta}_0 = \begin{pmatrix} y_{0,1} \exp(b/x_{0,1}) - a \\ y_{0,2} \exp(b/x_{0,2}) - a \\ \dots \\ y_{0,m} \exp(b/x_{0,m}) - a \end{pmatrix}.$$

Since

$$\boldsymbol{\eta}_0 = \boldsymbol{0} \tag{24.17}$$

we have

$$\boldsymbol{\beta}_0 = \boldsymbol{0} \tag{24.18}$$

as it should be.

24.4 Iteration

Given the estimators $\bar{a}$, $\bar{b}$ are closer to the true values a, b than the guessed starting values a^*, b^*, it stands to reason to cyclically repeat the adjustment until

$$\bar{\beta}_1 \approx 0 \quad \text{and} \quad \bar{\beta}_2 \approx 0 \, . \tag{24.19}$$

Clearly, if we were to feed in error free data, progressing cycles would steadily contract the estimators $\bar{\beta}_1, \bar{\beta}_2$ down to the numerical uncertainty of the computer. However, as the input data are erroneous, just after a few cycles, the numerical values of the estimators $\bar{\beta}_1, \bar{\beta}_2$ start to oscillate. To recall, the iteration, if convergent, is in no way apt to reduce the effect of the measuring errors—it may, however, reduce the effect of the linearization errors.

24.5 Uncertainties of the Components of the Solution Vector

As the respective starting values a^*, b^* are to be judged error-free, the uncertainties of the estimators $\bar{\beta}_1, \bar{\beta}_2$ pass over to the uncertainties of the estimated parameters $\bar{a}$ and $\bar{b}$,

$$u_{\bar{a}} = u_{\bar{\beta}_1} \,, \quad u_{\bar{b}} = u_{\bar{\beta}_2} \,. \tag{24.20}$$

Random Errors

As (24.14) provides us with repeated measurements, we are not reliant on the minimized sum of squared residuals. Inserting the $\bar{\eta}_i$ into (24.15) issues

$$\begin{aligned} \bar{\beta}_k &= \sum_{i=1}^{m} b_{ik} \bar{\eta}_i \\ &= \sum_{i=1}^{m} b_{ik} \left[\bar{y}_i \exp\left(b^*/x_{0,i} \right) - a^* \right] \\ &= \sum_{i=1}^{m} b_{ik}\, \bar{y}_i \exp\left(b^*/x_{0,i} \right) - a^* \sum_{i=1}^{m} b_{ik} \\ &= \frac{1}{n} \sum_{l=1}^{n} \sum_{i=1}^{m} b_{ik}\, y_{il} \exp\left(b^*/x_{0,i} \right) - a^* \sum_{i=1}^{m} b_{ik} \,; \qquad k = 1, 2 \,. \end{aligned} \tag{24.21}$$

Defining

$$\bar{\beta}_{kl} = \sum_{i=1}^{m} b_{ik}\, y_{il} \exp\left(b^*/x_{0,i} \right) - a^* \sum_{i=1}^{m} b_{ik} \,; \qquad k = 1, 2 \,. \tag{24.22}$$

we arrive at

$$\bar{\beta}_{kl} - \bar{\beta}_k = \sum_{i=1}^{m} b_{ik} \exp\left(b^*/x_{0,i} \right) \left(y_{il} - \bar{y}_i \right) ; \qquad k = 1, 2 \,. \tag{24.23}$$

Hence, the empirical variances and covariances of the solution vector are given by

$$s_{\bar{\beta}_k \bar{\beta}_{k'}} = \frac{1}{n-1} \sum_{l=1}^{n} \left(\bar{\beta}_{kl} - \bar{\beta}_k\right) \left(\bar{\beta}_{k'l} - \bar{\beta}_{k'}\right)$$

$$= \frac{1}{n-1} \sum_{l=1}^{n} \left[\sum_{i=1}^{m} b_{ik} \exp\left(b^*/x_{0,i}\right)(y_{il} - \bar{y}_i) \right] \left[\sum_{j=1}^{m} b_{jk'} \exp\left(b^*/x_{0,j}\right)(y_{jl} - \bar{y}_j) \right]$$

$$= \sum_{i,j=1}^{m} b_{ik} \exp\left(b^*/x_{0,i}\right) b_{jk'} \exp\left(b^*/x_{0,j}\right) s_{ij} \, ; \qquad k, k' = 1, 2, 3 \, .$$

We gather the

$$s_{ij} = \frac{1}{n-1} \sum_{l=1}^{n} (y_{il} - \bar{y}_i)(y_{jl} - \bar{y}_j) \tag{24.24}$$

within the empirical variance–covariance matrix of the input data

$$\boldsymbol{s} = (s_{ij}) \, ; \quad i, j = 1, \ldots, m \, . \tag{24.25}$$

For convenience, we further introduce an auxiliary matrix

$$\boldsymbol{H} = \begin{pmatrix} b_{11} \exp\left(b^*/x_{0,1}\right) & b_{12} \exp\left(b^*/x_{0,1}\right) \\ b_{21} \exp\left(b^*/x_{0,2}\right) & b_{22} \exp\left(b^*/x_{0,2}\right) \\ \ldots & \ldots \\ b_{m1} \exp\left(b^*/x_{0,m}\right) & b_{m2} \exp\left(b^*/x_{0,m}\right) \end{pmatrix}$$

so that

$$\boldsymbol{s}_{\bar{\beta}} = \begin{pmatrix} s_{\bar{\beta}_1 \bar{\beta}_1} & s_{\bar{\beta}_1 \bar{\beta}_2} \\ s_{\bar{\beta}_2 \bar{\beta}_1} & s_{\bar{\beta}_2 \bar{\beta}_2} \end{pmatrix} = \boldsymbol{H}^T \boldsymbol{s} \, \boldsymbol{H} , \qquad s_{\bar{\beta}_k \bar{\beta}_k} \equiv s_{\bar{\beta}_k}^2 \, . \tag{24.26}$$

Systematic Errors

Returning to (24.15), we observe

$$
\begin{aligned}
\bar{\beta}_k &= \sum_{i=1}^{m} b_{ik}\bar{\eta}_i \\
&= \sum_{i=1}^{m} b_{ik}\left[\bar{y}_i \exp\left(b^*/x_{0,i}\right) - a^*\right] \\
&= \sum_{i=1}^{m} b_{ik}\left[\left(y_{0,i} + (\bar{y}_i - \mu_{\bar{y}_i}) + f_{\bar{y}_i}\right) \exp\left(b^*/x_{0,i}\right) - a^*\right] \qquad (24.27)\\
&= \sum_{i=1}^{m} b_{ik}\left[y_{0,i} \exp\left(b^*/x_{0,i}\right) - a^*\right] \\
&+ \sum_{i=1}^{m} b_{ik} \exp\left(b^*/x_{0,i}\right)(\bar{y}_i - \mu_{\bar{y}_i}) + \sum_{i=1}^{m} b_{ik} \exp\left(b^*/x_{0,i}\right) f_{\bar{y}_i}\,; \quad k = 1, 2\,.
\end{aligned}
$$

Obviously, in practice, the terms

$$
\sum_{i=1}^{m} b_{ik}\left[y_{0,i} \exp\left(b^*/x_{0,i}\right) - a^*\right]; \quad k = 1, 2 \qquad (24.28)
$$

are inaccessible. However, should they be sufficiently close to zero, thus corresponding to the homologous true system, the propagated systematic error would be

$$
f_{\bar{\beta}_k} = \sum_{i=1}^{m} b_{ik} \exp\left(b^*/x_{0,i}\right) f_{\bar{y}_i}\,; \qquad k = 1, 2 \qquad (24.29)
$$

with the worst-case estimation

$$
f_{s,\bar{\beta}_k} = \sum_{i=1}^{m} \mid b_{ik} \mid \exp\left(b^*/x_{0,i}\right) f_{s,\bar{y}_i}\,; \qquad k = 1, 2\,. \qquad (24.30)
$$

The unknown, non-zero term (24.28) asks us to rate the uncertainties to come cautiously. For equal systematic errors

$$
f_{\bar{y}_i} = f_y\,, \qquad f_{s,\bar{y}_i} = f_{s,y}\,; \qquad i = 1, \dots m
$$

$$
-f_{s,y} \le f_y \le f_{s,y}
$$

(24.29) gives way to

$$
f_{s,\bar{\beta}_k} = f_{s,y} \sum_{i=1}^{m} \mid b_{ik} \mid \exp\left(b^*/x_{0,i}\right); \qquad k = 1, 2\,. \qquad (24.31)
$$

Overall Uncertainties

From (24.26) and (24.30) we take

$$u_{\bar{\beta}_k} = \frac{t_P(n-1)}{\sqrt{n}} s_{\bar{\beta}_k} + \sum_{i=1}^{m} \mid b_{ik} \mid f_{s,\bar{y}_i} \exp\left(b^*/x_{0,i}\right); \quad k = 1, 2 \tag{24.32}$$

In case of equal systematic errors, due to (24.31), this turns into

$$u_{\bar{\beta}_k} = \frac{t_P(n-1)}{\sqrt{n}} s_{\bar{\beta}_k} + f_{s,y} \sum_{i=1}^{m} \mid b_{ik} \mid \exp\left(b^*/x_{0,i}\right); \quad k = 1, 2 \tag{24.33}$$

Hence, the estimators (24.12) and the uncertainties (24.20) should localize the unknown parameters a and b according to

$$\bar{a} - u_{\bar{a}} \leq a \leq \bar{a} + u_{\bar{a}}\,, \qquad \bar{b} - u_{\bar{b}} \leq b \leq \bar{b} + u_{\bar{b}} \tag{24.34}$$

whereat the fitted function reads

$$y(x) = \bar{a} \exp\left(-\bar{b}/x\right); \quad x > 0\,. \tag{24.35}$$

Remarkably enough, given

$$\bar{a} > a_0\,, \quad \bar{b} > b_0 \qquad \text{or} \qquad \bar{a} < a_0\,, \quad \bar{b} < b_0$$

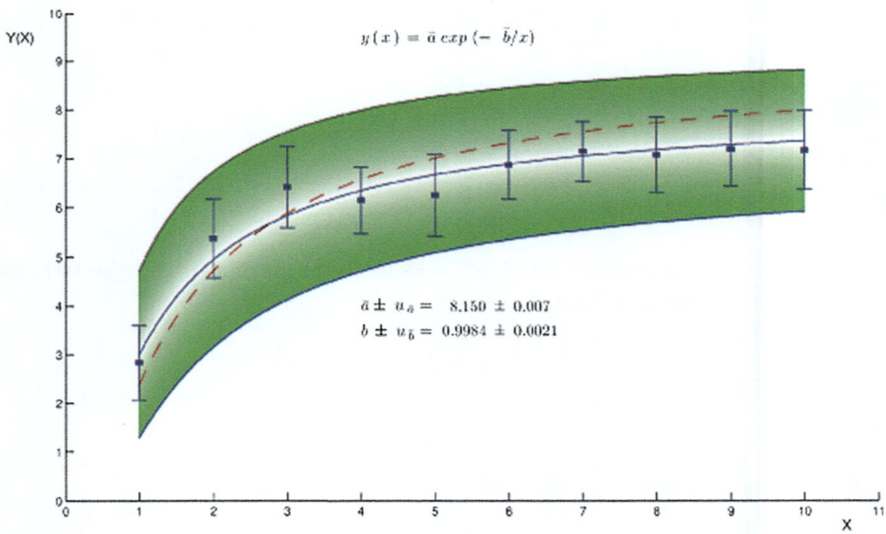

Fig. 24.1. Fitted function $y(x) = \bar{a} \exp(-\,\bar{b}/x)$, uncertainties of the input data, true function $y(x) = a \exp(-\,b/x)$ (*dashed line*), and uncertainty belt according to (24.36); arbitrarily chosen parameters $a = 3e$ and $b = 1$

the fitted curve is slightly revolved with respect to the true curve, either in the one or in the other sense of rotation. If, however,

$$\bar{a} > a_0\,, \quad \bar{b} < b_0 \qquad \text{or} \qquad \bar{a} < a_0\,, \quad \bar{b} > b_0$$

the fitted curve is at least in essence shifted parallel to itself.

In practice, we are unaware of what happens. In order to establish, say, an uncertainty belt which localizes the true function, we insert the pair of results

$$(\bar{a} + u_{\bar{a}}\,, \bar{b} - u_{\bar{b}}) \quad \text{and} \quad (\bar{a} - u_{\bar{a}}\,, \bar{b} + u_{\bar{b}}) \tag{24.36}$$

as implied in (24.34) into (24.35). This produces two boundary curves in-between which we should find the true non-linear function (24.13).

Figure 24.1 displays the true function (24.13), the fitted function (24.35), the uncertainties of the input data (24.7), and the suggested uncertainty belt following (24.36).

25 Transformation

Relationships non-linear with respect to the unknown parameters may be linearized via coordinate transformation. As a matter of course, the least squares estimators do not include linearization errors. However, they now depend on the properties of the transformation. Further, assigning measurement uncertainties, even a two step linearization is to be carried out.

25.1 Homologous True Function

We resume the example as discussed in Chap. 24,

$$\phi = a \exp(-b/T)\,; \quad a, b > 0\,; \quad T > 0\,. \tag{25.1}$$

To linearize, we put

$$y = \ln \phi\,, \quad \beta_1 = \ln a\,, \quad x = 1/T\,, \quad \beta_2 = -b \tag{25.2}$$

giving way to

$$\beta_1 + \beta_2\, x = y\,. \tag{25.3}$$

Considering m data pairs

$$(T_{0,1}, \phi_{0,1}), \quad (T_{0,2}, \phi_{0,2}), \quad \ldots, \quad (T_{0,m}, \phi_{0,m}) \tag{25.4}$$

satisfying

$$\phi_{0,i} = a \exp(-b/T_{0,i})\,; \quad i = 1, \ldots, m$$

the homologous true function, as introduced in the preceding section, reads

$$\beta_1 + \beta_2\, x_{0,i} = y_{0,i}\,; \quad i = 1, \ldots, m\,. \tag{25.5}$$

25.2 Fitting Conditions

Let there be $m > 2$ data pairs

$$(T_{0,1}, \bar{\phi}_1), \quad (T_{0,2}, \bar{\phi}_2), \quad \ldots, \quad (T_{0,m}, \bar{\phi}_m) \tag{25.6}$$

expressing n repeated measurements for each of the $\bar{\phi}_i$

$$\bar{\phi}_i = \frac{1}{n}\sum_{l=1}^{n}\phi_{il} = \phi_{0,i} + (\bar{\phi}_i - \mu_{\bar{\phi}_i}) + f_{\bar{\phi}_i}\,; \quad i = 1,\ldots,m$$

$$E\{\bar{\phi}_i\} = \mu_{\bar{\phi}_i}\,; \qquad -f_{s,\bar{\phi}_i} \le f_{\bar{\phi}_i} \le f_{s,\bar{\phi}_i}\,.$$

The uncertainties, as given by the empirical variances

$$s^2_{\bar{\phi}_i} = \frac{1}{n-1}\sum_{i=1}^{m}(\phi_{il} - \bar{\phi}_i)^2\,; \quad i = 1,\ldots,m \tag{25.7}$$

and the systematic errors $f_{\bar{\phi}_i}$, take the form

$$\bar{\phi}_i \pm u_{\bar{\phi}_i}\,, \quad u_{\bar{\phi}_i} = \frac{t_P(n-1)}{\sqrt{n}}\,s_{\bar{\phi}_i} + f_{s,\bar{\phi}_i}\,; \quad i = 1,\ldots,m\,. \tag{25.8}$$

Moreover, we shall need the uncertainties of the $\bar{y}_i$. We have

$$\begin{aligned} \bar{y}_i &= y_{0,i} + \frac{1}{\bar{\phi}_i}\,(\bar{\phi}_i - \mu_{\bar{\phi}_i}) + \frac{1}{\bar{\phi}_i}\,f_{\bar{\phi}_i}\,, \qquad i = 1,\ldots,m \\ y_{il} &= y_{0,i} + \frac{1}{\bar{\phi}_i}\,(\phi_{il} - \mu_{\bar{\phi}_i}) + \frac{1}{\bar{\phi}_i}\,f_{\bar{\phi}_i}\,, \qquad l = 1,\ldots,n\,. \end{aligned} \tag{25.9}$$

As $\bar{\phi}_i > 0$, we simply write

$$s_{\bar{y}_i} = \frac{1}{\bar{\phi}_i}\,s_{\bar{\phi}_i}\,; \qquad i = 1,\ldots,m\,. \tag{25.10}$$

Combined with the systematic errors $f_{\bar{y}_i} = \dfrac{1}{\bar{\phi}_i}\,f_{\bar{\phi}_i}$ the uncertainties are given by

$$\bar{y}_i \pm u_{\bar{y}_i}\,, \quad u_{\bar{y}_i} = \frac{t_P(n-1)}{\sqrt{n}}\,s_{\bar{y}_i} + f_{s,\bar{y}_i}\,; \quad i = 1,\ldots,m\,.$$

25.3 Orthogonal Projection

Written in matrices the inconsistent, over-determined, linear system

$$\beta_1 + \beta_2\,x_{0,i} \approx \bar{y}_i\,; \qquad i = 1,\ldots,m \tag{25.11}$$

reads

$$\boldsymbol{A\beta} \approx \bar{\boldsymbol{y}}$$

where

$$\boldsymbol{A} = \begin{pmatrix} 1 & x_{0,1} \\ 1 & x_{0,2} \\ \dots & \dots \\ 1 & x_{0,m} \end{pmatrix}, \qquad \boldsymbol{\beta} = \begin{pmatrix} \beta_1 \\ \beta_2 \end{pmatrix}, \qquad \bar{\boldsymbol{y}} = \begin{pmatrix} \bar{y}_1 \\ \bar{y}_2 \\ \dots \\ \bar{y}_m \end{pmatrix}.$$

The orthogonal projection produces

$$\bar{\boldsymbol{\beta}} = \boldsymbol{B}^{\mathrm{T}} \bar{\boldsymbol{y}}, \qquad \bar{\beta}_k = \sum_{i=1}^{m} b_{ik} \bar{y}_i ; \qquad k = 1, 2 \tag{25.12}$$

with

$$\boldsymbol{B} = \boldsymbol{A} \left(\boldsymbol{A}^{\mathrm{T}} \boldsymbol{A} \right)^{-1} = (b_{ik}) ; \quad i = 1, \dots, m; \quad k = 1, 2 .$$

The components $\bar{\beta}_k$; $k = 1, 2$ of the solution vector $\bar{\boldsymbol{\beta}}$ constitute the least squares line

$$y(x) = \bar{\beta}_1 + \bar{\beta}_2 \, x . \tag{25.13}$$

Ultimately, there are estimators

$$\bar{a} = e^{\bar{\beta}_1} , \quad \bar{b} = -\bar{\beta}_2 \tag{25.14}$$

for the unknown parameters a, b.

25.4 Uncertainties of the Components of the Solution Vector

As has been discussed in Sect.7.1, the uncertainties of the estimators $\bar{a}$ and $\bar{b}$ turn out to be

$$u_{\bar{a}} = \mathrm{e}^{\bar{\beta}_1} u_{\bar{\beta}_1} , \qquad u_{\bar{b}} = u_{\bar{\beta}_2} . \tag{25.15}$$

Random Errors

Again we are not reliant on the minimized sum of squared residuals in order to assess the influence of random errors. Rather, the components $\bar{\beta}_k$ of the solution vector (25.12)

$$\begin{aligned} \bar{\beta}_k &= \sum_{i=1}^{m} b_{ik} \left[\frac{1}{n} \sum_{l=1}^{n} y_{il} \right] = \frac{1}{n} \sum_{l=1}^{n} \left[\sum_{i=1}^{m} b_{ik} y_{il} \right] \\ &= \frac{1}{n} \sum_{l=1}^{n} \bar{\beta}_{kl} ; \quad k = 1, 2 \end{aligned} \tag{25.16}$$

produce the differences

$$\bar{\beta}_{kl} - \bar{\beta}_k = \sum_{i=1}^{m} b_{ik}(y_{il} - \bar{y}_i)$$
$$= \sum_{i=1}^{m} b_{ik}(\frac{\phi_{il} - \bar{\phi}_i}{\bar{\phi}_i})$$

which devise the empirical variance–covariance matrix

$$\boldsymbol{s}_{\bar{\beta}} = \begin{pmatrix} s_{\bar{\beta}_1\bar{\beta}_1} & s_{\bar{\beta}_1\bar{\beta}_2} \\ s_{\bar{\beta}_2\bar{\beta}_1} & s_{\bar{\beta}_2\bar{\beta}_2} \end{pmatrix}, \qquad s_{\bar{\beta}_k\bar{\beta}_k} \equiv s^2_{\bar{\beta}_k}\,.$$

With

$$s_{ij} = \frac{1}{n-1}\sum_{l=1}^{n}(y_{il} - \bar{y}_i)(y_{jl} - \bar{y}_j)\,; \qquad i,j = 1,\dots,m$$
$$= \frac{1}{n-1}\sum_{l=1}^{n}\frac{(\phi_{il} - \bar{\phi}_i)}{\bar{\phi}_i}\,\frac{(\phi_{jl} - \bar{\phi}_j)}{\bar{\phi}_j}$$

and

$$\boldsymbol{s} = (s_{ij})\,; \quad i,j = 1,\dots,m\,.$$

designating the empirical variance–covariance matrix of the input data, we have

$$\boldsymbol{s}_{\bar{\beta}} = \boldsymbol{B}^T \boldsymbol{s}\,\boldsymbol{B}\,. \tag{25.17}$$

Systematic Errors

The formal decompositions

$$\bar{\beta}_k = \sum_{i=1}^{m} b_{ik}\,[y_{0,i} + \frac{1}{\bar{\phi}_i}\,(\bar{\phi}_i - \mu_{\bar{\phi}_i}) + \frac{1}{\bar{\phi}_i}\,f_{\bar{\phi}_i}]\,; \quad k = 1,2 \tag{25.18}$$

issue the propagated systematic errors

$$f_{\bar{\beta}_k} = \sum_{i=1}^{m} b_{ik}\frac{1}{\bar{\phi}_i}\,f_{\bar{\phi}_i}\,; \quad k = 1,2 \tag{25.19}$$

their worst-case estimations being

$$f_{s,\bar{\beta}_k} = \sum_{i=1}^{m} \mid b_{ik} \mid \frac{1}{\bar{\phi}_i} f_{s,\bar{\phi}_i}\,; \quad k = 1,2\,. \tag{25.20}$$

In case of equal systematic errors

$$f_{\bar{\phi}_i} = f_\phi\,, \qquad f_{s,\bar{\phi}_i} = f_{s,\phi}\,; \qquad i = 1, \ldots m$$
$$-f_{s,\phi} \le f_\phi \le f_{s,\phi} \tag{25.21}$$

we have

$$f_{s,\bar{\beta}_k} = f_{s,\phi} \sum_{i=1}^{m} \mid b_{ik} \mid \frac{1}{\bar{\phi}_i} \qquad k = 1, 2\,. \tag{25.22}$$

Overall Uncertainties

Ultimately, combining the square roots of the diagonal elements of (25.17) either with (25.20) or with (25.22) brings forth

$$u_{\bar{\beta}_k} = \frac{t_P(n-1)}{\sqrt{n}}\, s_{\bar{\beta}_k} + f_{s,\bar{\beta}_k}\,; \qquad k = 1, 2\,. \tag{25.23}$$

Following (25.14) and (25.15) the results take the form

$$\bar{a} \pm u_{\bar{a}}\,; \qquad u_{\bar{a}} = \mathrm{e}^{\bar{\beta}_1} u_{\bar{\beta}_1}$$
$$\bar{b} \pm u_{\bar{b}}\,; \qquad u_{\bar{b}} = u_{\bar{\beta}_2} \tag{25.24}$$

where the fitted function itself is given by

$$\phi(T) = \bar{a}\; exp\,(-\,\bar{b}/T)\,. \tag{25.25}$$

Just to have an impression about the region which localizes the true function, we may consider a pair of functions deploying the set of estimators

$$(\bar{a} - u_{\bar{a}}\,,\, \bar{b} + u_{\bar{b}}) \quad \text{and} \quad (\bar{a} - u_{\bar{a}}\,,\, \bar{b} - u_{\bar{b}})\,. \tag{25.26}$$

In a sense, we might see them to constitute something like an uncertainty belt.

Figure 25.1 reverts to the data set used in Fig. 24.1 and displays the fitted function (25.25), the true function (25.1), the uncertainties of the input data and, finally, the uncertainty belt according to (25.26).

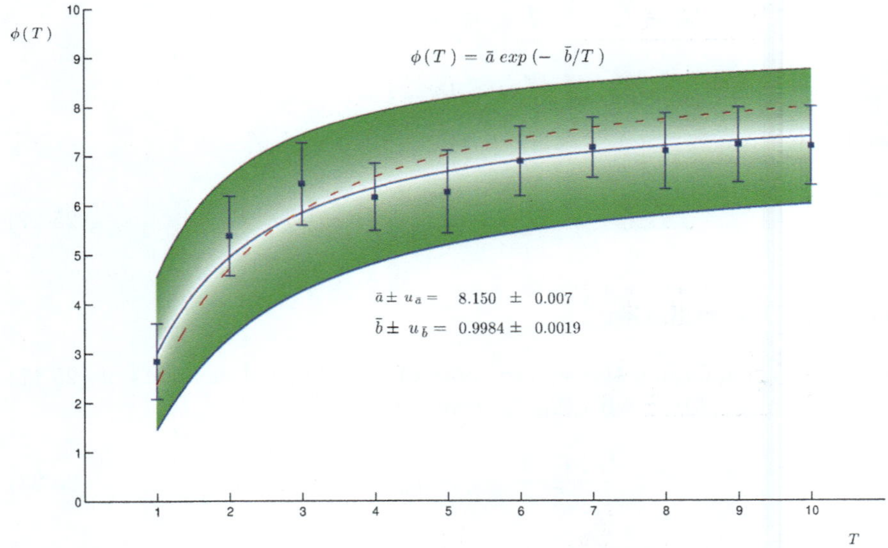

Fig. 25.1. Fitted function $\phi(T) = \bar{a}\,exp\,(-\,\bar{b}/T)$, true function $\phi(T) = a\,exp\,(-\,b/T)$ (*dashed line*), uncertainties of the input data, and uncertainty belt following (25.26)

Part IX

Appendices

A Graphical Scale Transformations

The graphical visualization of uncertainty statements present us with a hitch. As we commonly assume relative uncertainties of, say, 10^{-3} and less, our eyes are not in a position to resolve the details. A way out is to introduce graphical expansion factors.

Mean Values

Let us graphically expand the uncertainty $u_{\bar{x}}$ of the arithmetic mean $\bar{x}$

$$\bar{x} \pm u_{\bar{x}} \,.$$

To have a beneficial vista, instead of the metrologically defined result, we display

$$\bar{x} \pm V u_{\bar{x}} \,; \qquad V \gg 1 \,. \tag{A.1}$$

The factor V expands the primary uncertainty region with reference to the estimator $\bar{x}$. By means of data simulations, we constantly keep track of the true values of the measurands. Equation (A.1) tells us that any number lying within the expanded uncertainty region may symbolize the true value. Thus, in order to have a meaningful exposure the graph has, in a formal sense as a matter of course, to shift the true value x_0 of the mean $\bar{x}$ according to

$$x_0^* = \bar{x} + (x_0 - \bar{x})V \,; \qquad V \gg 1 \tag{A.2}$$

Putting $V = 1$, we retrace $x_0^* = x_0$.

A similar transformation applies to the expectation $E\{\bar{X}\} = \mu_{\bar{x}}$ of the random variable $\bar{X}$

$$\mu_{\bar{x}}^* = \bar{x} + (\mu_{\bar{x}} - \bar{x})V \,; \qquad V \gg 1 \,. \tag{A.3}$$

Again, $V = 1$ recovers $\mu_{\bar{x}}^* = \mu_{\bar{x}}$.

Straight Lines

Assume the graphical display of the uncertainty band of a least squares line to be too narrow to be visually resolved. Hence, instead of

$$\bar{y}(x) \pm u_{\bar{y}(x)}$$

we depict

$$\bar{y}(x) \pm V u_{\bar{y}(x)}\,; \qquad V \gg 1\,. \tag{A.4}$$

The expansion prevents the two branches of the uncertainty region to pretendedly match the fitted straight line

$$\bar{y}(x) = \bar{\beta}_1 + \bar{\beta}_2\, x\,. \tag{A.5}$$

The expansion of the uncertainty band asks us to formally readjust the position of the true straight line

$$y_0(x) = \beta_{0,1} + \beta_{0,2}\, x\,. \tag{A.6}$$

For any fixed x, any number lying in between $\bar{y}(x) - u_{\bar{y}(x)} \quad \ldots \quad \bar{y}(x) + u_{\bar{y}(x)}$ might be the true value $y_0(x)$. Hence, the true straight line is to be transformed according to

$$y_0^*(x) = \bar{y}(x) + (y_0(x) - \bar{y}(x))V\,. \tag{A.7}$$

Obviously, $V = 1$ yields $y_0^*(x) = y_0(x)$. Inserting (A.5) and (A.6), we have

$$\begin{aligned} y_0^*(x) &= \bar{\beta}_1 + \bar{\beta}_2\, x + (\beta_{0,1} + \beta_{0,2}\, x - \bar{\beta}_1 - \bar{\beta}_2\, x)V \\ &= [\,\bar{\beta}_1 + (\beta_{0,1} - \bar{\beta}_1)V\,] + [\,\bar{\beta}_2 + (\beta_{0,2} - \bar{\beta}_2)V\,]\, x\,. \end{aligned} \tag{A.8}$$

Hence, we display the true straight line following

$$y_0^*(x) = \beta_{0,1}^* + \beta_{0,2}^*\, x \tag{A.9}$$

with coefficients

$$\begin{aligned} \beta_{0,1}^* &= \bar{\beta}_1 + (\beta_{0,1} - \bar{\beta}_1)V \\ \beta_{0,2}^* &= \bar{\beta}_2 + (\beta_{0,2} - \bar{\beta}_2)V\,. \end{aligned} \tag{A.10}$$

To transform the expectations $E\{\bar{\beta}_1\} = \mu_{\bar{\beta}_1}$ and $E\{\bar{\beta}_2\} = \mu_{\bar{\beta}_2}$, we start from (A.5)

$$\mu_{\bar{y}(x)} = \mu_{\bar{\beta}_1} + \mu_{\bar{\beta}_2}\, x\,. \tag{A.11}$$

Equation (A.7) tells us

$$\mu_{\bar{y}(x)}^* = \bar{y}(x) + (\mu_{\bar{y}(x)} - \bar{y}(x))V\,. \tag{A.12}$$

Thus, inserting (A.5) and (A.11), we find

$$\begin{aligned} \mu_{\bar{\beta}_1}^* &= \bar{\beta}_1 + (\mu_{\bar{\beta}_1} - \bar{\beta}_1)V \\ \mu_{\bar{\beta}_2}^* &= \bar{\beta}_2 + (\mu_{\bar{\beta}_2} - \bar{\beta}_2)V\,. \end{aligned} \tag{A.13}$$

Planes

Instead of

$$\bar{z}(x,y) \pm u_{\bar{z}(x,y)}$$

the diagram is intended to display

$$\bar{z}(x,y) \pm V u_{\bar{z}(x,y)}\,; \qquad V \gg 1\,. \tag{A.14}$$

By necessity, the expansion has to take reference to the fitted plane

$$\bar{z}(x,y) = \bar{\beta}_1 + \bar{\beta}_2\, x - \bar{\beta}_3\, y\,. \tag{A.15}$$

Initially, we readjust the true plane

$$z_0(x,y) = \beta_{0,1} + \beta_{0,2}\, x + \beta_{0,3}\, y\,. \tag{A.16}$$

For any fixed point x, y, any number lying within $\bar{z}(x,y) - u_{\bar{z}(x,y)} \;\ldots\; \bar{z}(x,y) + u_{\bar{z}(x,y)}$ might symbolize the true value $z(x_0, y_0)$. Hence, we put

$$z_0^*(x,y) = \bar{z}(x,y) + (z_0(x,y) - \bar{z}(x,y))V\,. \tag{A.17}$$

$V = 1$ reproduces $z_0^*(x,y) = z_0(x,y)$. Inserting (A.15) and (A.16) we find

$$\begin{aligned} z_0^*(x,y) &= \bar{\beta}_1 + \bar{\beta}_2\, x + \bar{\beta}_3\, y \\ &\quad + (\beta_{0,1} + \beta_{0,2}\, x + \beta_{0,3}\, y - \bar{\beta}_1 - \bar{\beta}_2\, x - \bar{\beta}_3\, y)V \\ &= [\,\bar{\beta}_1 + (\beta_{0,1} - \bar{\beta}_1)V\,] + [\,\bar{\beta}_2 + (\beta_{0,2} - \bar{\beta}_2)V\,]\,x \\ &\quad + [\,\bar{\beta}_3 + (\beta_{0,3} - \bar{\beta}_3)V\,]\,y\,. \end{aligned} \tag{A.18}$$

Thus, for graphical ends the true plane reads

$$z_0^*(x,y) = \beta_{0,1}^* + \beta_{0,2}^*\, x + \beta_{0,2}^*\, y \tag{A.19}$$

with coefficients

$$\begin{aligned} \beta_{0,1}^* &= \bar{\beta}_1 + (\beta_{0,1} - \bar{\beta}_1)V \\ \beta_{0,2}^* &= \bar{\beta}_2 + (\beta_{0,2} - \bar{\beta}_2)V \\ \beta_{0,3}^* &= \bar{\beta}_3 + (\beta_{0,3} - \bar{\beta}_3)V\,. \end{aligned} \tag{A.20}$$

We also transform the expectations $E\{\bar{\beta}_1\} = \mu_{\bar{\beta}_1}$, $E\{\bar{\beta}_2\} = \mu_{\bar{\beta}_2}$ and $E\{\bar{\beta}_3\} = \mu_{\bar{\beta}_3}$. From (A.15) we take

$$\mu_{\bar{z}(x,y)} = \mu_{\bar{\beta}_1} + \mu_{\bar{\beta}_2}\, x + \mu_{\bar{\beta}_3}\, y\,. \tag{A.21}$$

But then

$$\mu^*_{\bar{z}(x,y)} = \bar{z}(x,y) + (\mu_{\bar{z}(x,y)} - \bar{z}(x,y))V \tag{A.22}$$

yields

$$\begin{aligned} \mu^*_{\bar{\beta}_1} &= \bar{\beta}_1 + (\mu_{\bar{\beta}_1} - \bar{\beta}_1)V \\ \mu^*_{\bar{\beta}_2} &= \bar{\beta}_2 + (\mu_{\bar{\beta}_2} - \bar{\beta}_2)V \\ \mu^*_{\bar{\beta}_3} &= \bar{\beta}_3 + (\mu_{\bar{\beta}_3} - \bar{\beta}_3)V\,. \end{aligned} \tag{A.23}$$

Parabolas

Instead of

$$\bar{y}(x) \pm u_{\bar{y}(x)}$$

we wish to display

$$\bar{y}(x) \pm V u_{\bar{y}(x)}\,; \qquad V \gg 1 \tag{A.24}$$

where the least squares parabola

$$\bar{y}(x) = \bar{\beta}_1 + \bar{\beta}_2\, x + \bar{\beta}_3\, x^2 \tag{A.25}$$

serves as a reference. The scale transformations shifts the true parabola

$$y_0(x) = \beta_{0,1} + \beta_{0,2}\, x + \beta_{0,3}\, x^2 \tag{A.26}$$

into

$$y^*_0(x) = \bar{y}(x) + (y_0(x) - \bar{y}(x))V \tag{A.27}$$

yielding

$$\begin{aligned} y^*_0(x) &= \bar{\beta}_1 + \bar{\beta}_2\, x + \bar{\beta}_3\, x^2 \\ &+ (\beta_{0,1} + \beta_{0,2}\, x + \beta_{0,3}\, x^2 - \bar{\beta}_1 - \bar{\beta}_2\, x - \bar{\beta}_3\, x^2)V \\ &= [\,\bar{\beta}_1 + (\beta_{0,1} - \bar{\beta}_1)V\,] + [\,\bar{\beta}_2 + (\beta_{0,2} - \bar{\beta}_2)V\,]\, x \\ &+ [\,\bar{\beta}_3 + (\beta_{0,3} - \bar{\beta}_3)V,]\, x^2\,. \end{aligned} \tag{A.28}$$

Hence, the coefficients of the readjusted true parabola

$$y^*_0(x) = \beta^*_{0,1} + \beta^*_{0,2}\, x + \beta^*_{0,3}\, x^2 \tag{A.29}$$

turn out to be

$$\begin{aligned}\beta_{0,1}^* &= \bar{\beta}_1 + (\beta_{0,1} - \bar{\beta}_1)V \\ \beta_{0,2}^* &= \bar{\beta}_2 + (\beta_{0,2} - \bar{\beta}_2)V \\ \beta_{0,3}^* &= \bar{\beta}_3 + (\beta_{0,3} - \bar{\beta}_3)V\,.\end{aligned} \tag{A.30}$$

Finally, we readjust the expectations $E\{\bar{\beta}_1\} = \mu_{\bar{\beta}_1}$, $E\{\bar{\beta}_2\} = \mu_{\bar{\beta}_2}$ and $E\{\bar{\beta}_3\} = \mu_{\bar{\beta}_3}$. (A.25) produces

$$\mu_{\bar{y}(x)} = \mu_{\bar{\beta}_1} + \mu_{\bar{\beta}_2}\,x + \mu_{\bar{\beta}_3}\,x^2 \tag{A.31}$$

As

$$\mu_{\bar{y}(x)}^* = \bar{y}(x) + (\mu_{\bar{y}(x)} - \bar{y}(x))V \tag{A.32}$$

we find, inserting (A.25) and (A.31),

$$\begin{aligned}\mu_{\bar{\beta}_1}^* &= \bar{\beta}_1 + (\mu_{\bar{\beta}_1} - \bar{\beta}_1)V \\ \mu_{\bar{\beta}_2}^* &= \bar{\beta}_2 + (\mu_{\bar{\beta}_2} - \bar{\beta}_2)V \\ \mu_{\bar{\beta}_3}^* &= \bar{\beta}_3 + (\mu_{\bar{\beta}_3} - \bar{\beta}_3)V\,.\end{aligned} \tag{A.33}$$

B Expansion of Solution Vectors

Given the abscissas as well as ordinates are considered erroneous, the fitting of geometrical objects implies erroneous design matrices. Then, error propagation requires us to carry out series expansions of the solution vectors.

Straight Lines – Section 15.3

The series expansions of the components $\bar{\beta}_1$, $\bar{\beta}_2$ produce coefficients

$$c_{i,1} = \frac{1}{D}\left[2\bar{x}_i\sum_{j=1}^{m}\bar{y}_j - \bar{y}_i\sum_{j=1}^{m}\bar{x}_j - \sum_{j=1}^{m}\bar{x}_j\bar{y}_j\right] - \frac{2\bar{\beta}_1}{D}\left[m\bar{x}_i - \sum_{j=1}^{m}\bar{x}_j\right]$$

$$c_{i+m,1} = \frac{1}{D}\left[\sum_{j=1}^{m}\bar{x}_j^2 - \bar{x}_i\sum_{j=1}^{m}\bar{x}_j\right]$$

$$c_{i,2} = \frac{1}{D}\left[-\sum_{j=1}^{m}\bar{y}_j + m\bar{y}_i\right] - \frac{2\bar{\beta}_2}{D}\left[m\bar{x}_i - \sum_{j=1}^{m}\bar{x}_j\right]$$

$$c_{i+m,2} = \frac{1}{D}\left[-\sum_{j=1}^{m}\bar{x}_j + m\bar{x}_i\right]\,; \qquad i = 1,\dots,m\,.$$

For convenience, the coefficients have been gathered within an auxiliary matrix as stated in (15.9).

Planes – Section 19.3

The design matrix

$$\boldsymbol{A} = \begin{pmatrix} 1 & \bar{x}_1 & \bar{y}_1 \\ 1 & \bar{x}_2 & \bar{y}_2 \\ \dots & \dots & \dots \\ 1 & \bar{x}_m & \bar{y}_m \end{pmatrix}$$

produces

$$\boldsymbol{A}^{\mathrm{T}}\boldsymbol{A} = \begin{pmatrix} H_{11} & H_{12} & H_{13} \\ H_{21} & H_{22} & H_{23} \\ H_{31} & H_{32} & H_{33} \end{pmatrix}$$

with elements

$$H_{11} = m\,, \qquad H_{22} = \sum_{j=1}^{m} \bar{x}_j^2\,, \qquad H_{33} = \sum_{j=1}^{m} \bar{y}_j^2$$

$$H_{12} = H_{21} = \sum_{j=1}^{m} \bar{x}_j\,, \quad H_{13} = H_{31} = \sum_{j=1}^{m} \bar{y}_j\,, \quad H_{23} = H_{32} = \sum_{j=1}^{m} \bar{x}_j \bar{y}_j\,.$$

Let

$$\begin{aligned} D = H_{11}\left(H_{22}H_{33} - H_{23}^2\right) + H_{12}\left(H_{31}H_{23} - H_{33}H_{12}\right) \\ +H_{13}\left(H_{21}H_{32} - H_{13}H_{22}\right)\,. \end{aligned}$$

denote the determinant of $\boldsymbol{A}^{\mathrm{T}}\boldsymbol{A}$. With this, the inverse of $\boldsymbol{A}^{\mathrm{T}}\boldsymbol{A}$ reads

$$\left(\boldsymbol{A}^{\mathrm{T}}\boldsymbol{A}\right)^{-1} = \frac{1}{D}\left(\begin{array}{c|c|c} H_{22}H_{33} - H_{23}^2 & H_{23}H_{13} - H_{12}H_{33} & H_{12}H_{23} - H_{22}H_{13} \\ H_{23}H_{13} - H_{12}H_{33} & H_{11}H_{33} - H_{13}^2 & H_{12}H_{13} - H_{11}H_{23} \\ H_{12}H_{23} - H_{22}H_{13} & H_{12}H_{13} - H_{11}H_{23} & H_{11}H_{22} - H_{12}^2 \end{array}\right)\,.$$

Putting $\boldsymbol{B} = \boldsymbol{A}\left(\boldsymbol{A}^{\mathrm{T}}\boldsymbol{A}\right)^{-1}$, the components of $\bar{\boldsymbol{\beta}} = \boldsymbol{B}^{\mathrm{T}}\bar{\boldsymbol{z}}$ take the form

$$\bar{\beta}_1 = \sum_{j=1}^{m} b_{j1}\bar{z}_j\,, \quad \bar{\beta}_2 = \sum_{j=1}^{m} b_{j2}\bar{z}_j\,, \quad \bar{\beta}_3 = \sum_{j=1}^{m} b_{j3}\bar{z}_j\,,$$

more detailed

$$\begin{aligned} \bar{\beta}_1 = \frac{1}{D}\sum_{j=1}^{m} \Big\{ &\left[H_{22}H_{33} - H_{23}^2\right] + \left[H_{23}H_{13} - H_{12}H_{33}\right]\bar{x}_j \\ &+ \left[H_{12}H_{23} - H_{22}H_{13}\right]\bar{y}_j \Big\}\,\bar{z}_j \end{aligned}$$

$$\begin{aligned} \bar{\beta}_2 = \frac{1}{D}\sum_{j=1}^{m} \Big\{ &\left[H_{23}H_{13} - H_{12}H_{33}\right] + \left[H_{11}H_{33} - H_{13}^2\right]\bar{x}_j \\ &+ \left[H_{12}H_{13} - H_{11}H_{23}\right]\bar{y}_j \Big\}\,\bar{z}_j \end{aligned}$$

$$\begin{aligned} \bar{\beta}_3 = \frac{1}{D}\sum_{j=1}^{m} \Big\{ &\left[H_{12}H_{23} - H_{22}H_{13}\right] + \left[H_{12}H_{13} - H_{11}H_{23}\right]\bar{x}_j \\ &+ \left[H_{11}H_{22} - H_{12}^2\right]\bar{y}_j \Big\}\,\bar{z}_j\,. \end{aligned}$$

The partial derivatives of the determinant are issued as

$$\begin{aligned}
\frac{\partial D}{\partial \bar{x}_i} &= 2\left[H_{11}H_{33} - H_{13}^2\right]\bar{x}_i + 2\left[H_{12}H_{13} - H_{11}H_{23}\right]\bar{y}_i \\
&\quad +2\left[H_{13}H_{23} - H_{12}H_{33}\right] \\
\frac{\partial D}{\partial \bar{y}_i} &= 2\left[H_{12}H_{13} - H_{11}H_{23}\right]\bar{x}_i + 2\left[H_{11}H_{22} - H_{12}^2\right]\bar{y}_i \\
&\quad +2\left[H_{12}H_{23} - H_{13}H_{22}\right] \\
\frac{\partial D}{\partial \bar{z}_i} &= 0\,.
\end{aligned}$$

Differentiating each of the components $\bar{\beta}_k$; $k = 1, 2, 3$ with respect to x, y, and z, we find

$$\begin{aligned}
c_{i1} = \frac{\partial \bar{\beta}_1}{\partial \bar{x}_i} &= -\frac{\bar{\beta}_1}{D}\frac{\partial D}{\partial \bar{x}_i} \\
&+\frac{1}{D}\left\{2\left(H_{33}\bar{x}_i - H_{23}\bar{y}_i\right)\sum_{j=1}^{m}\bar{z}_j + \left(H_{13}H_{23} - H_{12}H_{33}\right)\bar{z}_i\right. \\
&\left. +\left(H_{13}\bar{y}_i - H_{33}\right)\sum_{j=1}^{m}\bar{x}_j\bar{z}_j + \left(H_{23} + H_{12}\bar{y}_i - 2H_{13}\bar{x}_i\right)\sum_{j=1}^{m}\bar{y}_j\bar{z}_j\right\}
\end{aligned}$$

$$\begin{aligned}
c_{i+m,1} = \frac{\partial \bar{\beta}_1}{\partial \bar{y}_i} &= -\frac{\bar{\beta}_1}{D}\frac{\partial D}{\partial \bar{y}_i} \\
&+\frac{1}{D}\left\{2\left(H_{22}\bar{y}_i - H_{23}\bar{x}_i\right)\sum_{j=1}^{m}\bar{z}_j + \left(H_{12}H_{23} - H_{13}H_{22}\right)\bar{z}_i\right. \\
&\left. +\left(H_{23} + H_{13}\bar{x}_i - 2H_{12}\bar{y}_i\right)\sum_{j=1}^{m}\bar{x}_j\bar{z}_j + \left(H_{12}\bar{x}_i - H_{22}\right)\sum_{j=1}^{m}\bar{y}_j\bar{z}_j\right\} \\
c_{i+2m,1} = \frac{\partial \bar{\beta}_1}{\partial \bar{z}_i} &= b_{i1} \\
c_{i2} = \frac{\partial \bar{\beta}_2}{\partial \bar{x}_i} &= -\frac{\bar{\beta}_2}{D}\frac{\partial D}{\partial \bar{x}_i} + \frac{1}{D}\left\{\left(H_{13}\bar{y}_i - H_{33}\right)\sum_{j=1}^{m}\bar{z}_j\right. \\
&\left. +\left(H_{11}H_{33} - H_{13}^2\right)\bar{z}_i + \left(H_{13} - H_{11}\bar{y}_i\right)\sum_{j=1}^{m}\bar{y}_j\bar{z}_j\right\}
\end{aligned}$$

$$c_{i+m,2} = \frac{\partial \bar{\beta}_2}{\partial \bar{y}_i} = -\frac{\bar{\beta}_2}{D}\frac{\partial D}{\partial \bar{y}_i} + \frac{1}{D}\Bigg\{(H_{23} + H_{13}\bar{x}_i - 2H_{12}\bar{y}_i)\sum_{j=1}^{m}\bar{z}_j$$

$$+ (H_{12}H_{13} - H_{11}H_{23})\,\bar{z}_i + 2\,(H_{11}\bar{y}_i - H_{13})\sum_{j=1}^{m}\bar{x}_j\bar{z}_j$$

$$+ (H_{12} - H_{11}\bar{x}_i)\sum_{j=1}^{m}\bar{y}_j\bar{z}_j\Bigg\}$$

$$c_{i+2m,2} = \frac{\partial \bar{\beta}_2}{\partial \bar{z}_i} = b_{i2}$$

$$c_{i3} = \frac{\partial \bar{\beta}_3}{\partial \bar{x}_i} = -\frac{\bar{\beta}_3}{D}\frac{\partial D}{\partial \bar{x}_i} + \frac{1}{D}\Bigg\{(H_{23} + H_{12}\bar{y}_i - 2H_{13}\bar{x}_i)\sum_{j=1}^{m}\bar{z}_j$$

$$+ (H_{12}H_{13} - H_{11}H_{23})\,\bar{z}_i$$

$$+ (H_{13} - H_{11}\bar{y}_i)\sum_{j=1}^{m}\bar{x}_j\bar{z}_j + 2\,(H_{11}\bar{x}_i - H_{12})\sum_{j=1}^{m}\bar{y}_j\bar{z}_j\Bigg\}$$

$$c_{i+m,3} = \frac{\partial \bar{\beta}_3}{\partial \bar{y}_i} = -\frac{\bar{\beta}_3}{D}\frac{\partial D}{\partial \bar{y}_i} + \frac{1}{D}\Bigg\{(H_{12}\bar{x}_i - H_{22})\sum_{j=1}^{m}\bar{z}_j$$

$$+ \left(H_{11}H_{22} - H_{12}^2\right)\bar{z}_i + (H_{12} - H_{11}\bar{x}_i)\sum_{j=1}^{m}\bar{x}_j\bar{z}_j\Bigg\}$$

$$c_{i+2m,3} = \frac{\partial \bar{\beta}_3}{\partial \bar{z}_i} = b_{i3}\,.$$

Parabolas – Section 23.4

The design matrix

$$\boldsymbol{A} = \begin{pmatrix} 1 & \bar{x}_1 & \bar{x}_1^2 \\ 1 & \bar{x}_2 & \bar{x}_2^2 \\ \dots & \dots & \dots \\ 1 & \bar{x}_m & \bar{x}_m^2 \end{pmatrix}$$

produces

$$\boldsymbol{A}^{\mathrm{T}}\boldsymbol{A} = \begin{pmatrix} H_{11} & H_{12} & H_{13} \\ H_{21} & H_{22} & H_{23} \\ H_{31} & H_{32} & H_{33} \end{pmatrix}$$

with elements

$$H_{11} = m\,, \qquad H_{22} = \sum_{j=1}^{m} \bar{x}_j^2\,, \qquad H_{33} = \sum_{j=1}^{m} \bar{x}_j^4$$

$$H_{12} = H_{21} = \sum_{j=1}^{m} \bar{x}_j\,, \quad H_{13} = H_{31} = \sum_{j=1}^{m} \bar{x}_j^2\,, \quad H_{23} = H_{32} = \sum_{j=1}^{m} \bar{x}_j^3\,.$$

Denoting the determinant of $\boldsymbol{A}^{\mathrm{T}}\boldsymbol{A}$ by

$$\begin{aligned} D = {} & H_{11}\left(H_{22}H_{33} - H_{23}^2\right) + H_{12}\left(H_{31}H_{23} - H_{33}H_{12}\right) \\ & + H_{13}\left(H_{21}H_{32} - H_{13}H_{22}\right) \end{aligned}$$

we have

$$\left(\boldsymbol{A}^{\mathrm{T}}\boldsymbol{A}\right)^{-1} = \frac{1}{D}\left(\begin{array}{c|c|c} H_{22}H_{33} - H_{23}^2 & H_{23}H_{13} - H_{12}H_{33} & H_{12}H_{23} - H_{22}H_{13} \\ H_{23}H_{13} - H_{12}H_{33} & H_{11}H_{33} - H_{13}^2 & H_{12}H_{13} - H_{11}H_{23} \\ H_{12}H_{23} - H_{22}H_{13} & H_{12}H_{13} - H_{11}H_{23} & H_{11}H_{22} - H_{12}^2 \end{array}\right).$$

Putting $\boldsymbol{B} = \boldsymbol{A}(\boldsymbol{A}^{\mathrm{T}}\boldsymbol{A})^{-1}$ the components of $\bar{\boldsymbol{\beta}} = \boldsymbol{B}^{\mathrm{T}}\bar{\boldsymbol{y}}$ take the form

$$\bar{\beta}_1 = \sum_{j=1}^{m} b_{j1}\bar{y}_j\,, \quad \bar{\beta}_2 = \sum_{j=1}^{m} b_{j2}\bar{y}_j\,, \quad \bar{\beta}_3 = \sum_{j=1}^{m} b_{j3}\bar{y}_j$$

i.e.

$$\begin{aligned} \bar{\beta}_1 = {} & \frac{1}{D}\sum_{j=1}^{m}\left\{\left[H_{22}H_{33} - H_{23}^2\right] + \left[H_{23}H_{13} - H_{12}H_{33}\right]\bar{x}_j \right. \\ & \left. + \left[H_{12}H_{23} - H_{22}H_{13}\right]\bar{x}_j^2\right\}\bar{y}_j \\ \bar{\beta}_2 = {} & \frac{1}{D}\sum_{j=1}^{m}\left\{\left[H_{23}H_{13} - H_{12}H_{33}\right] + \left[H_{11}H_{33} - H_{13}^2\right]\bar{x}_j \right. \\ & \left. + \left[H_{12}H_{13} - H_{11}H_{23}\right]\bar{x}_j^2\right\}\bar{y}_j \\ \bar{\beta}_3 = {} & \frac{1}{D}\sum_{j=1}^{m}\left\{\left[H_{12}H_{23} - H_{22}H_{13}\right] + \left[H_{12}H_{13} - H_{11}H_{23}\right]\bar{x}_j \right. \\ & \left. + \left[H_{11}H_{22} - H_{12}^2\right]\bar{x}_j^2\right\}\bar{y}_j\,. \end{aligned}$$

The partial derivatives of the determinant turn out to be

$$\begin{aligned} \frac{\partial D}{\partial \bar{x}_i} = {} & 2\left[H_{13}H_{23} - H_{12}H_{33}\right] \\ & + 2\left[H_{11}H_{33} + 2\left(H_{12}H_{23} - H_{13}H_{22}\right) - H_{13}^2\right]\bar{x}_i \\ & + 6\left[H_{12}H_{13} - H_{11}H_{23}\right]\bar{x}_i^2 + 4\left(H_{11}H_{22} - H_{12}^2\right)\bar{x}_i^3 \\ \frac{\partial D}{\partial \bar{y}_i} = {} & 0\,. \end{aligned}$$

Those of the components of the solution vector read

$$
\begin{aligned}
c_{i1} = \frac{\partial \bar{\beta}_1}{\partial \bar{x}_i} = -\frac{\bar{\beta}_1}{D}\frac{\partial D}{\partial \bar{x}_i} + \frac{1}{D}\Bigg\{ & 2\left(H_{33}\bar{x}_i - 3H_{23}\bar{x}_i^2 + 2H_{22}\bar{x}_i^3\right)\sum_{j=1}^{m}\bar{y}_j \\
& + \left(H_{23}H_{13} - H_{12}H_{33}\right)\bar{y}_i \\
& + \left(-H_{33} + 2H_{23}\bar{x}_i + 3H_{13}\bar{x}_i^2 - 4H_{12}\bar{x}_i^3\right)\sum_{j=1}^{m}\bar{x}_j\bar{y}_j \\
& + 2\left(H_{12}H_{23} - H_{22}H_{13}\right)\bar{x}_i\bar{y}_i \\
& + \left[H_{23} - 2\left(H_{13} + H_{22}\right)\bar{x}_i + 3H_{12}\bar{x}_i^2\right]\sum_{j=1}^{m}\bar{x}_j^2\bar{y}_j\Bigg\}
\end{aligned}
$$

$$
c_{i+m,1} = \frac{\partial \bar{\beta}_1}{\partial \bar{y}_i} = b_{i1}
$$

$$
\begin{aligned}
c_{i2} = \frac{\partial \bar{\beta}_2}{\partial \bar{x}_i} = -\frac{\bar{\beta}_2}{D}\frac{\partial D}{\partial \bar{x}_i} + \frac{1}{D}\Bigg\{ & \left(-H_{33} + 2H_{23}\bar{x}_i + 3H_{13}\bar{x}_i^2 - 4H_{12}\bar{x}_i^3\right)\sum_{j=1}^{m}\bar{y}_j \\
& + 4\left(-H_{13}\bar{x}_i + H_{11}x_i^3\right)\sum_{j=1}^{m}\bar{x}_j\bar{y}_j \\
& + \left(H_{11}H_{33} - H_{13}^2\right)\bar{y}_i + 2\left(H_{12}H_{13} - H_{11}H_{23}\right)\bar{x}_i\bar{y}_i \\
& + \left[H_{13} + 2H_{12}\bar{x}_i - 3H_{11}\bar{x}_i^2\right]\sum_{j=1}^{m}\bar{x}_j^2\bar{y}_j\Bigg\}
\end{aligned}
$$

$$
c_{i+m,2} = \frac{\partial \bar{\beta}_2}{\partial \bar{y}_i} = b_{i2}
$$

$$
\begin{aligned}
c_{i3} = \frac{\partial \bar{\beta}_3}{\partial \bar{x}_i} = -\frac{\bar{\beta}_3}{D}\frac{\partial D}{\partial \bar{x}_i} + \frac{1}{D}\Bigg\{ & \left[H_{23} - 2\left(H_{13} + H_{22}\right)\bar{x}_i + 3H_{12}\bar{x}_i^2\right]\sum_{j=1}^{m}\bar{y}_j \\
& + \left(H_{13} + 2H_{12}\bar{x}_i - 3H_{11}\bar{x}_i^2\right)\sum_{j=1}^{m}\bar{x}_j\bar{y}_j + \left(H_{12}H_{13} - H_{11}H_{23}\right)\bar{y}_i \\
& + 2\left(H_{11}H_{22} - H_{12}^2\right)\bar{x}_i\bar{y}_i + 2\left(-H_{12} + H_{11}\bar{x}_i\right)\sum_{j=1}^{m}\bar{x}_j^2\bar{y}_j\Bigg\}
\end{aligned}
$$

$$
c_{i+m,3} = \frac{\partial \bar{\beta}_3}{\partial \bar{y}_i} = b_{i3}\,.
$$

C Special Confidence Ellipses and Ellipsoids

In case repeated measurements are lacking, confidence ellipses and ellipsoids are not ready to hand. In the following, a heuristic approach is discussed.

Ellipses

Chapter 13 relates the adjustment of a straight line to individual measurements. In order to design a confidence ellipse, we start from Hotelling's ellipse as given, e.g., in (3.71), [28]

$$
\begin{aligned}
& t^2(2,n-1) = n(\bar{\boldsymbol{\zeta}} - \boldsymbol{\mu})^{\mathrm{T}} \boldsymbol{s}^{-1} (\bar{\boldsymbol{\zeta}} - \boldsymbol{\mu}) \\
& = \frac{n}{|\boldsymbol{s}|} \left[s_{yy}(\bar{x} - \mu_x)^2 - 2 s_{xy} (\bar{x} - \mu_x)(\bar{y} - \mu_y) + s_{xx}(\bar{y} - \mu_y)^2 \right] .
\end{aligned}
\tag{C.1}
$$

Putting

$$
\bar{\boldsymbol{s}}^{-1} = n \boldsymbol{s}^{-1}
$$

where

$$
\boldsymbol{s} = \begin{pmatrix} s_{xx} & s_{xy} \\ s_{yx} & s_{yy} \end{pmatrix} ; \qquad \bar{\boldsymbol{s}} = \begin{pmatrix} \dfrac{s_{xx}}{n} & \dfrac{s_{xy}}{n} \\ \dfrac{s_{yx}}{n} & \dfrac{s_{yy}}{n} \end{pmatrix}
\tag{C.2}
$$

we have

$$
\begin{aligned}
& t^2(2,n-1) = (\bar{\boldsymbol{\zeta}} - \boldsymbol{\mu})^{\mathrm{T}} \bar{\boldsymbol{s}}^{-1} (\bar{\boldsymbol{\zeta}} - \boldsymbol{\mu}) \\
& = \frac{1}{|\bar{\boldsymbol{s}}|} \left[\frac{s_{yy}}{n}(\bar{x} - \mu_x)^2 - 2 \frac{s_{xy}}{n} (\bar{x} - \mu_x)(\bar{y} - \mu_y) + \frac{s_{xx}}{n}(\bar{y} - \mu_y)^2 \right] .
\end{aligned}
\tag{C.3}
$$

To recall, the ellipse refers to arithmetic means $\bar{x}$, $\bar{y}$ with expectations μ_x, μ_y and an empirical variance–covariance matrix $\bar{\boldsymbol{s}}$, the elements of which have degrees of freedom $n-1$.

By contrast, Sect. 13.6 considers estimators $\bar{\beta}_1$, $\bar{\beta}_2$ with expectations $\mu_{\bar{\beta}_1}$ and $\mu_{\bar{\beta}_2}$ and empirical variance–covariance matrix

$$s_{\bar{\boldsymbol{\beta}}} = \begin{pmatrix} s_{\bar{\beta}_1\bar{\beta}_1} & s_{\bar{\beta}_1\bar{\beta}_2} \\ s_{\bar{\beta}_2\bar{\beta}_1} & s_{\bar{\beta}_2\bar{\beta}_2} \end{pmatrix} \tag{C.4}$$

the elements of which have degrees of freedom $m - 2$.

Let us tentatively cast Hotelling's ellipse into

$$\begin{aligned} t^2(2,\ldots) &= (\bar{\boldsymbol{\beta}} - \boldsymbol{\mu}_{\bar{\beta}})^{\mathrm{T}} s_{\bar{\boldsymbol{\beta}}}^{-1} (\bar{\boldsymbol{\beta}} - \boldsymbol{\mu}_{\bar{\beta}}) \\ &= \frac{1}{|s_{\bar{\boldsymbol{\beta}}}|} \left[s_{\bar{\beta}_2\bar{\beta}_2} (\bar{\beta}_1 - \mu_{\bar{\beta}_1})^2 - 2 s_{\bar{\beta}_1\bar{\beta}_2} (\bar{\beta}_1 - \mu_{\bar{\beta}_1})(\bar{\beta}_2 - \mu_{\bar{\beta}_2}) \right. \\ &\quad \left. + \; s_{\bar{\beta}_1\bar{\beta}_1} (\bar{\beta}_2 - \mu_{\bar{\beta}_2})^2 \right] . \end{aligned} \tag{C.5}$$

While (C.3) includes repeated measurements (C.5) does not. Nevertheless, in both cases, the statistical fluctuations per se stand on an equal footing. To invoke the pertaining density a formal n is needed.

Hotelling's density reads

$$\begin{aligned} &p_T(t; m, n-1) \\ &= \frac{2\Gamma(n/2)}{(n-1)^{m/2}\Gamma[(n-m)/2]\Gamma(m/2)} \frac{t^{m-1}}{[1 + t^2/(n-1)]^{n/2}} \\ &t > 0\,, \quad n > m \end{aligned} \tag{C.6}$$

Let us firstly substitute the number 2 for m as we are considering two variables,

$$p_T(t; 2, n-1) = \frac{2\Gamma(n/2)}{(n-1)\Gamma[(n-2)/2]} \frac{t}{[1 + t^2/(n-1)]^{n/2}} \,.$$

Next we replace Hotelling's degrees of freedom $n-1$ by the degrees of freedom of the elements of the matrix (C.4), i.e. by $m - 2$. Hence

$$p_T(t; 2, m-2) = \frac{2\Gamma((m-1)/2)}{(m-2)\Gamma[(m-3)/2]} \frac{t}{[1 + t^2/(m-2)]^{(m-1)/2}} \,. \tag{C.7}$$

After all, the proceeding suggests to assign the confidence ellipse

$$\begin{aligned} s_{\bar{\beta}_2\bar{\beta}_2}(\beta_1 - \bar{\beta}_1)^2 - 2 s_{\bar{\beta}_1\bar{\beta}_2}(\beta_1 - \bar{\beta}_1)(\beta_2 - \bar{\beta}_2) + s_{\bar{\beta}_1\bar{\beta}_1}(\beta_2 - \bar{\beta}_2)^2 \\ = t_P^2(2, m-2)|s_{\bar{\boldsymbol{\beta}}}| \end{aligned} \tag{C.8}$$

to the empirical estimators $\bar{\beta}_1, \bar{\beta}_2$ of the adjustment of a straight line as discussed in Chap. 13. Consulting a table for the quantiles of Hotelling's density, we look for the entry which refers to 2 variables and $m-2$ degrees of freedom. After all, we expect (C.8) to localize the point

$$\boldsymbol{\mu}_{\bar{\boldsymbol{\beta}}} = \begin{pmatrix} \mu_{\bar{\beta}_1} \\ \\ \mu_{\bar{\beta}_2} \end{pmatrix} \tag{C.9}$$

with probability P.

Finally, we hint that the ellipse's angle of rotation

$$\tan(2\varphi) = \frac{-s_{\bar{\beta}_1\bar{\beta}_2}}{s_{\bar{\beta}_2\bar{\beta}_2} - s_{\bar{\beta}_1\bar{\beta}_1}}\,, \tag{C.10}$$

measured counterclockwise against the $\beta_{\bar{}}$-axis of a rectangular β_1, β_2-coordinate system, proves sample-independent as the empirical variance s_y^2 cancels out.

Ellipsoids

Chapter 17 relates the adjustment of a plane to individual measurements. Following similar arguments we arrive at

$$(\boldsymbol{\beta} - \bar{\boldsymbol{\beta}})^{\mathrm{T}} s_{\bar{\boldsymbol{\beta}}}^{-1} (\boldsymbol{\beta} - \bar{\boldsymbol{\beta}}) = t_P^2(3, m-3) \tag{C.11}$$

as here the elements of the empirical variance–covariance matrix $s_{\bar{\boldsymbol{\beta}}}$ have degrees of freedom $m-3$. Hence, we have to refer to the quantiles of the density

$$p_T(t; 3, m-3)$$

$$= \frac{2\Gamma[(m-2)/2]}{(m-3)^{3/2}\Gamma[(m-5)/2]\,\Gamma(3/2)} \frac{t^2}{\left[1 + \dfrac{t^2}{m-3}\right]^{(m-2)/2}} \tag{C.12}$$

for 3 variables and $m-3$ degrees of freedom.

D Extreme Points of Ellipses and Ellipsoids

To facilitate graphical representations of ellipses and ellipsoids we draw upon their extreme points.

Confidence Ellipses

The extreme point of the confidence ellipse (13.27) with respect to the β_1- and β_2-directions are given by

$$\begin{pmatrix} \bar{\beta}_1 \\ \bar{\beta}_2 \end{pmatrix}_{\beta_1\text{-direction}} = \pm \frac{t}{\sqrt{s_{xx}}} \begin{pmatrix} s_{xx} \\ s_{xy} \end{pmatrix} \tag{D.1}$$

and

$$\begin{pmatrix} \bar{\beta}_1 \\ \bar{\beta}_2 \end{pmatrix}_{\beta_2\text{-direction}} = \pm \frac{t}{\sqrt{s_{yy}}} \begin{pmatrix} s_{yx} \\ s_{yy} \end{pmatrix}, \tag{D.2}$$

respectively. Correspondingly, for (14.26) we have

$$\begin{pmatrix} \bar{\beta}_1 \\ \bar{\beta}_2 \end{pmatrix}_{\beta_1\text{-direction}} = \pm \frac{t}{\sqrt{n}\sqrt{s_{xx}}} \begin{pmatrix} s_{xx} \\ s_{xy} \end{pmatrix} \tag{D.3}$$

and

$$\begin{pmatrix} \bar{\beta}_1 \\ \bar{\beta}_2 \end{pmatrix}_{\beta_2\text{-direction}} = \pm \frac{t}{\sqrt{n}\sqrt{s_{yy}}} \begin{pmatrix} s_{yx} \\ s_{yy} \end{pmatrix}. \tag{D.4}$$

Confidence Ellipsoids

With respect to ellipsoids, let us start out considering

$$\boldsymbol{x}^{\mathrm{T}} \boldsymbol{s}^{-1} \boldsymbol{x} = t^2/n \tag{D.5}$$

where $\boldsymbol{x} = (x \quad y \quad z)^{\mathrm{T}}$ marks any point on the ellipsoid's skin. Furthermore, let

$$\boldsymbol{s} = \begin{pmatrix} s_{11} & s_{12} & s_{13} \\ s_{21} & s_{22} & s_{23} \\ s_{31} & s_{32} & s_{33} \end{pmatrix} \tag{D.6}$$

denote the empirical variance–covariance matrix and

$$\text{adj } \boldsymbol{s} = \begin{pmatrix} \rho_{11} & \rho_{21} & \rho_{31} \\ \rho_{12} & \rho_{22} & \rho_{32} \\ \rho_{13} & \rho_{23} & \rho_{33} \end{pmatrix} \tag{D.7}$$

its adjoint, with ρ_{ij} the cofactors of s_{ij}, both matrices being symmetrical. As is well known, we have

$$\boldsymbol{s}^{-1} = \frac{\text{adj } \boldsymbol{s}}{|\boldsymbol{s}\,|} \tag{D.8}$$

with $|\boldsymbol{s}\,|$ the determinant of $\boldsymbol{s}$. In the following we shall refer to the relations

$$\begin{aligned} |\boldsymbol{s}\,|s_{11} &= \rho_{22}\rho_{33} - \rho_{32}\rho_{23}\,, & |\boldsymbol{s}\,|s_{22} &= \rho_{11}\rho_{33} - \rho_{13}\rho_{31} \\ |\boldsymbol{s}\,|s_{12} &= \rho_{31}\rho_{23} - \rho_{21}\rho_{33}\,, & |\boldsymbol{s}\,|s_{23} &= \rho_{12}\rho_{31} - \rho_{11}\rho_{32} \\ |\boldsymbol{s}\,|s_{13} &= \rho_{21}\rho_{32} - \rho_{22}\rho_{13}\,, & |\boldsymbol{s}\,|s_{33} &= \rho_{11}\rho_{22} - \rho_{12}\rho_{21} \end{aligned} \tag{D.9}$$

and

$$\begin{aligned} \rho_{11}s_{31} + \rho_{21}s_{32} + \rho_{31}s_{33} &= 0 \\ \rho_{12}s_{31} + \rho_{22}s_{32} + \rho_{32}s_{33} &= 0 \\ \rho_{13}s_{31} + \rho_{23}s_{32} + \rho_{33}s_{33} &= |\boldsymbol{s}\,|\,. \end{aligned} \tag{D.10}$$

From (D.5) and (D.8) we draw

$$\begin{aligned} &f(x, y, z) = \\ &\rho_{11}x^2 + \rho_{22}y^2 + \rho_{33}z^2 + 2\rho_{12}xy + 2\rho_{23}yz + 2\rho_{31}zx - \frac{t^2}{n}\,|\,\boldsymbol{s}\,| = 0 \end{aligned} \tag{D.11}$$

Let $(x_1 \quad y_1 \quad z_1)^{\mathrm{T}}$ denote one of the two extreme points in the z-direction. The respective normal vector is given by $(0 \quad 0 \quad \partial f/\partial z)^{\mathrm{T}}$. Hence

$$\begin{aligned} \rho_{11}x_1 + \rho_{12}y_1 + \rho_{31}z_1 &= 0 \\ \rho_{12}x_1 + \rho_{22}y_1 + \rho_{23}z_1 &= 0 \\ \rho_{31}x_1 + \rho_{23}y_1 + \rho_{33}z_1 &= \frac{\partial f}{\partial z}\,. \end{aligned} \tag{D.12}$$

Solving for x_1, y_1, z_1 yields

$$
\begin{aligned}
x_1 &= \frac{1}{|\boldsymbol{s}|^2} \frac{\partial f}{\partial z} (\rho_{12}\rho_{23} - \rho_{31}\rho_{22}) = \frac{1}{|\boldsymbol{s}|} \frac{\partial f}{\partial z}\, s_{13} \\
y_1 &= \frac{1}{|\boldsymbol{s}|^2} \frac{\partial f}{\partial z} (\rho_{12}\rho_{31} - \rho_{23}\rho_{11}) = \frac{1}{|\boldsymbol{s}|} \frac{\partial f}{\partial z}\, s_{23} \\
z_1 &= \frac{1}{|\boldsymbol{s}|^2} \frac{\partial f}{\partial z} (\rho_{11}\rho_{22} - \rho_{12}\rho_{12}) = \frac{1}{|\boldsymbol{s}|} \frac{\partial f}{\partial z}\, s_{33}\,.
\end{aligned}
\tag{D.13}
$$

Inserting this into (D.11) we find

$$
\frac{\partial f}{\partial z} = \pm \frac{1}{\sqrt{s_{33}}}\, |\boldsymbol{s}|\, \frac{t}{\sqrt{n}}\,. \tag{D.14}
$$

Analogue procedures lead to the extreme points in the x- and y-direction, respectively. Returning to the center coordinates $\bar{\beta}_1$, $\bar{\beta}_2$, and $\bar{\beta}_3$ we have

in the β_1-direction

$$
\begin{pmatrix} \bar{\beta}_1 \\ \bar{\beta}_2 \\ \bar{\beta}_3 \end{pmatrix}_{\beta_1\text{-direction}} = \pm \frac{t}{\sqrt{n}\sqrt{s_{11}}} \begin{pmatrix} s_{11} \\ s_{12} \\ s_{13} \end{pmatrix}, \tag{D.15}
$$

in the β_2-direction

$$
\begin{pmatrix} \bar{\beta}_1 \\ \bar{\beta}_2 \\ \bar{\beta}_3 \end{pmatrix}_{\beta_2\text{-direction}} = \pm \frac{t}{\sqrt{n}\sqrt{s_{22}}} \begin{pmatrix} s_{21} \\ s_{22} \\ s_{23} \end{pmatrix} \tag{D.16}
$$

and, finally, in the β_3-direction

$$
\begin{pmatrix} \bar{\beta}_1 \\ \bar{\beta}_2 \\ \bar{\beta}_3 \end{pmatrix}_{\beta_3\text{-direction}} = \pm \frac{t}{\sqrt{n}\sqrt{s_{33}}} \begin{pmatrix} s_{31} \\ s_{32} \\ s_{33} \end{pmatrix}. \tag{D.17}
$$

E Drawing Ellipses and Ellipsoids

Just for convenience, we summarize the basic formulas to be used to draw ellipses and ellipsoids.

Ellipses

To have an example, we refer to (13.27),

$$
\begin{aligned}
s_{\bar{\beta}_2\bar{\beta}_2}(\beta_1 - \bar{\beta}_1)^2 - 2s_{\bar{\beta}_1\bar{\beta}_2}(\beta_1 - \bar{\beta}_1)(\beta_2 - \bar{\beta}_2) + s_{\bar{\beta}_1\bar{\beta}_1}(\beta_2 - \bar{\beta}_2)^2 \\
= t_P^2(2, m-2)|s_{\bar{\boldsymbol{\beta}}}|\,.
\end{aligned}
$$

Putting

$$
\beta_1 - \bar{\beta}_1 = r\cos\varphi; \qquad \beta_2 - \bar{\beta}_2 = r\sin\varphi
$$

we have

$$
r = \frac{t_P(2, m-2)\sqrt{|s_{\bar{\boldsymbol{\beta}}}|}}{\sqrt{s_{\bar{\beta}_2\bar{\beta}_2}\cos^2\varphi - 2s_{\bar{\beta}_1\bar{\beta}_2}\sin\varphi\cos\varphi + s_{\bar{\beta}_1\bar{\beta}_1}\sin^2\varphi}}\,.
$$

In case of (14.26)

$$
\begin{aligned}
s_{\bar{\beta}_2\bar{\beta}_2}\left(\beta_1 - \bar{\beta}_1\right)^2 - 2s_{\bar{\beta}_1\bar{\beta}_2}\left(\beta_1 - \bar{\beta}_1\right)\left(\beta_2 - \bar{\beta}_2\right) + s_{\bar{\beta}_1\bar{\beta}_1}\left(\beta_2 - \bar{\beta}_2\right)^2 \\
= |\, s_{\bar{\boldsymbol{\beta}}}\, |\, \frac{t_P^2(2, n-1)}{n}
\end{aligned}
$$

we find

$$
r = \frac{\dfrac{t_P(2, n-1)}{\sqrt{n}}\sqrt{|s_{\bar{\boldsymbol{\beta}}}|}}{\sqrt{s_{\bar{\beta}_2\bar{\beta}_2}\cos^2\varphi - 2s_{\bar{\beta}_1\bar{\beta}_2}\sin\varphi\cos\varphi + s_{\bar{\beta}_1\bar{\beta}_1}\sin^2\varphi}}\,.
$$

Ellipsoids

We refer to (17.22),

$$(\boldsymbol{\beta} - \bar{\boldsymbol{\beta}})^{\mathrm{T}} s_{\bar{\boldsymbol{\beta}}}^{-1} (\boldsymbol{\beta} - \bar{\boldsymbol{\beta}}) = t_P^2(3, m-3)\,, \tag{E.1}$$

with $s_{\bar{\boldsymbol{\beta}}}$ as given in (17.14),

$$\boldsymbol{s}_{\bar{\beta}} = \begin{pmatrix} s_{\bar{\beta}_1\bar{\beta}_1} & s_{\bar{\beta}_1\bar{\beta}_2} & s_{\bar{\beta}_1\bar{\beta}_3} \\ s_{\bar{\beta}_2\bar{\beta}_1} & s_{\bar{\beta}_2\bar{\beta}_2} & s_{\bar{\beta}_2\bar{\beta}_3} \\ s_{\bar{\beta}_3\bar{\beta}_1} & s_{\bar{\beta}_3\bar{\beta}_2} & s_{\bar{\beta}_3\bar{\beta}_3} \end{pmatrix}.$$

To shorten the notation we put

$$x = \beta_1 - \bar{\beta}_1\,, \qquad y = \beta_2 - \bar{\beta}_2\,, \qquad z = \beta_3 - \bar{\beta}_3$$

and, in a formal sense,

$$s_{\bar{\boldsymbol{\beta}}}^{-1} = \begin{pmatrix} \gamma_{11} & \gamma_{12} & \gamma_{13} \\ \gamma_{21} & \gamma_{22} & \gamma_{23} \\ \gamma_{31} & \gamma_{32} & \gamma_{33} \end{pmatrix} \tag{E.2}$$

so that (E.1) turns into

$$\gamma_{11}x^2 + \gamma_{22}y^2 + \gamma_{33}z^2 + 2\gamma_{12}xy + 2\gamma_{23}yz + 2\gamma_{31}zx = t_P^2(3, m-2)\,. \tag{E.3}$$

The contour lines $z =$ const. are obviously ellipses. Following (D.17), we have to consider the interval

$$-t_P(3, m-3)\sqrt{s_{\bar{\beta}_3\bar{\beta}_3}} \le z \le t_P(3, m-3)\sqrt{s_{\bar{\beta}_3\bar{\beta}_3}}\,. \tag{E.4}$$

As the centers of the ellipses are shifted with respect to the origin $x = y = 0$, we put

$$x = \xi + x_M; \qquad y = \eta + y_M \tag{E.5}$$

where

$$x_M = \frac{\gamma_{22}\gamma_{31} - \gamma_{12}\gamma_{23}}{\gamma_{12}^2 - \gamma_{11}\gamma_{22}}\,z\,; \qquad y_M = \frac{\gamma_{11}\gamma_{23} - \gamma_{12}\gamma_{31}}{\gamma_{12}^2 - \gamma_{11}\gamma_{22}}\,z\,. \tag{E.6}$$

For any z=const. we have

$$\begin{aligned} \gamma_{11}\xi^2 + 2\gamma_{12}\xi\eta + \gamma_{22}\eta^2 &= t_P^2(3, m-3) \\ -(\gamma_{33}z^2 + 2(\gamma_{31}x_M + \gamma_{23}y_M)z &+ \gamma_{11}x_M^2 + 2\gamma_{12}x_M\,y_M + \gamma_{22}y_M^2)\,. \end{aligned} \tag{E.7}$$

After all, the problem to draw a spatially rotated ellipsoid has been reduced to the problem to draw a sufficiently dense set of ellipses perpendicular to the z-axis.

F Security Polygons and Polyhedra

Unknown systematic errors cause a new, in error calculus hitherto unknown species of geometrical objects to go on stage [8]. These are polygons in case of two measurands, polyhedra in case of three and abstract polytopes for more than three measurands. At any rate, the objects are convex and point-symmetric [28]. The naming security polygon, security polyhedron and security polytope establishes a correspondence to the terms confidence ellipse and confidence ellipsoid. We shall confine ourselves to polygons and polyhedra.

Security Polygons

Edges

Let us reconsider (14.10). For $r = 2$ the components of the propagated systematic error

$$f_{\bar{\beta}_k} = \sum_{i=1}^{m} b_{ik} f_{\bar{y}_i}\,; \quad k = 1, 2$$

span a polygon. In a way, the polygon comes into being while the $f_{\bar{y}_1}, f_{\bar{y}_1}, \ldots, f_{\bar{y}_m}$ "scan" the set of points lying within or on the faces of the m-dimensional hypercuboid

$$-f_{s,\bar{y}_i} \le f_{\bar{y}_i} \le f_{s,\bar{y}_i}\,; \quad i = 1, \ldots, m\,.$$

To simplify matters, let us put $f_{\bar{y}_i} = f_i$, $f_{s\,\bar{y}_i} = f_{s,i}$ and $m = 5$ so that

$$\begin{aligned} f_{\bar{\beta}_1} &= b_{11} f_1 + b_{21} f_2 + b_{31} f_3 + b_{41} f_4 + b_{51} f_5 \\ f_{\bar{\beta}_2} &= b_{12} f_1 + b_{22} f_2 + b_{32} f_3 + b_{42} f_4 + b_{52} f_5 \end{aligned} \tag{F.1}$$

with

$$-f_{s,i} \le f_i \le f_{s,i}\,; \quad i = 1, \ldots, 5\,. \tag{F.2}$$

With reference to a rectangular $f_{\bar{\beta}_1}, f_{\bar{\beta}_2}$ coordinate system, the polygon is obviously enclosed within a rectangle

$$-f_{s,\bar{\beta}_k} \le f_{\bar{\beta}_k} \le f_{s,\bar{\beta}_k}\,; \quad f_{s,\bar{\beta}_k} = \sum_{i=1}^{5} |b_{ik}|\, f_{s,i}\,; \quad k = 1,2\,.$$

Let us assume $b_{11} \neq 0$. Eliminating f_1, we find a straight line

$$f_{\bar{\beta}_2} = \frac{b_{12}}{b_{11}} f_{\bar{\beta}_1} + c_1 \tag{F.3}$$

where

$$c_1 = h_{12} f_2 + h_{13} f_3 + h_{14} f_4 + h_{15} f_5$$

and

$$h_{12} = \frac{1}{b_{11}}(b_{11}b_{22} - b_{12}b_{21})\,, \qquad h_{13} = \frac{1}{b_{11}}(b_{11}b_{32} - b_{12}b_{31})$$

$$h_{14} = \frac{1}{b_{11}}(b_{11}b_{42} - b_{12}b_{41})\,, \qquad h_{15} = \frac{1}{b_{11}}(b_{11}b_{52} - b_{12}b_{51})\,.$$

Any variation of c_1 shifts the line (F.3) parallel to itself; c_1 is maximal if

$$\begin{aligned} f_2 &= f_2^* = \operatorname{sign}(h_{12})\, f_{s,2}\,, \quad f_3 = f_3^* = \operatorname{sign}(h_{13})\, f_{s,3} \\ f_4 &= f_4^* = \operatorname{sign}(h_{14})\, f_{s,4}\,, \quad f_5 = f_5^* = \operatorname{sign}(h_{15})\, f_{s,5}\,. \end{aligned} \tag{F.4}$$

We obviously have

$$c_{s,1} = h_{12} f_2^* + h_{13} f_3^* + h_{14} f_4^* + h_{15} f_5^*$$

and

$$-c_{s,1} \le c_1 \le c_{s,1}\,. \tag{F.5}$$

Hence, the lines

$$f_{\bar{\beta}_2} = \frac{b_{12}}{b_{11}} f_{\bar{\beta}_1} + c_{s,1}\,, \qquad f_{\bar{\beta}_2} = \frac{b_{12}}{b_{11}} f_{\bar{\beta}_1} - c_{s,1} \tag{F.6}$$

hold maximum distance thus pegging two edges of the polygon. Shifting f_1 in $-f_{s,1}, \ldots, f_{s,1}$, the point

$$\begin{aligned} f_{\bar{\beta}_1} &= b_{11} f_1 + b_{21} f_2^* + b_{31} f_3^* + b_{41} f_4^* + b_{51} f_5^* \\ f_{\bar{\beta}_2} &= b_{12} f_1 + b_{22} f_2^* + b_{32} f_3^* + b_{42} f_4^* + b_{52} f_5^*\,, \end{aligned} \tag{F.7}$$

slides along the polygon's $c_{s,1}$ edge and something similar applies to the $-c_{s,1}$ edge.

Vertices

When it comes to specify the vertices of the two edges obtained so far, the $c_{s,1}$ edge is marked off by

$$\text{Vertex} \quad V_1$$

$$f_{\bar{\beta}_1,1} = b_{11} f_{s,1} + b_{21} f_2^* + b_{31} f_3^* + b_{41} f_4^* + b_{51} f_5^*$$

$$f_{\bar{\beta}_2,1} = b_{12} f_{s,1} + b_{22} f_2^* + b_{32} f_3^* + b_{42} f_4^* + b_{52} f_5^*$$

and

$$\text{Vertex} \quad V_2$$

$$f_{\bar{\beta}_1,2} = -b_{11} f_{s,1} + b_{21} f_2^* + b_{31} f_3^* + b_{41} f_4^* + b_{51} f_5^*$$

$$f_{\bar{\beta}_2,2} = -b_{12} f_{s,1} + b_{22} f_2^* + b_{32} f_3^* + b_{42} f_4^* + b_{52} f_5^* \,.$$

For the $-c_{s,1}$ edge we have

$$\text{Vertex} \quad V_3$$

$$f_{\bar{\beta}_1,3} = b_{11} f_{s,1} - b_{21} f_2^* - b_{31} f_3^* - b_{41} f_4^* - b_{51} f_5^*$$

$$f_{\bar{\beta}_2,3} = b_{12} f_{s,1} - b_{22} f_2^* - b_{32} f_3^* - b_{42} f_4^* - b_{52} f_5^*$$

and

$$\text{Vertex} \quad V_4$$

$$f_{\bar{\beta}_1,4} = -b_{11} f_{s,1} - b_{21} f_2^* - b_{31} f_3^* - b_{41} f_4^* - b_{51} f_5^*$$

$$f_{\bar{\beta}_2,4} = -b_{12} f_{s,1} - b_{22} f_2^* - b_{32} f_3^* - b_{42} f_4^* - b_{52} f_5^* \,.$$

This suggests a point symmetry with respect to the vertices V_1 and V_4 and the vertices V_2 and V_3, respectively. After all, we have found two of the polygon's edges including their vertices. We conclude that the polygon holds at most m pairs of parallel edges.

Given $b_{11} = 0$, we cannot eliminate f_1. Instead, (F.1) produces two abscissas

$$\begin{aligned} f_{\bar{\beta}_1} = \pm \; & [\operatorname{sign}(b_{21})\, b_{21} f_{s,2} + \operatorname{sign}(b_{31})\, b_{31} f_{s,3} \\ & + \operatorname{sign}(b_{41})\, b_{41} f_{s,4} + \operatorname{sign}(b_{51})\, b_{51} f_{s,5}\,] \end{aligned} \tag{F.8}$$

through each of which a line is to be drawn parallel to the $f_{\bar{\beta}_2}$-axis. The coordinate $f_{\bar{\beta}_2}$ slides along these verticals following

$$\begin{aligned} f_{\bar{\beta}_2} = b_{12} f_1 \pm \; & [\operatorname{sign}(b_{21})\, b_{22} f_{s,2} + \operatorname{sign}(b_{31})\, b_{32} f_{s,3} \\ & + \operatorname{sign}(b_{41})\, b_{42} f_{s,4} + \operatorname{sign}(b_{51})\, b_{52} f_{s,5}\,] \,. \end{aligned} \tag{F.9}$$

Equal systematic errors

Assuming $f_{\bar{y}_i} = f_y\,;\ f_{s,\bar{y}_i} = f_{s,y};\ \ i = 1, \ \ldots\ , m$, due to (14.13), the security polygon degenerates into an interval

$$-f_{s,y} \le f_{s,\bar{\beta}_1} \le f_{s,y}\,; \qquad f_{\bar{\beta}_2} = 0\,. \tag{F.10}$$

Security Polyhedra

Faces

To examine the geometrical properties of the convex, point-symmetric hull spanned by three propagated systematic errors,

$$f_{\bar{\beta}_k} = \sum_{i=1}^{m} b_{ik} f_i\,, \quad -f_{s,i} \le f_i \le f_{s,i}\,; \qquad k = 1, 2, 3$$

we again confine ourselves to $m = 5$

$$\begin{aligned} f_{\bar{\beta}_1} &= b_{11}f_1 + b_{21}f_2 + b_{31}f_3 + b_{41}f_4 + b_{51}f_5 \\ f_{\bar{\beta}_2} &= b_{12}f_1 + b_{22}f_2 + b_{32}f_3 + b_{42}f_4 + b_{52}f_5 \\ f_{\bar{\beta}_3} &= b_{13}f_1 + b_{23}f_2 + b_{33}f_3 + b_{43}f_4 + b_{53}f_5\,. \end{aligned} \tag{F.11}$$

With respect to a rectangular $f_{\bar{\beta}_1}, f_{\bar{\beta}_2}, f_{\bar{\beta}_3}$ coordinate system, the polyhedron is enclosed within a cuboid given by

$$-f_{s,\bar{\beta}_k} \le f_{\bar{\beta}_k} \le f_{s,\bar{\beta}_k}\,; \quad f_{s,\bar{\beta}_k} = \sum_{i=1}^{5} |b_{ik}|\, f_{s,i}\,; \quad k = 1, 2, 3\,. \tag{F.12}$$

Rewriting (F.11) according to

$$\begin{aligned} b_{11}f_1 + b_{21}f_2 + b_{31}f_3 &= f_{\bar{\beta}_1} - b_{41}f_4 - b_{51}f_5 = \xi \\ b_{12}f_1 + b_{22}f_2 + b_{32}f_3 &= f_{\bar{\beta}_2} - b_{42}f_4 - b_{52}f_5 = \eta \\ b_{13}f_1 + b_{23}f_2 + b_{33}f_3 &= f_{\bar{\beta}_3} - b_{43}f_4 - b_{53}f_5 = \zeta \end{aligned} \tag{F.13}$$

we may eliminate f_1 and f_2 thus solving for f_3

$$f_3 \begin{vmatrix} b_{11} & b_{21} & b_{31} \\ b_{12} & b_{22} & b_{32} \\ b_{13} & b_{23} & b_{33} \end{vmatrix} = \begin{vmatrix} b_{11} & b_{21} & \xi \\ b_{12} & b_{22} & \eta \\ b_{13} & b_{23} & \zeta \end{vmatrix}\,.$$

Defining

$$\lambda_{12} = b_{12}b_{23} - b_{22}b_{13}\,, \quad \mu_{12} = b_{21}b_{13} - b_{11}b_{23}\,, \quad \nu_{12} = b_{11}b_{22} - b_{21}b_{12}$$

we have

$$\lambda_{12} f_{\bar{\beta}_1} + \mu_{12} f_{\bar{\beta}_2} + \nu_{12} f_{\bar{\beta}_3} = c_{12} \tag{F.14}$$

where

$$c_{12} = h_{12,3} f_3 + h_{12,4} f_4 + h_{12,5} f_5 \tag{F.15}$$

and

$$\begin{aligned} h_{12,3} &= (\lambda_{12} b_{31} + \mu_{12} b_{32} + \nu_{12} b_{33}) \\ h_{12,4} &= (\lambda_{12} b_{41} + \mu_{12} b_{42} + \nu_{12} b_{43}) \\ h_{12,5} &= (\lambda_{12} b_{51} + \mu_{12} b_{52} + \nu_{12} b_{53}) \, . \end{aligned}$$

Varying c_{12}, which we can do via f_3, f_4, f_5, the plane (F.14) gets shifted parallel to itself. The choices

$$\begin{aligned} f_3 &= f_3^* = \operatorname{sign}(h_{12,3})\, f_{s,3} \\ f_4 &= f_4^* = \operatorname{sign}(h_{12,4})\, f_{s,4} \\ f_5 &= f_5^* = \operatorname{sign}(h_{12,5})\, f_{s,5} \end{aligned}$$

assign a maximum value to c_{12}

$$c_{s,12} = h_{12,3} f_3^* + h_{12,4} f_4^* + h_{12,5} f_5^*$$

whereat

$$-c_{s,12} \le c_{12} \le c_{s,12} \, .$$

After all, we have found two of the solid's faces. Calling them F1 and F2, we observe

$$\begin{aligned} \lambda_{12} f_{\bar{\beta}_1} + \mu_{12} f_{\bar{\beta}_2} + \nu_{12} f_{\bar{\beta}_3} &= -c_{s,12} \qquad \text{F1} \\ \lambda_{12} f_{\bar{\beta}_1} + \mu_{12} f_{\bar{\beta}_2} + \nu_{12} f_{\bar{\beta}_3} &= c_{s,12} \qquad \text{F2}\, . \end{aligned} \tag{F.16}$$

In

$$\boldsymbol{f} = (f_1 \quad f_2 \quad f_3 \quad f_4 \quad f_5)^{\mathrm{T}}$$

the variables f_1, f_2 are free to vary, while the variables f_3, f_4, f_5 are to be kept fixed. If f_1, f_2 vary, the vector $\boldsymbol{f}_{\bar{\beta}} = \boldsymbol{B}^{\mathrm{T}} \boldsymbol{f}$ slides along the faces F1 and F2. Cyclically swapping the variables, we successively find the polyhedron's remaining faces. As Table F.1 indicates, the maximum number of faces is given by $m(m-1)$. Table F.2 visualizes the swapping procedure.

Given in (F.11), say, b_{11} and b_{12} vanish, we cannot eliminate f_1. Moreover, as ν_{12} becomes zero, (F.16) represents a pair of planes parallel to the $f_{\bar{\beta}_3}$-axis,

$$\lambda_{12} f_{\bar{\beta}_1} + \mu_{12} f_{\bar{\beta}_2} = \pm c_{s,12}\, ; \qquad \nu_{12} = 0\, .$$

The cases $\lambda_{12} = 0$ and $\mu_{12} = 0$ are to be treated correspondingly.

Table F.1. Maximum Number of Faces

×	$f_1 f_2$	$f_1 f_3$	$f_1 f_4$	$f_1 f_5$
×	×	$f_2 f_3$	$f_2 f_4$	$f_2 f_5$
×	×	×	$f_3 f_4$	$f_3 f_5$
×	×	×	×	$f_4 f_5$
×	×	×	×	×

Table F.2. Cyclic Elimination, $r = 3$ and $m = 5$

	Eliminated variable			Variables remaining within the system		
1st step	f_1	f_2	$\Rightarrow$	f_3	f_4	f_5
2nd step	f_1	f_3	$\Rightarrow$	f_4	f_5	f_2
3rd step	f_1	f_4	$\Rightarrow$	f_5	f_2	f_3
4th step	f_1	f_5	$\Rightarrow$	f_2	f_3	f_4
5th step	f_2	f_3	$\Rightarrow$	f_4	f_5	f_1
6th step	f_2	f_4	$\Rightarrow$	f_5	f_1	f_3
7th step	f_2	f_5	$\Rightarrow$	f_1	f_3	f_4
8th step	f_3	f_4	$\Rightarrow$	f_5	f_1	f_2
9th step	f_3	f_5	$\Rightarrow$	f_1	f_2	f_4
10th step	f_4	f_5	$\Rightarrow$	f_1	f_2	f_3

Vertices

Given $h_{12,i} \neq 0;\ i = 3, 4, 5$, we may consider two sets of $\boldsymbol{f}$-vectors with respect to the faces F1 and F2, namely

$$\begin{matrix} [\,-f_{s,1} & -f_{s,2} & -f_3^* & -f_4^* & -f_5^*\,]^{\mathrm{T}} \\ [\;\;\, f_{s,1} & -f_{s,2} & -f_3^* & -f_4^* & -f_5^*\,]^{\mathrm{T}} \\ [\,-f_{s,1} & f_{s,2} & -f_3^* & -f_4^* & -f_5^*\,]^{\mathrm{T}} \\ [\;\;\, f_{s,1} & f_{s,2} & -f_3^* & -f_4^* & -f_5^*\,]^{\mathrm{T}} \end{matrix} \qquad \text{F1}$$

and

$$\begin{matrix} [\;\;\, f_{s,1} & f_{s,2} & f_3^* & f_4^* & f_5^*\,]^{\mathrm{T}} \\ [\,-f_{s,1} & f_{s,2} & f_3^* & f_4^* & f_5^*\,]^{\mathrm{T}} \\ [\;\;\, f_{s,1} & -f_{s,2} & f_3^* & f_4^* & f_5^*\,]^{\mathrm{T}} \\ [\,-f_{s,1} & -f_{s,2} & f_3^* & f_4^* & f_5^*\,]^{\mathrm{T}} \end{matrix} \qquad \text{F2}$$

On each of the faces, these vectors appoint four vertices via $\boldsymbol{f}_{\bar{\boldsymbol{\beta}}} = \boldsymbol{B}^{\mathrm{T}}\boldsymbol{f}$. For F1, we find

Vertex V_1

$$f_{\bar{\beta}_1,1} = -b_{11}f_{s,1} - b_{21}f_{s,2} - b_{31}f_3^* - b_{41}f_4^* - b_{51}f_5^*$$

$$f_{\bar{\beta}_2,1} = -b_{12}f_{s,1} - b_{22}f_{s,2} - b_{32}f_3^* - b_{42}f_4^* - b_{52}f_5^*$$

$$f_{\bar{\beta}_3,1} = -b_{13}f_{s,1} - b_{23}f_2 - b_{33}f_3^* - b_{43}f_4^* - b_{52}f_5^*$$

Vertex V_2

$$f_{\bar{\beta}_1,1} = b_{11}f_{s,1} - b_{21}f_{s,2} - b_{31}f_3^* - b_{41}f_4^* - b_{51}f_5^*$$

$$f_{\bar{\beta}_2,1} = b_{12}f_{s,1} - b_{22}f_{s,2} - b_{32}f_3^* - b_{42}f_4^* - b_{52}f_5^*$$

$$f_{\bar{\beta}_3,1} = b_{13}f_{s,1} - b_{23}f_{s,2} - b_{33}f_3^* - b_{43}f_4^* - b_{52}f_5^*$$

Vertex V_3

$$f_{\bar{\beta}_1,1} = -b_{11}f_{s,1} + b_{21}f_{s,2} - b_{31}f_3^* - b_{41}f_4^* - b_{51}f_5^*$$

$$f_{\bar{\beta}_2,1} = -b_{12}f_{s,1} + b_{22}f_{s,2} - b_{32}f_3^* - b_{42}f_4^* - b_{52}f_5^*$$

$$f_{\bar{\beta}_3,1} = -b_{13}f_{s,1} + b_{23}f_{s,2} - b_{33}f_3^* - b_{43}f_4^* - b_{52}f_5^*$$

Vertex V_4

$$f_{\bar{\beta}_1,1} = b_{11}f_{s,1} + b_{21}f_{s,2} - b_{31}f_3^* - b_{41}f_4^* - b_{51}f_5^*$$

$$f_{\bar{\beta}_2,1} = b_{12}f_{s,1} + b_{22}f_{s,2} - b_{32}f_3^* - b_{42}f_4^* - b_{52}f_5^*$$

$$f_{\bar{\beta}_3,1} = b_{13}f_{s,1} + b_{23}f_{s,2} - b_{33}f_3^* - b_{43}f_4^* - b_{52}f_5^*$$

As the polyhedra are point-symmetric, the vertices on F2 are to be obtained from those on F1 by mirroring.

Now assume $h_{12,5}$ to vanish. Then, as (F.11) and (F.15) suggest, each of the faces carries eight vertices instead of four. Should additionally $h_{12,4}$ vanish, each face would hold 16 vertices. However, at least one of the coefficients $h_{12,i}$, $i = 3, 4, 5$ must be unequal to zero as otherwise the planes F1 and F2 would pass through the origin.

Example

We shall find the surfaces and vertices of the polyhedron

$$\boldsymbol{f}_{\bar\beta} = \boldsymbol{B}^{\mathrm{T}}\boldsymbol{f}\,, \quad \boldsymbol{B}^{\mathrm{T}} = \begin{pmatrix} 1 & 2 & 3 & -1 & 2 \\ -1 & 3 & 2 & 2 & -1 \\ 2 & -1 & -3 & 3 & -3 \end{pmatrix}, \quad -1 \le f_i \le 1\,; \quad i = 1, \ldots, 4\,.$$

As suggested by

$$\begin{aligned} f_{\bar\beta_1} &= f_1 + 2f_2 + 3f_3 - f_4 + 2f_5 \\ f_{\bar\beta_2} &= -f_1 + 3f_2 + 2f_3 + 2f_4 - f_5 \\ f_{\bar\beta_3} &= 2f_1 - f_2 - 3f_3 + 3f_4 - 3f_5\,, \end{aligned}$$

the polyhedron is enclosed within the cuboid

$$-9 \le f_{\bar\beta_1} \le 9\,, \quad -9 \le f_{\bar\beta_2} \le 9\,, \quad -12 \le f_{\bar\beta_3} \le 12\,.$$

Cyclic elimination yields the coefficients λ, μ, ν, the constants $h_{12,3}, \ldots$ and $\pm c_{s,12}, \ldots$ and finally the vertices and faces. The polyhedron is depicted in Fig. F.1 and shows 20 faces, each face holding four vertices.

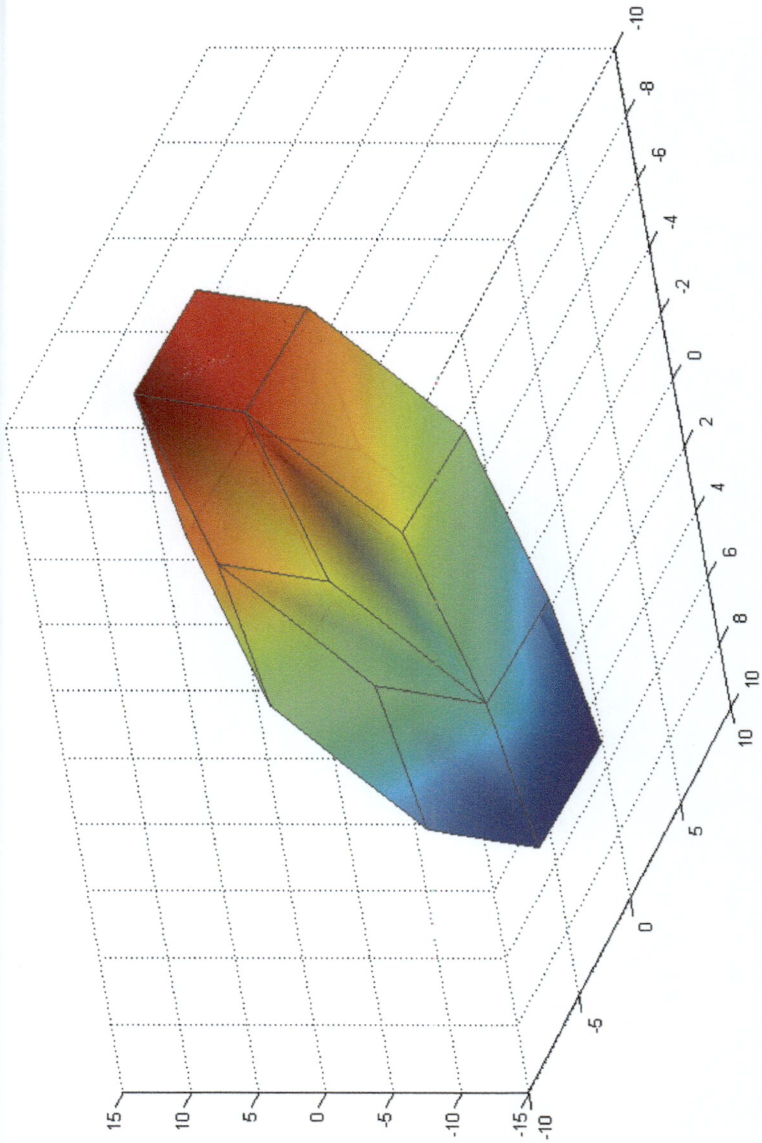

Fig. F.1. Security polyhedron as specified in the example: there are 20 faces, each face having four vertices

G EP Boundaries and EPC Hulls

The configuration of confidence ellipses and ellipsoids on the one hand with security polygons and polyhedra on the other gives rise to EP boundaries and EPC hulls as termed by the author. An EP boundary is intended to localize a 2-tuple of true values with respect to the related 2-tuple of estimators. An EPC hull does, mutatis mutandis, the same for a 3-tuple. We shall abstain from considering tuples with dimensionality greater than 3. To recall [8, 29, 28]:

- An EP boundary denotes a convex, continuously differentiable closed line composed of the segments of a confidence Ellipse and the edges of a Polygon.
- An EPC hull devises a convex, continuously differentiable closed surface composed of the segments of a confidence Ellipsoid, the faces of a Polyhedron and certain segments of elliptical Cylinders providing smooth transitions between the flat and elliptically curved parts of the hull.

While an EP boundary resembles a convex slice of a potato, an EPC hull takes after a convex potato as a whole.

EP Boundary

Ellipse and Interval

Let us consider a rectangular β_1, β_2 coordinate system and estimators $\bar{\beta}_1$, $\bar{\beta}_2$ with expectations $\mu_{\bar{\beta}_1}$, $\mu_{\bar{\beta}_2}$ and true values $\beta_{0,1}$, $\beta_{0,2}$, respectively. Any point of the ellipse's circumference may coincide with the point

$$\boldsymbol{\mu}_{\bar{\beta}} = \begin{pmatrix} \mu_{\bar{\beta}_1} \\ \mu_{\bar{\beta}_2} \end{pmatrix} .$$

Let the interval limiting the propagated systematic error be stretched along the β_1- axes and notionally be taken as a "stick."

To span the EP region, we move the midpoint of the "stick" along the circumference of the ellipse, keeping the stick's orientation steadily horizontal, i.e. parallel to the β_1-axis. The end of the stick pointing away from the ellipse produces a point of the EP boundary. For a practical approach, we observe that there are two points of the ellipse in which the "stick" coincides with the local tangent. Here, we decompose the ellipse shifting one of the partial arcs to the left and the other to right, in each case parallel to the β_1-axis. Reconnecting the arcs on either side by copies of the horizontally kept "stick" produces the EP boundary as a whole.

We expect the tuple $(\beta_{0,1},\ \beta_{0,2})$ of true values to be an element of the set of $(\beta_1,\ \beta_2)$-tuples lying within or on the borderline of the EP region.

Ellipse and Polygon

Again, we refer to a rectangular β_1, β_2 coordinate system and estimators $\bar{\beta}_1$, $\bar{\beta}_2$ with expectations $\mu_{\bar{\beta}_1}$, $\mu_{\bar{\beta}_2}$ and true values $\beta_{0,1}$, $\beta_{0,2}$, respectively. Any point of the ellipse's circumference may coincide with the point

$$\boldsymbol{\mu}_{\bar{\boldsymbol{\beta}}} = \begin{pmatrix} \mu_{\bar{\beta}_1} \\ \\ \mu_{\bar{\beta}_2} \end{pmatrix} .$$

Let us now move the center of the polygon in discrete, sufficiently narrow steps along the circumference of the ellipse, keeping the polygon's orientation steadily constant. In each of the breakpoints we draw a tangent to the ellipse. Shifting the tangent away from its osculation point, parallel to itself and to the ellipse's outside, it either strikes a vertex of the polygon or coincides with one of its edges at any rate marking the largest possible distance from the osculation point. In the first case we produce a point, in the latter a line segment of the EP boundary. After all, the boundary of the EP region consists of the segments of a contiguous decomposition of the ellipse's circumference and the edges of the polygon, where arcs and edges follow in succession [29].

We expect the tuple $(\beta_{0,1},\ \beta_{0,2})$ of true values to be an element of the set of $(\beta_1,\ \beta_2)$-tuples lying within or on the borderline of the EP region.

EPC Hull

Ellipsoid and Interval

We refer to a rectangular β_1, β_2, β_3 coordinate system and estimators $\bar{\beta}_1$, $\bar{\beta}_2$, $\bar{\beta}_3$ with expectations $\mu_{\bar{\beta}_1}$, $\mu_{\bar{\beta}_2}$, $\mu_{\bar{\beta}_3}$ and true values $\beta_{0,1}$, $\beta_{0,2}$, $\beta_{0,3}$, respectively. Any point of the ellipsoid's skin may coincide with the point

$$\boldsymbol{\mu}_{\bar{\boldsymbol{\beta}}} = \begin{pmatrix} \mu_{\bar{\beta}_1} \\ \\ \mu_{\bar{\beta}_2} \\ \\ \mu_{\bar{\beta}_3} \end{pmatrix} .$$

Let the interval limiting the propagated systematic error be stretched along the β_1-axis and notionally taken as a "stick."

For convenience, we cover the ellipsoid by an ensemble of sufficiently dense contour lines $\beta_3 = \text{const.}$ being on their part ellipses, obviously. We then move the center of the "stick" along any one of these ellipses—keeping the stick's spatial orientation steadily horizontal and parallel to the $\bar{\beta}_1$-axis. Again, there are two points in which the "stick" coincides with the local tangent. Here, we decompose the ellipse shifting one of the partial arcs to the left and the other to the right, respectively, in each case parallel to the β_1-axis. Reconnecting the arcs on either side by inserting copies of the "stick" produces a first continuous and differentiable contour line of the EPC hull. Repeating the procedure with the ellipsoid's other contour lines eventually yields the EPC hull as a whole.

We expect the triple $(\beta_{0,1}, \beta_{0,2}, \beta_{0,3})$ of true values to be an element of the set of $(\beta_1, \beta_2, \beta_3)$-tuples lying within or on the border of the EPC hull.

Ellipsoid and Polygon

We conceive a rectangular β_1, β_2, β_3 coordinate system and estimators $\bar{\beta}_1$, $\bar{\beta}_2$, $\bar{\beta}_3$ with expectations $\mu_{\bar{\beta}_1}$, $\mu_{\bar{\beta}_2}$, $\mu_{\bar{\beta}_3}$ and true values $\beta_{0,1}$, $\beta_{0,2}$, $\beta_{0,3}$, respectively. Any point of the ellipsoid's skin may coincide with the point

$$\boldsymbol{\mu}_{\bar{\boldsymbol{\beta}}} = \begin{pmatrix} \mu_{\bar{\beta}_1} \\ \mu_{\bar{\beta}_2} \\ \mu_{\bar{\beta}_3} \end{pmatrix} .$$

Let $f_{\bar{\beta}_3} = 0$ so that the polygon confining the pair $(f_{\bar{\beta}_1}, f_{\bar{\beta}_2})$ of propagated systematic error is aligned parallel to β_1, β_2-plane.

Again, we cover the ellipsoid by an ensemble of contour lines $\beta_3 = \text{const.}$ We then move the center of the polygon in discrete, sufficiently narrow steps along any such contour line $\beta_3 = \text{const.}$, keeping the polygon's orientation steadily constant and parallel to the β_1, β_2-plane. In each of the breakpoints we conceive a local tangent. Shifting the tangent away from its osculation point, parallel to itself and to the ellipse's outside, it either strikes a vertex of the polygon or coincides with one of its edges, at any rate marking the largest possible distance from the osculation point. In the first case we get a single point, in the latter a line segment of a contour line of the EPC hull. Repeating the procedure with the other contour lines of the ellipsoid successively produces the EPC hull in its entirety.

We expect the triple $(\beta_{0,1}, \beta_{0,2}, \beta_{0,3})$ of true values to be an element of the set of $(\beta_1, \beta_2, \beta_3)$-tuples lying within or on the border of the EPC hull.

Ellipsoid and Polyhedron

We refer to a rectangular β_1, β_2, β_3 coordinate system and estimators $\bar{\beta}_1$, $\bar{\beta}_2$, $\bar{\beta}_3$ with expectations $\mu_{\bar{\beta}_1}$, $\mu_{\bar{\beta}_2}$, $\mu_{\bar{\beta}_3}$ and true values $\beta_{0,1}$, $\beta_{0,2}$, $\beta_{0,3}$, respectively. Any point of the ellipsoid's skin may coincide with the point

$$\boldsymbol{\mu}_{\bar{\boldsymbol{\beta}}} = \begin{pmatrix} \mu_{\bar{\beta}_1} \\ \mu_{\bar{\beta}_2} \\ \mu_{\bar{\beta}_3} \end{pmatrix} .$$

To span the EPC hull, we move the center of the polyhedron along any contour line β_3 = const. of the ellipsoid, keeping the spatial orientation of the polyhedron steadily constant. In each of the breakpoints we conceive a local tangent plane. We shift the plane away from its osculation point parallel to itself and to the ellipsoid's outside so that, finally, it strikes either a vertex of the polyhedron or coincides with one of the polyhedron's edges or faces. In the first case we get a point, in the second a line segment and in the third a plane segment of the EPC hull. Repeating the procedure with the ellipsoid's other contour lines eventually produces the EPC hull as a whole (Fig. G.1).

We expect the triple $(\beta_{0,1}, \beta_{0,2}, \beta_{0,3})$ of true values to be an element of the set of $(\beta_1, \beta_2, \beta_3)$-tuples lying within or on the border of the EPC hull.

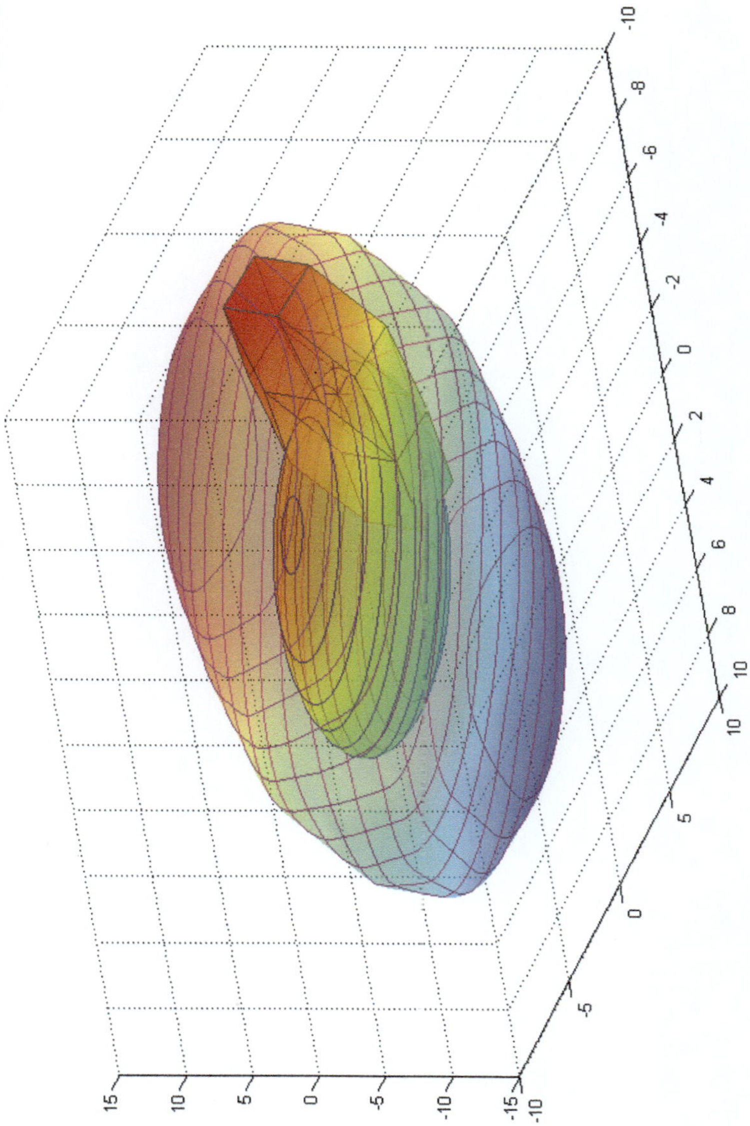

Fig. G.1. Exemplary configuration of an ellipsoid with a polyhedron

H Student's Density

We consider it beneficial to conceptualize two versions of the variable of Student's density.

Variable $T(\nu) = \dfrac{X - \mu}{S}$

Let the random variable X be $N(\mu, \sigma^2)$-distributed,

$$P(X \le x) = \frac{1}{\sigma\sqrt{2\pi}} \int\limits_{-\infty}^{x} \exp\left(-\frac{(x-\mu)^2}{2\sigma^2}\right) \mathrm{d}x\,. \tag{H.1}$$

Furthermore, let the

$$x_1, x_2, \ldots, x_n \tag{H.2}$$

be n realization of the random variable X. The sample variance is given by

$$s^2 = \frac{1}{n-1} \sum_{l=1}^{n} (x_l - \bar{x})^2 \; ; \quad \nu = n - 1\,. \tag{H.3}$$

For a fixed t we define

$$x = \mu + ts \tag{H.4}$$

and ask for

$$P(X \le \mu + ts) = \frac{1}{\sigma\sqrt{2\pi}} \int\limits_{-\infty}^{\mu+ts} \exp\left(-\frac{(x-\mu)^2}{2\sigma^2}\right) \mathrm{d}x\,. \tag{H.5}$$

Substituting $\eta = (x - \mu)/\sigma$ for x produces

$$P(X \le \mu + ts) = P(t, s, \nu) = \frac{1}{\sqrt{2\pi}} \int\limits_{-\infty}^{ts/\sigma} \mathrm{e}^{-\eta^2/2} \mathrm{d}\eta\,. \tag{H.6}$$

For this to become statistically representative, we average[1] by means of

$$p_{S^2}\left(s^2\right) = \frac{\nu^{\nu/2}}{2^{\nu/2}\Gamma\left(\nu/2\right)\sigma^{\nu}} \exp\left(-\frac{\nu s^2}{2\sigma^2}\right) s^{\nu-2} \tag{H.7}$$

so that

$$P\left(t,\nu\right) = E\left\{P\left(t,s,\nu\right)\right\} = \int_0^{\infty} \left\{ \frac{1}{\sqrt{2\pi}} \int_{-\infty}^{ts/\sigma} \mathrm{e}^{-\eta^2/2} \mathrm{d}\eta \right\} p_{S^2}\left(s^2\right) \mathrm{d}s^2 . \tag{H.8}$$

But this is a distribution function with a variable t. Hence, differentiating with respect to t

$$\frac{\mathrm{d}P\left(t,\nu\right)}{\mathrm{d}t} = \int_0^{\infty} \left\{ \frac{1}{\sigma\sqrt{2\pi}} \exp\left(-\frac{t^2 s^2}{2\sigma^2}\right) s \right\} p_{S^2}\left(s^2\right) \mathrm{d}s^2 \tag{H.9}$$

and integrating out the right-hand side issues a density, namely

$$p_T\left(t,\nu\right) = \frac{\Gamma\left(\frac{\nu+1}{2}\right)}{\sqrt{\pi\nu}\,\Gamma\left(\frac{\nu}{2}\right)} \frac{1}{\left(1+\frac{t^2}{\nu}\right)^{(\nu+1)/2}} . \tag{H.10}$$

This, obviously, is Student's or Gosset's density for the random variable

$$T(\nu) = \frac{X-\mu}{S} . \tag{H.11}$$

Due to (H.3) and (H.7), the number of degrees of freedom is $\nu = n-1$.

Variable $T(\nu) = \dfrac{\bar{X}-\mu}{S/\sqrt{n}}$

Let us replace (H.4) by

$$\bar{x} = \mu + ts/\sqrt{n} \tag{H.12}$$

in which

$$\bar{x} = \frac{1}{n}\sum_{l=1}^{n} \bar{x}_l \tag{H.13}$$

denotes the sample mean. As

[1]Averaging by $p_S(s)$ leads to the same result as $p_S(s)ds = p_{S^2}(S^2)dS^2$.

$$P\left(\bar{X} \le \bar{x}\right) = \frac{1}{\sigma/\sqrt{n}\sqrt{2\pi}} \int\limits_{-\infty}^{\bar{x}} \exp\left(-\frac{(\bar{x} - mu)^2}{2\sigma^2/n}\right) \mathrm{d}\bar{x} \tag{H.14}$$

we have

$$P\left(\bar{X} \le \mu + ts/\sqrt{n}\right) = \frac{1}{\sigma/\sqrt{n}\sqrt{2\pi}} \int\limits_{-\infty}^{\mu+ts/\sqrt{n}} \exp\left(-\frac{(\bar{x} - \mu)^2}{2\sigma^2/n}\right) \mathrm{d}\bar{x}\,. \tag{H.15}$$

Substituting $\eta = (\bar{x} - \mu)/(\sigma/\sqrt{n})$ leads us back to (H.6). Hence, the random variable

$$T(\nu) = \frac{\bar{X} - \mu}{S/\sqrt{n}} \tag{H.16}$$

is t-distributed with degrees of freedom $\nu = n - 1$.

I Uncertainty Band Versus EP-Region

The fitting of straight lines asks for two-dimensional uncertainty regions. Remarkably enough, there are two different, however, equivalent approaches [29].

Set of Straight Lines Versus Set of 2-Tuples

Consider a straight line as defined by a 2-tuple of true values

$$\begin{pmatrix} \beta_{0,1} \\ \beta_{0,2} \end{pmatrix}$$

pegging its ordinate intercept $\beta_{0,1}$ and slope $\beta_{0,2}$. The least squares adjustment provides a pair of estimators

$$\begin{pmatrix} \bar{\beta}_1 \\ \bar{\beta}_2 \end{pmatrix}$$

devising a straight line

$$y(x) = \bar{\beta}_1 + \bar{\beta}_2 x\,.$$

However, there are still other straight lines compatible with the error model and the input data. The idea is to engird this bunch by an uncertainty band

$$\bar{y}(x) \pm u_{\bar{y}(x)}\,.$$

We expect the upper and lower borderlines

$$\bar{y}(x) + u_{\bar{y}(x)} \quad \text{and} \quad \bar{y}(x) - u_{\bar{y}(x)}$$

to localize the true straight line.

At the same time, the fitted straight line may be endowed by a confidence ellipse and a security polygon. Their piecing together devises an EP-region. The latter marks a set of tuples (β_1, β_2) being compatible with the input data and the error model. We expect the 2-tuple of true values, $(\beta_{0,1}, \beta_{0,2})$, to be an element of this set.

The approaches, though ostensibly different, should put forth concurrent results: The set of straight lines, as devised by the uncertainty band $\bar{y}(x) \pm u_{\bar{y}(x)}$ should reproduce the set of 2-tuples $(\beta_1,\, \beta_2)$ as confined by the EP-region, and vice versa.

We shall establish this complementarity. To this end, we step back and decompose the uncertainty band and the EP-region into its components due to random and systematic errors.

Transformation Procedures

Initially, we transform the random share of the uncertainty band into the confidence ellipsoid and, thereafter, the systematic share into the security polygon.

Random Errors

Consider case (i) as quoted in Table 12.1. Abstaining from systematic errors, the uncertainty band is given by (13.26) disregarding the term $f_{s,y}$. The confidence ellipse is given in (13.27).

We envisage the set of tangents to the upper and lower boundaries of the uncertainty band. Each tangent specifies a certain y-intercept and slope, i.e a tuple

$$\begin{pmatrix} \beta_1 \\ \beta_2 \end{pmatrix} .$$

Considering Fig. I.1, we see the parameters β_1, β_2 of the tangent to P_1 to reproduce the point P_1^* on the ellipse. More general, a tangent gliding from P_1 via P_2 to P_3 is seen to establish the right arc P_1^*, P_2^*, P_3^* of the confidence ellipse. Similarly, a tangent gliding from P_5 via P_6 to P_7 is seen to reproduce the ellipse's left arc P_5^*, P_6^*, P_7^* .

Let us specify the differences $\beta_1 - \bar{\beta}_1$ and $\beta_2 - \bar{\beta}_2$. For this, we denote the upper boundary of the uncertainty band by

$$\begin{aligned} y_2(x) &= \bar{y}(x) + u_{\bar{y}(x)} \\ &= \bar{\beta}_1 + \bar{\beta}_2\, x + t_P(m-2)\, s_y \sqrt{\sum_{i=1}^{m} (b_{i1} + b_{i2}\, x)^2} \, . \end{aligned} \tag{I.1}$$

Due to (13.17) we observe

$$y_2(x) = \bar{\beta}_1 + \bar{\beta}_2\, x + t_P(m-2) \sqrt{s_{\bar{\beta}_1\bar{\beta}_1} + 2 s_{\bar{\beta}_1\bar{\beta}_2}\, x + s_{\bar{\beta}_2\bar{\beta}_2} x^2} \, . \tag{I.2}$$

The tangent to any point $(x, y_2(x))$

$$y(\xi) = y_2(x) + y_2'(x)(\xi - x) = \beta_1 + \beta_2 \xi \tag{I.3}$$

implies

$$\beta_1 = y_2(x) - y_2'(x)\, x \quad \text{and} \quad \beta_2 = y_2'(x)\,.$$

With

$$y_2'(x) = \bar{\beta}_2 + t_P(m-2) \frac{s_{\bar{\beta}_1 \bar{\beta}_2} + s_{\bar{\beta}_2 \bar{\beta}_2}\, x}{\sqrt{s_{\bar{\beta}_1 \bar{\beta}_1} + 2 s_{\bar{\beta}_1 \bar{\beta}_2}\, x + s_{\bar{\beta}_2 \bar{\beta}_2} x^2}} \tag{I.4}$$

we have

$$\beta_1 - \bar{\beta}_1 = t_P(m-2) \frac{s_{\bar{\beta}_1 \bar{\beta}_1} + s_{\bar{\beta}_1 \bar{\beta}_2}\, x}{\sqrt{s_{\bar{\beta}_1 \bar{\beta}_1} + 2 s_{\bar{\beta}_1 \bar{\beta}_2}\, x + s_{\bar{\beta}_2 \bar{\beta}_2} x^2}}\,. \tag{I.5}$$

Similarly, we find

$$\beta_2 - \bar{\beta}_2 = t_P(m-2) \frac{s_{\bar{\beta}_1 \bar{\beta}_2} + s_{\bar{\beta}_2 \bar{\beta}_2}\, x}{\sqrt{s_{\bar{\beta}_1 \bar{\beta}_1} + 2 s_{\bar{\beta}_1 \bar{\beta}_2}\, x + s_{\bar{\beta}_2 \bar{\beta}_2} x^2}}\,. \tag{I.6}$$

Ultimately, inserting (I.5) and (I.6) into (13.27) and gathering terms in powers of x^0, x^1, and x^2 issues an identity, provided that

$$t_P(m-2) = t_P(2, m-2)\,.$$

This, however, ascertains differing probabilities as

$$P(m-2) \neq P(2, m-2)\,. \tag{I.7}$$

The disparity is to be attributed to the observation that the straight line's uncertainty band is established by Student's density, which is one-dimensional, while the confidence ellipse is issued by Hotelling's density, which is two-dimensional. To have an example, we assume $m = 14$ and choose $P(2, m-2) = 95\%$ for Hotelling's density. This produces $t_P(2, m-2) = 3.0$. On the other hand, for $t_P(m-2) = 3.0$, Student's density issues a probability of $P(m-2) = 99\%$.

Letting x tend to infinity, we get the tangent to $y_2(x)$ with y-intercept and slope

$$\begin{aligned} \lim_{x \to \infty} \{\beta_1\} &= \bar{\beta}_1 + t_P(m-2) \frac{s_{\bar{\beta}_1 \bar{\beta}_2}}{\sqrt{s_{\bar{\beta}_2 \bar{\beta}_2}}} \\ \lim_{x \to \infty} \{\beta_2\} &= \bar{\beta}_2 + t_P(m-2) \frac{s_{\bar{\beta}_2 \bar{\beta}_2}}{\sqrt{s_{\bar{\beta}_2 \bar{\beta}_2}}} \end{aligned} \tag{I.8}$$

producing the point P_4^* of Fig. I.1. Putting $x = 0$, we find the tangent to $y_2(x)$ with y-intercept and slope

$$\begin{aligned}\beta_1 - \bar{\beta}_1 &= t_P(m-2)\frac{s_{\bar{\beta}_1\bar{\beta}_1}}{\sqrt{s_{\bar{\beta}_1\bar{\beta}_1}}}\\ \beta_2 - \bar{\beta}_2 &= t_P(m-2)\frac{s_{\bar{\beta}_1\bar{\beta}_2}}{\sqrt{s_{\bar{\beta}_1\bar{\beta}_1}}}\,.\end{aligned} \tag{I.9}$$

We observe (I.8) to coincide with (D.2) and (I.9) with (D.1). Deriving corresponding expressions for the lower boundary of the uncertainty band

$$\begin{aligned}y_1(x) &= \bar{y}(x) - u_{\bar{y}(x)}\\ &= \bar{\beta}_1 + \bar{\beta}_2\,x - t_P(m-2)\,s_y\sqrt{\sum_{i=1}^{m}(b_{i1} + b_{i2}\,x)^2}\end{aligned} \tag{I.10}$$

produces

$$\begin{aligned}\beta_1 - \bar{\beta}_1 &= -t_P(m-2)\frac{s_{\beta_1\beta_1} + s_{\beta_1\beta_2}\,x}{\sqrt{s_{\beta_1\beta_1} + 2s_{\beta_1\beta_2}\,x + s_{\beta_2\beta_2}x^2}}\\ \beta_2 - \bar{\beta}_2 &= -t_P(m-2)\frac{s_{\beta_1\beta_2} + s_{\beta_2\beta_2}\,x}{\sqrt{s_{\beta_1\beta_1} + 2s_{\beta_1\beta_2}\,x + s_{\beta_2\beta_2}x^2}}\end{aligned} \tag{I.11}$$

For $x \to \infty$, we obtain the tangent to $y_1(x)$ with y-intercept and slope

$$\begin{aligned}\lim_{x\to\infty}\{\beta_1\} &= \bar{\beta}_1 - t_P(m-2)\frac{s_{\bar{\beta}_1\bar{\beta}_2}}{\sqrt{s_{\bar{\beta}_2\bar{\beta}_2}}}\\ \lim_{x\to\infty}\{\beta_2\} &= \bar{\beta}_2 - t_P(m-2)\frac{s_{\bar{\beta}_2\bar{\beta}_2}}{\sqrt{s_{\bar{\beta}_2\bar{\beta}_2}}}\end{aligned} \tag{I.12}$$

establishing the point P_8^* of Fig. I.1.

Vice versa, we might wish to deploy the coordinates β_1 and β_2 of the ellipse's circumference directly as given in (13.27) and draw a bunch of straight lines

$$y(x) = \beta_1 + \beta_2\,x$$

in-between the upper and lower borders of the straight line's uncertainty band. We in fact find the bunch to fit the uncertainty band, given we boost Student's t from $t_P(m-2)$ up to Hotelling's $t_P(2, m-2)$ putting $t_P(m-2) = t_P(2, m-2)$. Otherwise, the bunch would exceed the borderlines of the uncertainty band.

Case (ii), as specified in Table 12.1, leads to a corresponding result, here, however, we have to put

$$t_P(n-1) = t_P(2, n-1)\,.$$

Systematic Errors

To illustrate the proceeding, we address case (ii) of Table 12.1 ignoring random errors. To simplify the discussion, the number of measuring points is reduced to just three. On its left, Fig. I.2 displays the least squares line and its uncertainty band. On the right we observe the associated security polygon. The symbols LS_1 to LS_6 designate the line segments of the upper and lower borderlines of the uncertainty band.

Intuitively, we expect the y-intercept and slope of line segment LS_1 to reproduce the polygon's vertex V_1; furthermore, segment LS_2 to reproduce vertex V_2 and, finally, segment LS_3 vertex to issue V_3. Corresponding relations should hold with respect to the segments of the upper borderline and the polygon's remaining three vertices. To keep track of the supposed relationships, we resort to a numerical example,

$$\boldsymbol{B} = \begin{pmatrix} b_{11} & b_{12} \\ b_{21} & b_{22} \\ b_{31} & b_{32} \end{pmatrix} = \begin{pmatrix} 1.33 & -0.50 \\ 0.33 & 0.00 \\ -0.67 & 0.50 \end{pmatrix} .$$

Referring to (14.21)

$$f_{s,\bar{y}(x)} = \mid b_{11} + b_{12}x \mid f_{s,\bar{y}_1} + \mid b_{21} + b_{22}x \mid f_{s,\bar{y}_2} + \mid b_{31} + b_{32}x \mid f_{s,\bar{y}_3} \,,$$

we have

$$LS_1 : x < -\frac{b_{31}}{b_{32}}; \quad LS_2 : -\frac{b_{31}}{b_{32}} \le x < -\frac{b_{11}}{b_{12}}; \quad LS_3 : x \ge -\frac{b_{11}}{b_{12}} . \tag{I.13}$$

Regarding LS_1, the propagated systematic error is given by

$$\begin{aligned} f_{s,\bar{y}(x)} &= (b_{11} + b_{12}x) f_{s,\bar{y}_1} + b_{21} f_{s,\bar{y}_2} - (b_{31} + b_{32}x) f_{s,\bar{y}_3} \\ &= (b_{11} f_{s,\bar{y}_1} + b_{21} f_{s,\bar{y}_2} - b_{31} f_{s,\bar{y}_3}) + (b_{12} f_{s,\bar{y}_1} - b_{32} f_{s,\bar{y}_3})\, x . \end{aligned} \tag{I.14}$$

Hence, segment LS_1

$$\begin{gathered} \bar{y}(x) - f_{s,\bar{y}(x)} = \bar{\beta}_1 + \bar{\beta}_2\, x \\ -(b_{11} f_{s,\bar{y}_1} + b_{21} f_{s,\bar{y}_2} - b_{31} f_{s,\bar{y}_3}) - (b_{12} f_{s,\bar{y}_1} - b_{32} f_{s,\bar{y}_3})\, x \end{gathered} \tag{I.15}$$

has y-intercept and slope

$$\begin{aligned} \beta_1 &= \bar{\beta}_1 - (b_{11} f_{s,\bar{y}_1} + b_{21} f_{s,\bar{y}_2} - b_{31} f_{s,\bar{y}_3}) \\ \beta_2 &= \bar{\beta}_2 - (b_{12} f_{s,\bar{y}_1} - b_{32} f_{s,\bar{y}_3}) . \end{aligned} \tag{I.16}$$

We suppose the pair β_1, β_2 to reproduce the polygon's upper leftmost vertex V_1. To show this, we turn to Appendix F. Indeed, rewriting (14.10),

$$\begin{aligned} f_{\bar{\beta}_1} &= b_{11} f_{\bar{y}_1} + b_{21} f_{\bar{y}_2} + b_{31} f_{\bar{y}_3} \\ f_{\bar{\beta}_2} &= b_{12} f_{\bar{y}_1} \qquad\qquad + b_{32} f_{\bar{y}_3} \,, \end{aligned} \tag{I.17}$$

and putting

$$f^*_{\bar{y}_1} = \text{sign}(b_{12}) f_{s,\bar{y}_1}; \quad f^*_{\bar{y}_3} = \text{sign}(b_{32}) f_{s,\bar{y}_3}$$

we find

$$\begin{aligned} f_{s,\bar{\beta}_1} &= b_{11}\text{sign}(b_{12}) f_{s,\bar{y}_1} - b_{21} f_{s,\bar{y}_2} + b_{31}\text{sign}(b_{32}) f_{s,\bar{y}_3} \\ f_{s,\bar{\beta}_2} &= b_{12}\text{sign}(b_{12}) f_{s,\bar{y}_1} + b_{32}\text{sign}(b_{32}) f_{s,\bar{y}_3} \end{aligned}$$

or

$$\begin{aligned} f_{s,\bar{\beta}_1} &= -b_{11} f_{s,\bar{y}_1} - b_{21} f_{s,\bar{y}_2} + b_{31} f_{s,\bar{y}_3} \\ f_{s,\bar{\beta}_2} &= -b_{12} f_{s,\bar{y}_1} + b_{32} f_{s,\bar{y}_3} \end{aligned} \tag{I.18}$$

As the polygon is centered in $(\bar{\beta}_1, \bar{\beta}_2)$, (I.18) coincides with (I.16) as it should be.

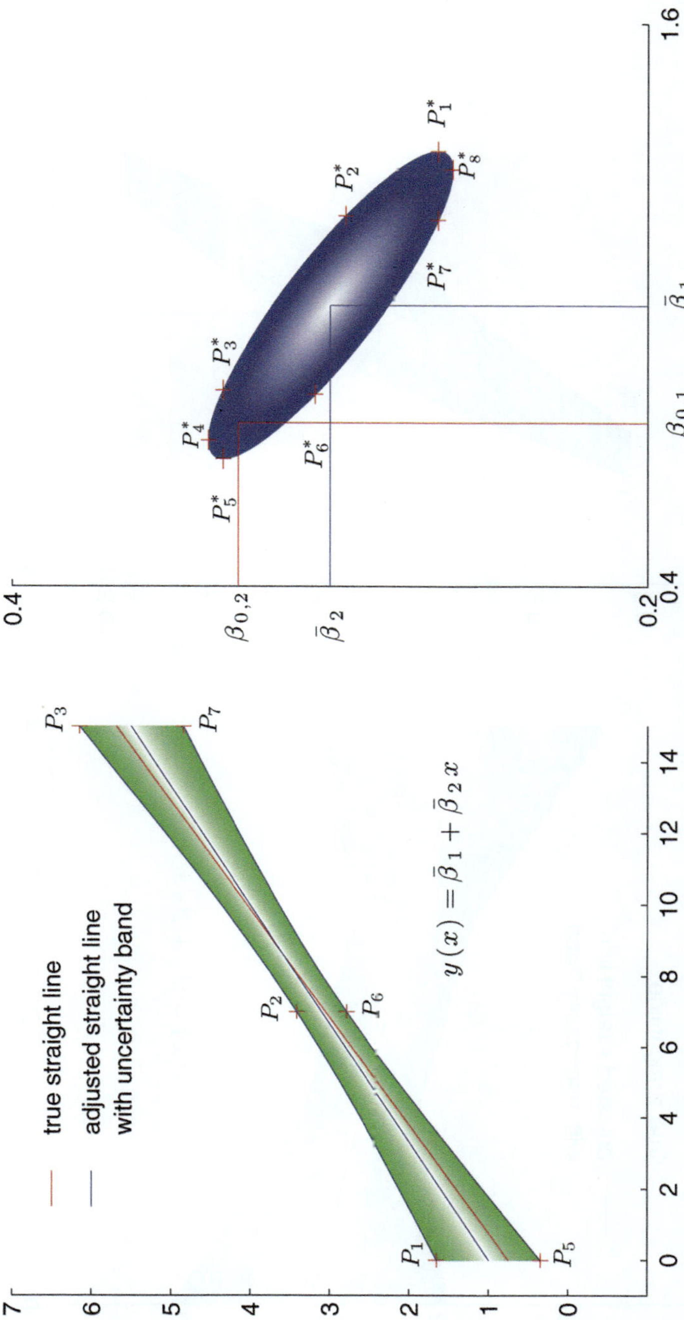

Fig. I.1. Random share of uncertainty band versus confidence ellipse

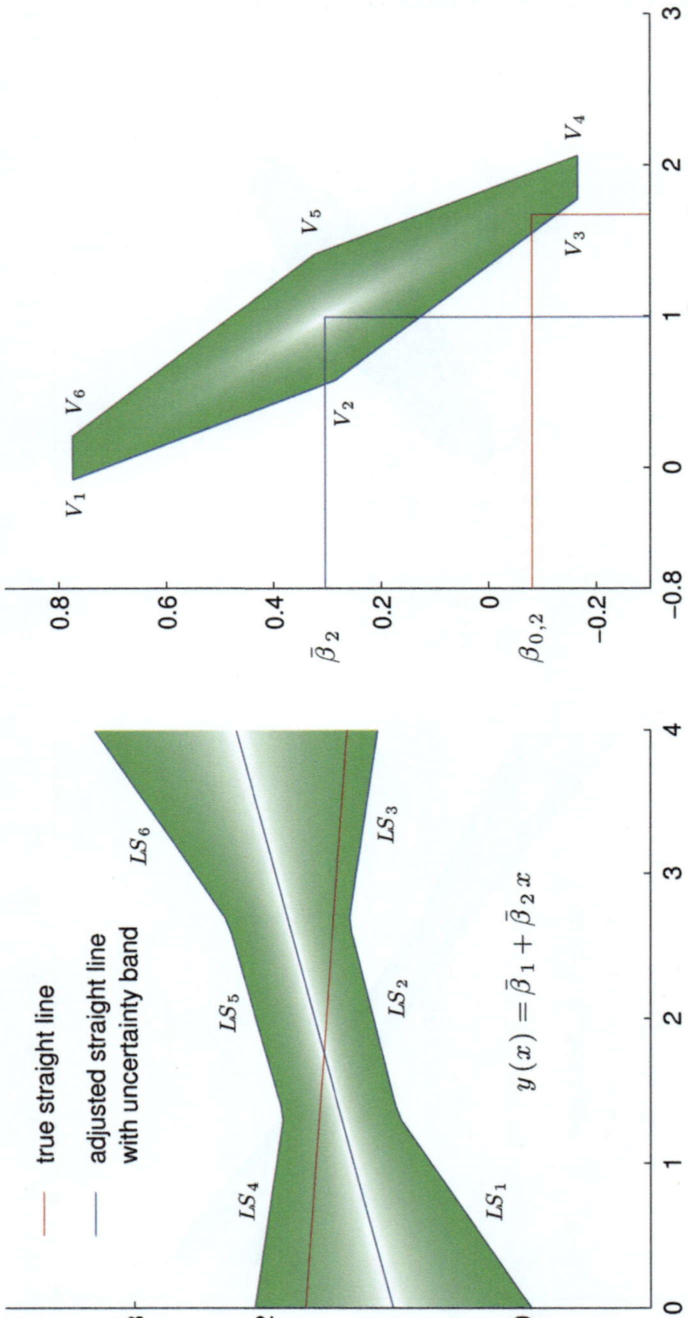

Fig. I.2. Systematic proportion of uncertainty band versus security polygon

J Quantiles of Hotelling's Density

$t_{0.95}$; P = 95%

m $n-1$	1	2	3	4	5	6	7	8	9	10
1	13									
2	4.3	28								
3	3.2	7.6	44							
4	2.8	5.1	10.7	60						
5	2.6	4.2	6.8	13.9	76					
6	2.5	3.7	5.4	8.5	17.0	92				
7	2.4	3.5	4.8	6.7	10.3	20.1	108			
8	2.3	3.3	4.4	5.8	7.9	12.0	23.3	124		
9	2.3	3.2	4.1	5.2	6.7	9.1	13.7	26.4	140	
10	2.2	3.1	3.9	4.9	6.0	7.7	10.3	15.4	29.5	156
11	2.2	3.0	3.8	4.6	5.6	6.9	8.7	11.5	17.1	32.7
12	2.2	3.0	3.7	4.4	5.3	6.3	7.7	9.6	12.7	18.8
13	2.2	2.9	3.6	4.3	5.0	5.9	7.0	8.5	10.6	13.9
14	2.1	2.9	3.5	4.1	4.8	5.6	6.6	7.7	9.3	11.2
15	2.1	2.8	3.4	4.0	4.7	5.4	6.2	7.2	8.4	10.1
16	2.1	2.8	3.4	4.0	4.6	5.2	5.9	6.8	7.8	9.1
17	2.1	2.8	3.3	3.9	4.5	5.1	5.7	6.5	7.4	8.4
18	2.1	2.8	3.3	3.8	4.4	4.9	5.5	6.2	7.0	7.9
19	2.1	2.7	3.3	3.8	4.3	4.8	5.4	6.0	6.7	7.5
20	2.1	2.7	3.3	3.7	4.2	4.7	5.3	5.8	6.5	7.2
21	2.1	2.7	3.2	3.7	4.2	4.7	5.2	5.7	6.3	6.9
22	2.1	2.7	3.2	3.7	4.1	4.6	5.1	5.6	6.1	6.7
23	2.1	2.7	3.2	3.6	4.1	4.5	5.0	5.5	6.0	6.5
24	2.1	2.7	3.2	3.6	4.0	4.5	4.9	5.4	5.9	6.4
25	2.1	2.7	3.1	3.6	4.0	4.4	4.8	5.3	5.7	6.2
26	2.1	2.7	3.1	3.6	4.0	4.4	4.8	5.2	5.7	6.1
27	2.1	2.7	3.1	3.5	3.9	4.3	4.7	5.1	5.6	6.0
28	2.1	2.6	3.1	3.5	3.9	4.3	4.7	5.1	5.5	5.9
29	2.0	2.6	3.1	3.5	3.9	4.3	4.6	5.0	5.4	5.8
30	2.0	2.6	3.1	3.5	3.9	4.2	4.6	5.0	5.4	5.8
31		2.6	3.1	3.5	3.8	4.2	4.6	4.9	5.3	5.7
32			3.1	3.5	3.8	4.2	4.5	4.9	5.3	5.6
33				3.4	3.8	4.2	4.5	4.9	5.2	5.6
34					3.8	4.1	4.5	4.8	5.2	5.5
35						4.1	4.5	4.8	5.1	5.4
36							4.4	4.7	5.1	5.4
37								4.7	5.1	5.4
38									5.0	5.3
39										5.3

References

Journals and Reports

1. BIPM, Draft recommendation on the statement of uncertainties. Metrologia, **17**, 69–74 (1981).
2. Bureau International des Poids et Mesures (BIPM), Sèvres, France.
3. E.R. Cohen, B.R. Taylor, The 1986 Adjustement of the Fundamental Physical Constants, Codata Bulletin, No. 63 (1986).
4. Deutsches Institut für Normung e.V., Grundbegriffe der Meßtechnik, DIN 1319-1 (1995), DIN 1319-2 (1980), DIN 1319-3 (1996), DIN 1319-4 (1999), Beuth Verlag GmbH, Berlin 30.
5. Deutsches Institut für Normung e.V., Zahlenangaben, DIN 1333 (1992), Beuth Verlag GmbH, Berlin 30.
6. C. Eisenhart, The reliability of measured values – Part I fundamental concepts. Photogramm. Eng. **18**, 543–561 (1952)
7. Engineering Statistics Handbook, NIST/SEMATECH e-Handbook of Statistical Methods, http://www.itl.nist.gov/div898/handbook/
8. M. Grabe, Anpassung eines Kreises nach kleinsten Quadraten, Jahresbericht der Physikalisch-Technischen Bundesanstalt, (1989) 210–211 (Least Squares Adjustment of a Circle, Annual Report of the Physikalisch-Technische Bundesanstalt, (1989) 210–211).
9. M. Grabe, Estimation of measurement uncertainties – an alternative to the *ISO Guide*. Metrologia **38**, 97–106 (2001).
10. M. Grabe, On the Assignment of Uncertainties Within the Method of Least Squares, Poster Paper, Second International Conference on Precision Measurement and Fundamental Constants, Washington, D.C., USA, 8–12 June 1981.
11. M. Grabe, On the Estimation of One- and Multidimensional Uncertainties, National Conference of Standards Laboratories, Proceedings 569-576, Albuquerque, USA, 25–29 July 1993.
12. M. Grabe, Principles of "Metrological Statistics". Metrologia **23**, 213–219 (1986/87).
13. M. Grabe, The Alternative Error Model and its Impact on Traceability and Key Comparison, Poster Paper, BIPM-NPL Workshop on the Evaluation of Interlaborators Comparison Data, NPL Teddington, 19 September 2002.
14. M. Grabe, Über die Fortpflanzung zufälliger und systematischer Fehler, Seminar über die Angabe der Meßunsicherheit, 20. und 21. Februar 1978, Physikalisch-Technische Bundesanstalt Braunschweig. (On the Propagation of Random and Systematic Errors, Seminar on the Statement of the Measurement

Uncertainty, 20 und 21 February 1978, Physikalisch-Technische Bundesanstalt Braunschweig).
15. Guide to the Expression of Uncertainty in Measurement (GUM), 1 Rue Varambé, Case Postale 56, CH 1221, Geneva 20, Switzerland.
16. R.L. Plackett, Some theorems in least squares. Biometrika **37**, 149–157 (1950).
17. S. Wagner, Zur behandlung systematischer Fehler bei der Angabe von Meunsicherheiten. PTB-Mitt. **79**, 343–347 (1969).

Monographs

18. F. Ayres, *Matrices, Schaum's Outline Series* (McGraw-Hill, New York, 1974).
19. P. Beckmann, *Elements of Applied Probability Theory* (Harcourt, Brace & World Inc., New York, 1968).
20. P.L. Bender, B.N. Taylor, E.R. Cohen, J.S. Thomsen, P. Franken, C. Eisenhart, Should Least Squares Adjustment of the Fundamental Constants be Abolished? In Precision Measurement and Calibration, NBS Special Publication **343** (United States Department of Commerce, Washington D.C., 197[illegible]).
21. L.L. Chao, *Statistics Methods and Analyses* (McGraw-Hill Kogakusha Ltd., Tokyo, 1974).
22. A.A. Clifford, *Multivariate Error Analysis* (Applied Science Publisher Ltd., London, 1973).
23. H. Cramér, *Mathematical Methods of Statistics* (Princeton University Press, Princeton, 1954).
24. N.R. Draper, H. Smith, *Applied Regression Analysis* (John Wiley & Sons, New York, 1981).
25. W.T. Eadie et al., *Statistical Methods in Experimental Physics* (North Holland, Amsterdam, 1971).
26. M. Fisz, *Wahrscheinlichkeitsrechnung und Mathematische Statistik* (VEB Deutscher Verlag der Wissenschaften, Berlin, 1978).
27. C.F. Gauss, Abhandlungen zur Methode der kleinsten Quadrate, Physica, Würzburg, 1964.
28. M. Grabe, *Measurement Uncertainties in Science and Technology* (Springer, Berlin, 2005).
29. M. Grabe, *Schätzen von Meßunsicherheiten in Wissenschaft und Technik*, (Books on Demand GmbH Braunschweig, 2000).
30. F.A. Graybill, *An Introduction to Linear Statistical Models* (McGraw-Hill, New York, 1961).
31. H.H. Ku (ed.), Precision Measurement and Calibration, NBS Special Publication 300, vol 1 (United States Department of Commerce, Washington D.C., 1969).
32. A. Papoulis, *Probability, Random Variables and Stochastic Processes* (McGraw-Hill Kogakusha Ltd, Tokyo, 1965).
33. G.A.F. Seber, *Linear Regression Analysis* (John Wiley & Sons, New York, 1977).
34. G. Strang, *Linear Algebra and Its Applications* (Harcourt Brace Jovanovich College Publishers, New York, 1988).

Index

GPSR Compliance

The European Union's (EU) General Product Safety Regulation (GPSR) is a set of rules that requires consumer products to be safe and our obligations to ensure this.

If you have any concerns about our products, you can contact us on ProductSafety@springernature.com

In case Publisher is established outside the EU, the EU authorized representative is:

Springer Nature Customer Service Center GmbH
Europaplatz 3
69115 Heidelberg, Germany

Batch number: 10614906

Printed by Printforce, the Netherlands